BIOASSAYS OF ENTOMOPATHOGENIC MICROBES AND NEMATODES

Bioassays of Entomopathogenic Microbes and Nematodes

Edited by

A. Navon

and

K.R.S. Ascher

Department of Entomology
Agricultural Research Organization
Bet Dagan
Israel

CABI *Publishing*

CABI *Publishing* is a division of CAB *International*

CABI Publishing
CAB International
Wallingford
Oxon OX10 8DE
UK

Tel: +44 (0)1491 832111
Fax: +44 (0)1491 833508
Email: cabi@cabi.org
Web site: http://www.cabi.org

CABI Publishing
10 E 40th Street
Suite 3203
New York, NY 10016
USA

Tel: +1 212 481 7018
Fax: +1 212 686 7993
Email: cabi-nao@cabi.org

A catalogue record for this book is available from the British Library,
London, UK.

Library of Congress Cataloging-in-Publication Data
Bioassays of entomopathogenic microbes and nematodes / edited by A. Navon and
K.R.S. Ascher.
      p. ; cm.
  Includes bibliographical references and index.
  ISBN 0-85199-422-9 (alk. paper)
    1. Nematoda as biological pest control agents--Research--Technique. 2. Microbial
pesticides--Research--Technique. 3. Biological assay. I. Navon, Amos. II. Ascher, K. R.
S.
    [DNLM: 1. Nematoda--parasitology. 2. Biological assay--methods. 3.
Parasites--parasitology. 4. Pest control, Biological--methods. QX 203 B615 2000]
SB933.334.B56 2000
632′.96--dc21
                                         99–046507
ISBN 0 85199 422 9

Typeset by Columns Design Ltd, Reading.
Printed and bound in the UK by Biddles Ltd, Guildford and King's Lynn.

Contents

Contributors

K.R.S. Ascher, Department of Entomology, Agricultural Research Organization, The Volcani Center, Bet Dagan 50250, Israel

N. Becker, German Mosquito Control Association, (KABS), Ludwigstrasse 99, D-67165, Waldsee, Germany

S. Braun, The Biotechnology Laboratory, Institute of Life Sciences, The Hebrew University of Jerusalem, Jerusalem 91904, Israel

W.M. Brooks, Department of Entomology, 2315 Gardner Hall, North Carolina State University, Raleigh, NC 27695, USA

T.M. Butt, School of Biological Sciences, University of Wales, Singleton Park, Swansea, SA2 8PP, UK

D.M. Eaves, Department of Mathematics and Statistics, Simon Fraser University, Burnaby, British Columbia, V5A 1S6, Canada

I. Glazer, Department of Nematology, Agricultural Research Organization, The Volcani Center, Bet Dagan 50250, Israel

M.S. Goettel, Lethbridge Research Centre, Agriculture and Agri-Food Canada, PO Box 3000, Alberta, T1J 4B1, Canada

T.A. Jackson, AgResearch, PO Box 60, Lincoln, New Zealand

K.A. Jones, Natural Resources Institute, University of Greenwich, Central Avenue, Chatham Maritime, Chatham, Kent ME4 4TB, UK

E.E. Lewis, Department of Entomology, Virginia Tech, Price Hall, Blacksburg, VA 24061, USA

J.V. Maddox, Center for Economic Entomology, Illinois Natural History Survey and Illinois Agricultural Experiment Station, 607 Peabody Drive, Champaign, IL 61820, USA

R. Marcus, Department of Statistics, Agricultural Research Organization, The Volcani Center, Bet Dagan 50250, Israel

A. Navon, Department of Entomology, Agricultural Research Organization, The Volcani Center, Bet Dagan 50250, Israel

D.J. Saville, AgResearch, PO Box 60, Lincoln, New Zealand

S.R. Sims, Whitmire Micro-Gen, 3568 Tree Court Industrial Blvd, St Louis, MO 63122, USA

O. Skovmand, Intelligent Insect Control, 80 Rue Pauline Ramart, F-34070, Montpellier, France

D. Smith, CABI Bioscience UK Centre (Egham), Bakeham Lane, Egham, Surrey TW20 9TY, UK

L.F. Solter, Center for Economic Entomology, Illinois Natural History Survey and Illinois Agricultural Experiment Station, 607 Peabody Drive, Champaign, IL 61820, USA

Foreword

Since World War II, pest control methods have been dominated by the use of synthetic chemical insecticides. These insecticides have been very successful in the marketplace and have contributed greatly to the production of food and fibre, as well as to the control of many diseases of medical and veterinary importance transmitted by insects, especially mosquitoes and blackflies. Despite these benefits, the use of chemical insecticides has not been without problems. Among the most important of these are the development of high levels of resistance in target pest and vector populations, high mortality rates in non-target beneficial insect, mite and spider populations, and contamination of the environment, particularly water supplies, with chemicals that the public perceives as being harmful. Due to these problems, scientists in academia, government and industry have placed a great deal of effort over the past few decades in developing more environmentally compatible pest control products and methods. The best of these include safer chemical insecticides, microbial insecticides based on pathogens that cause disease in insects, parasitic nematodes, and during the last decade, insect-resistant transgenic plants based primarily on insecticidal proteins produced by *Bacillus thuringiensis*. Products based on these new technologies are increasingly being accepted in the marketplace, and legislated into use, either by the banning of certain chemical insecticides or by legal restrictions placed on their use by environmental protection agencies around the world. The need for new products and pest control technologies will continue to accelerate due not only to demands for safer pest control technologies, but to the continuing increase in the world's population.

Many of the new technologies noted above offer excellent promise for the development of safe and effective pest control methods. However, each of the various new types of agents being developed requires special expertise and methods for accurate determination of efficacy through bioassays. Though many methods have been published in the scientific literature, these are typically found in a variety of disparate journals and technical manuals. In the current text, Dr A. Navon and Dr K.R.S. Ascher have enlisted the assistance of experts on all of the new and developing pest control technologies based on viruses, microorganisms and nematodes, to bring together the most pertinent methods in a single volume.

The volume begins with a series of subchapters on *B. thuringiensis*, the most successful of the microbial insecticides. These subchapters cover a wide range of topics ranging from how to isolate and grow new bacterial strains, assess their efficacy and evaluate the potential of wild-type and recombinant bacterial strains and proteins for commercial development. Methods are provided for assessing the activity of strains, proteins and formulations against numerous insect pests of crops, forests, and the larvae of pest and vector mosquitoes and blackflies. Perhaps the most important advance of the last decade emerging from the field of microbial control is the development of insect-resistant transgenic crops based on the insecticidal crystal proteins of *B. thuringiensis* (*Bt*). Millions of acres of *Bt*-transgenic cotton and maize are already being planted in the USA. Yet methods for assessing the efficacy of these are not widely known or available. The current text fulfils this need by providing bioassay methods for different plant tissues including leaves, roots and pollen. Whereas most of the focus on bacterial insecticides has been on *B. thuringiensis*, other bacteria continue to show potential for control of soil-dwelling pests. To treat this area, a chapter has been included on the recent development of a species of *Serratia* to control soil-dwelling beetle grubs.

Though *Bt* has been the most successful commercial microbial insecticide, insect-pathogenic viruses, fungi, microsporidia and nematodes have been used or are being considered for use against many crop pests, especially against pests not controlled easily by bacteria. These insects include locusts, sucking insects such as whiteflies and aphids, termites, ants, and even lepidopterous pests insensitive to *Bt*. Viruses must be eaten to be effective, whereas the fungi typically invade the insect target through the cuticle. Both of these pathogen types can cause acute diseases. Microsporidia, which are commonly transmitted horizontally by feeding or transovarially on or within eggs, cause chronic diseases. The nematodes are minute parasitic worms which, unlike most pathogens, can actually seek out their host, invade the body through various orifices (the mouth, anus, or spiracles), and cause death within a day or two. The methods to assess the efficacy of these agents differ markedly owing to these different modes of action. To treat this diversity, four separate chapters, one dealing with each of these agent types, are included here that cover a wide range of methods which can be used to evaluate efficacy.

Critical to the accurate assessment of efficacy are methods for designing bioassays and analysing the resultant data. Whereas by necessity many of the chapters deal with this subject, its importance is emphasized by the inclusion of a separate chapter that provides comprehensive coverage of the statistical problems encountered by biologists conducting bioassays with biopesticides. Statistical methods and solutions to problems often encountered in designing bioassays for pathogens and nematodes are covered in detail. The last chapter deals with regulatory issues and the topic of intellectual property rights. For the purpose of increasing international trade, many countries are attempting to harmonize regulations and laws that regulate the development and use of pest control agents and methods. This chapter covers the most critical issues, including patents and evolving governmental regulations in these expanding and increasingly important areas.

Though the new technology of insect-resistant transgenic crops provides an important pest control method, only a few insect-resistant transgenic crops are currently on the market. Moreover, many minor crops may not be engineered to be insect-resistant, at least in the near future. Thus there will be an ongoing if not expanding need for new types of pest control technologies which employ microorganisms and nematodes. This volume provides a comprehensive series of methods to assess the efficacy and commercial potential of these organisms as well as transgenic plants, and to move them into the marketplace. It should find wide use in the field of pest control in both highly industrialized and developing countries well into the 21st century.

Brian A. Federici
University of California, Riverside

Bioassays of *Bacillus thuringiensis* | 1

1A. Bioassays of *Bacillus thuringiensis* Products Used Against Agricultural Pests

A. Navon

Department of Entomology, Agricultural Research Organization, Bet Dagan, Israel

Introduction

Bacillus thuringiensis (*Bt*) has become the leading biopesticide since the beginning of the 1970s, due to the lethality of the toxin to insects. It has attracted industry to use it worldwide as an effective weapon against agricultural pests and insect vectors of human diseases. Originally, *B. thuringiensis* was considered an entomopathogen. Within the last four decades the complexity and diversity of *B. thuringiensis* as an insecticidal microbe have been elucidated.

The first report on the crystalline parasporal body in the bacterium that might be associated with the insecticidal activity appeared by 1953 (Hannay, 1953). Angus (1954) demonstrated that this crystal contains an alkaline-soluble toxin for insects. *B. thuringiensis* produces a β-exotoxin known also as the fly-toxin, thermostable toxin, or thuringiensin, but this toxin was not approved for use in agriculture because its toxicity was not limited to insect pests (Sebesta *et al.*, 1981). The δ-endotoxin showed the most promising characteristics of an insect-specific bioinsecticide. By the end of the 1950s, the toxicity of the spore–crystal complex was classified into 'insect types' (Heimpel and Angus, 1959) – an early indication that the spore contributes to the insecticidal power of *B. thuringiensis*.

In the 1960s, initial efforts were made to quantify the spore–crystal mixtures by means of standardized bioassays (Bonnefoi *et al.*, 1958; Burgerjon, 1959; Menn, 1960; Mechalas and Anderson, 1964). The introduction of commercial *B. thuringiensis* products by the agrochemical industry and the growing knowledge of the insecticidal toxin led to an urgent need to replace the spore count that had labelled the commercial products with a standardized value of insecticidal power. The idea of using a standard microbe for potency determination of the bioinsecticide product was accepted at the 1966 Colloquium of Insect Pathology and Microbial Control at Wageningen, The Netherlands. The E-61 formulation from the Institute Pasteur, Paris, France, with an assigned potency of 1000 IU mg^{-1}, was first adopted as the standard (Burges, 1967), but was later replaced in 1971 by the HD-1-S-1971 with an assigned potency of 18,000 IU mg^{-1}. As from 1980 the HD-1-S-80 standard with a potency of 16,000 IU mg^{-1} has been available from international *B. thuringiensis* cultures. The standardized bioassay procedures proposed by Dulmage *et al.* (1971) were reviewed (Dulmage *et al.*, 1981; Beegle, 1989) and further standardized for several insects species (Navon *et al.*, 1990). The 'activity ratio' in which a microbe powder is assayed against two insect species in parallel (Dulmage *et al.*, 1981) is useful for measuring differences between the tested microbe powder and the standard. Their potencies may be identical in one insect but different in another insect species. This ratio for any two insect species can be compared between laboratories, provided that the microbe powder is the same.

So far, however, the possibility of using new international standardization of *Bt* has not been accepted after discussing this issue in a recent Society for Invertebrate Pathology (SIP) meeting (Navon and Gelernter, 1996), mainly because: (i) insect strains and qualities differ between laboratories, and therefore mortalities are not comparable; and (ii) each of the *Bt* manufacturers has developed its own bioassay procedures for labelling the *Bt* products and changing the protocols will not be feasible for commercial considerations. Moreover, in the last two decades, additional bioassay protocols had to be designed in view of the following new developments and discoveries: (i) identification of new Cry proteins and their use in genetically engineered products (Baum *et al.*, 1999); (ii) development of new conventional formulations (Burges and Jones, 1999); (iii) an accumulating evidence of insect resistance to δ-endotoxin and Cry proteins (see also S. Sims in Subchapter 1B); and (iv) analytical assays for the δ-endotoxin, Cry proteins and insecticidal crystal protein (ICP) genes (Cannon, 1993; Plimmer, 1999), as complementary information on the activity of *Bt* proteins in the microbe isolates.

These developments and the new knowledge acquired challenged us to describe a comprehensive collection of bioassay protocols and procedures for quantifying activities of conventional *B. thuringiensis* products. Bioassays of genetically engineered *B. thuringiensis* plant products are also described in this chapter (see S. Sims, Subchapter 1B).

Bioassays of *B. thuringiensis* Against Lepidopterous Pests

Artificial diets

The aim of using bioassays based on an artificial diet was to provide the worker with a rapid, standardized and simple procedure for estimating the activity of a microbial strain.

The nutrients in the diet are a substitute for the natural food and the agar gel provides a texture similar to that of plant tissues but devoid of their undesirable side-effects due to plant allelochemicals and microorganisms. Initially, a diet was proposed for the single bioassay insect, *Trichoplusia ni* (Dulmage *et al.*, 1971), or *Anagasta kuehniella* for the French E-61 standard (*Bt* subsp. *thuringiensis*). However, with the growing international interest in *B. thuringiensis* as a useful substitute for chemical insecticides, efforts were made to select the most effective *B. thuringiensis* strains against specific insect pests. This change required the use of more than one bioassay diet. Navon *et al.* (1990) proposed a standardized diet that, with additions of feeding stimulants available from processed food fractions, would be suitable for almost any lepidopterous species. In addition, a wide choice of diets that could be adapted for a bioassay diet are available from rearing manuals (Singh and Moore, 1985). Even a diet based on a calcium alginate gel that is prepared without a heating step has been developed (Navon *et al.*, 1983). In this diet, heat-labile components of *B. thuringiensis* and enzymes can be used.

A standardized bioassay diet (Navon *et al.*, 1990) can be used for any

Table 1A.1.   Composition of standardized bioassay diets for lepidopterous insects.

No.	Ingredient	Neonate bioassay	3rd-instar bioassay
1	Seed products/processed food[a]	16.00	16.00
2	B vitamin solution[b]	10.50	10.50
3	Cholesterol	0.05	0.05
4	Choline chloride	0.09	0.09
5	L-ascorbic acid	0.10	0.38
6	Cellulose powder	1.67	1.67
7	Agar	1.67	1.67
8	Methyl-*p*-hydroxybenzoate	0.19	0.19
9	Sorbic acid	0.03	0.03
10	Formaldehyde	–	0.22
11	Deionized water	69.70	69.20
	Total	100.00	100.00

[a]Examples for nutrient ratios in insect diets:beans:whole milk powder – 6:1 for bioassays with *Spodoptera littoralis* and *Helicoverpa armigera*; beans:whole milk:cotton seed protein – 4.5:1:1.5 for bioassays with *Earias insulana*.
[b]*i*-Inositol – 4600 mg; pantothenic acid – 850 mg; niacin – 300 mg; *p*-amino-benzoic acid – 154 mg; riboflavin – 24 mg; pyridoxine – 38 mg; thiamine – 28 mg; folic acid – 12 mg; biotin – 4 mg; vitamin B_{12} – 2 mg; all dissolved in 1 litre distilled water (according to Levinson and Navon, 1969).

target insect provided that the specific phagostimulants are included for specialist feeders. Also, to preserve spore activity, inclusion of antibiotics in the diet should be avoided. However, diet preservatives with bacteriostatic effects, such as methyl-*p*-hydroxybenzoate (Nipagin), can be used. One standardized diet has been made suitable for rearing neonate larvae. The other diet formula is for third instars, as detailed in Table 1A.1. Changes in the seed product/condensed food source renders the diet suitable for several lepidopterous species. Other food sources with similar nutritional ratios can be used for any lepidopterous species, provided that the nutrients contain the necessary phagostimulants to induce insect feeding.

To prepare the diet dissolve the agar and the methyl-*p*-hydroxybenzoate in half of the water by heating in an autoclave. Use the other half of the water to homogenize the remaining ingredients, except the ascorbic acid, in a blender. Mix together the hot agar/methyl-*p*-hydroxybenzoate solution and the nutrient homogenate. Add the L-ascorbic acid at 50°C. Keep the diet at this temperature by holding it in a container in a hot water bath at approximately 70°C or in an electric heating basket. Weigh diet portions from this

(a)

Fig. 1A.1. (*and opposite*) Dietary larval rearing trays for bioassays of *B. thuringiensis*. (a) Disposable gel trays for bioassays with 3rd-instar larvae. Each cell is filled with 1 g of diet. The lid is an empty gel tray placed upside down. (b) A 24-well tray. Well dimensions: 15 mm diameter × 17 mm depth. (c) A grid of 21 cells each sized 13 × 13 × 6 mm cut from a fluorescent light shading, made of plastic (cell size: 11 mm²; wall thickness: 2.5 mm) and fitted into a 9-cm disposable Petri dish.

(b)

(c)

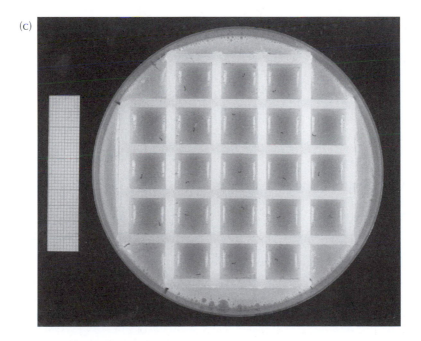

Fig. 1A.1. *Continued.*

container for each of the microbe dilution series, mix in the aqueous microbe mixture, stir with a laboratory stirrer for 30 s and pour the diet into the rearing cavities or cells (Fig. 1A.1). The microbial mixture should not exceed 5% of the diet's volume; larger microbial portions may dilute the nutrients' concentration and cool the diet so that it will set prematurely. Mix the dilute microbe mixtures first and then the concentrated ones.

Use the standard reference microbes obtained from the international *B. thuringiensis* collections (see Appendix in Subchapter 1D). HD-1-S-80 is the most recent reference standard. Store all the standard microbes at −10°C. If you use the same insect species in most of your bioassays and you have small amounts of the microbial standard, select an internal standard based on the HD-1 or another strain of subsp. *kurstaki*. Then calibrate the insecticidal power of the internal (secondary) standard against the international standard (HD-1-S-80) to express the microbial activity in potency units. This procedure of selecting an internal standard can also be used for insects susceptible to subsp. *aizawai*. Calibration of an internal standard has to be determined for each species of bioassay insect. Since there is no international standard based on a subsp. *aizawai* strain, companies created their own internal standard using *Spodoptera exigua* as the bioassay insect and expressing the activity of their product in *Spodoptera* Units. Also, Abbott Laboratories (1992) used a 'Diamondback Moth Unit' to determine the activity of the product XenTari based on an *aizawai* strain bioassayed with *Plutella xylostella*. This modification in the original bioassay is justified since insects such as *Spodoptera littoralis* have an unacceptably high LC_{50} against the HD-1-S-80 standard (Navon *et al.*, 1993).

The tested *B. thuringiensis* products are available mostly as wettable powders and liquid concentrates. The dilution solution used for the microbe preparations is saline buffer solution (8.5 g NaCl, 6.0 g K_2HPO_4 and 3.0 g KH_2PO_4 per litre, pH 7.0) together with 0.05% w/v polysorbitan monooleate (Tween 80) as a surfactant. To test potencies of unknown *B. thuringiensis* preparations, conduct a preliminary assay with a tenfold dilution series of the experimental microbial powders and the standard. The results of these assays are used to select a narrower dilution in which the LC_{50} will fall approximately midway in the series.

Mixing the *B. thuringiensis* insecticidal components (spores, crystals, protoxin and more) with the diet renders the microbe available not only to defoliators but also to larvae that penetrate into the diet and feed on inner layers of the medium. In addition, by using this mixing procedure undesired effects of the adjuvants of the *B. thuringiensis* product, mostly fermentation residues, on larval feeding will be minimized. In contrast to this, if commercial products are applied to the diet's surface, the fermentation adjuvants will accumulate there and may introduce dose-dependent errors in the bioassay at high *B. thuringiensis* concentrations.

Many types of larval rearing trays are available from commercial equipment and insect rearing catalogues. Three examples of rearing trays for single larvae are: (i) disposable J-2 50-cavity trays (Corrigan and Company

Inc., Jacksonville, FL 32203, USA) with 50 cavities for 3rd-instar larvae (Fig. 1A.1a); (ii) 24-well trays (Fig. 1A.1b); and (iii) a rearing unit of 21 cells produced by a plastic grid inserted into Petri dishes (Navon *et al.*, 1990) (Fig. 1A.1c). Collective larval rearing in the bioassay is not recommended, since feeding interference and cannibalism among the larvae can introduce erroneous variations in mortality.

In bioassays with 3rd-instar larvae, diet portions in the cells should be about 1 ml or 1 g. In a 1st-instar bioassay less than 1 g can be used, provided the diet does not dehydrate within the bioassay period. Weigh diet portions for a single *B. thuringiensis* concentration, mix with the aqueous microbe mixture in a blender and pour into the cavities. In the grid cells (Fig. 1A.1c), pour the diet into the Petri dish first and then fit the grid.

For neonate bioassays, keep the diet at room temperature in a hood for 1 h to evaporate any condensed water. Put a single larva in each of the cavities. To transfer neonates use a camel-hair brush. Avoid touching the larvae with the brush; instead, let the larva 'parachute' on its spinning thread. Hold the thread with the brush and let the larva touch the diet. You may hold several larvae together with the brush and save inoculation time. Special attention should be paid to closing the rearing units. In the grid cells, a filter paper of 9-cm diameter is placed on the grid and a 5–10-mm thick plastic sponge is placed on top of the filter paper. The Petri dish lid is put on top of the plastic sponge and rubber bands are used to close the cells tightly. In this way, the sponge is pressed against the grid so that neonate larvae cannot escape from the cells. Neonate larvae are used instead of 3rd instars for several reasons: (i) this instar is available from the insect colony in much larger numbers than any other instar and with less input of labour and materials; (ii) the bioassay period is shorter; and (iii) precision is higher because larval mortality is more uniform and confidence intervals of the LC_{50} are smaller.

In 'official bioassays' (Dulmage *et al.*, 1981) take measurements of larval weight, size and head capsule to describe the instar. In 3rd-instar bioassays, postecdysed larvae (after moult) are preferable as 'standardized larvae'. The length of this bioassay is based on 3rd-instar larvae surviving for 7 days. Bioassays starting on Tuesday instead of Monday would save termination of the experiment during weekends. Touch the larvae with a needle to confirm mortality. For the neonate bioassays use larvae that are 0–12 h old and deprived of food; mortality is counted after 48 or 96 h.

Natural food

Leaf and greenhouse-plant bioassays are an intermediate step between dietary (artificial diet) bioassays and field assays. Whereas dietary bioassays were designed to determine the activity of the spores and crystals accurately in an artificial medium, plant bioassays have two purposes. First, they consist of plant tissues with most of the chemical and physical barriers presented by the agricultural crop. For example, the alkaline leaf surface produced by the epidermal glands is present in both greenhouse and field plants (Navon *et al.*, 1988). However, other barriers may not be expressed in the sheltered plants; the trichome density in greenhouse seedlings is significantly lower than in field plants (Navon, unpublished). Second, they allow the evaluation of effects of formulation adjuvants on the phylloplane. Feeding stimulants, fermentation residues, surfactants, rain-fasting materials and stickers may affect the larval feeding behaviour and thereby the ingestion of the spore–crystal mixture. Cotton, tomato, maize and cruciferous species are among the most common crops raised in the greenhouse for bioassays. Cotton is a useful bioassay plant because it is a host of several major lepidopterous insect pests. Also, within 2–3 weeks, the seedlings already have eight to ten leaves that can be used either for detached-leaf or for potted plant bioassays. However, leaf bioassays with bollworm and borer larvae cannot substitute for assays with flower buds and fruit that are the natural target organs of these insects in the field.

Leaf bioassays

One of the common leaf bioassays uses leaf discs. The disc test is based on a standardized size of leaf tissue and therefore application of the *B. thuringiensis* in aqueous mixture per unit area is simple and accurate. However, a leaf disc cannot be infested in bioassays with more than one neonate larva of lepidopterous species with cannibalistic tendencies. An alternative method is to use the entire leaf with a water supply (Fig. 1A.2a). The whole leaf provides hiding places for the larvae, so that the physical contact among larvae that occurs in the disc bioassay is reduced. In addition, the water balance of the plant tissue is maintained better when the leaf has a water supply. The method is as follows: pour a 2-cm layer of 1% agar solution in a 15- or 20-cm^3 glass vial. Dry the condensation water by exposing the vial to reduced air pressure in a hood for 1–2 h. Pipette 50 µl of the test solution on each 10–15 cm^2 sized leaf side. Let the mixture dry on the leaves. Cut the petiole at 2-cm distance from the leaf. Hold the petiole with forceps and insert it into the agar layer. Put 5 neonate larvae in each vial. Close the vial with a cotton cloth held tightly with a rubber band. Use five vials per treatment (25 neonates). In this bioassay, leaf freshness is preserved for 3 days whereas the bioassay period is 1–2 days only. Flower bud bioassays are conducted in a similar manner with one or two larvae; the bud petiole is inserted into the agar layer as for the leaves.

A conventional mixing of two *B. thuringiensis* subspecies in the formulation is one of the means to widen the insect host range of the microbial product. When combining subspecies *kurstaki* and *aizawai* in an aqueous mixture or in the tank mix, different insect species should be used in parallel, for example *Helicoverpa armigera* and others which are susceptible to subsp. *kurstaki* strains and *Spodoptera* species for subsp. *aizawai* strains. Such a combination of *B. thuringiensis* strains has been developed in a granular feeding bait formulation (Navon *et al.*, 1997).

The leaf and flower bud bioassays are also suitable for assaying granular formulations. An accurate quantitative application of granules on leaves and flower buds was developed using a dispersion tower (Navon *et al.*, 1997). With this tower, effects of granular sizes on the larvae were determined. In the agar vials, granules larger than 250 µm were not suitable for the bioassay because they dropped off the leaf and were not available to the larvae, whereas granules of less than 150 µm adequately adhered to the leaf surface.

Potted-plant bioassays

Potted-plant bioassays have several uses that cannot be provided with the leaf bioassays. These include: (i) testing the activity of *B. thuringiensis* with intact plant organs; (ii) applying the microbial product by spraying or dusting the plant, where an accuracy of dosing exceeding that of a field application can be achieved; (iii) extending the bioassay time for neonates exposed to the intact leaves to more than 3 days (until the leaf cage area is totally consumed); and (iv) assaying the residual effect of the microbial preparation. Potted-plants bioassays, conducted by caging 1st-, 2nd- or 3rd-instar larvae on leaves of potted plants (Navon *et al.*, 1987), are useful for assaying mortality, leaf consumption and inhibition of larval weight gain under greenhouse conditions. The cages consist of two plastic cylinders (Figs 1A.2b, c), one attached to each side of the leaf. Each half cage is closed on its outer side by a 150–224 µm mesh metal screen. The screen is attached by heating it and pressing it against the plastic cylinder until sealed. The two half-cage units are held together by an uncoiled wire paper clip. The leaf area consumed within the cage can be measured with a leaf area meter (e.g. Li-COR, Lincoln, Nebraska, USA) as a parameter for assessing the efficacy of the *B. thuringiensis* preparation against 3rd-instar larvae. This cage is also used in bioassays with bollworms on cotton flower buds and fruits in potted plants. For this purpose, a nick measuring 3–4 mm^2 is cut out from the half-cage surface as a space for the insertion of the petiole of the flower bud or fruit (Figs 1A.2d, e) and an elastic filler is used to prevent larval migration through this hole in the cage. For neonate bioassays, a ring of polyethylene sponge is glued to the perimeter of each of the half-cage units. In this way, when the cage is closed, the sponge is pressed against the leaf surface preventing the 1st-instar larvae from escaping.

In order to determine the EC_{50}, the larvae are weighed every 1–2 days. In a 7–8-day bioassay, larvae have to be transferred to new leaves because 3rd-instar larvae will consume the whole leaf area in the cage within 3–4 days.

(a)

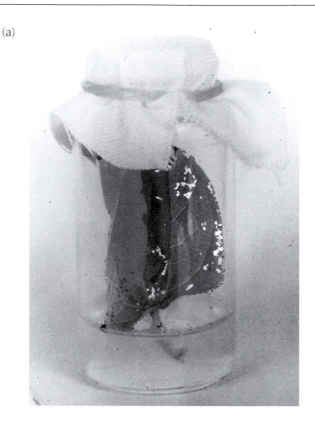

(b)

Fig. 1A.2.   (*and opposite*) Leaf and plant bioassays of *B. thuringiensis.* (a) Whole leaf bioassay. The petiole is inserted in agar gel to preserve leaf freshness. (b) Leaf cages for larvae in potted-cotton leaf bioassays. Cage dimensions: 30 mm inner diameter × 18 mm height. (c) Single larval cage. The two halves of the cage, the metal wire for holding them and the leaf arena infested by a *S. littoralis* larva. (d) Flower bud cages in cotton in the field. (e) The flower bud cage with an *H. armigera* larva.

(c)

(d)

(e)

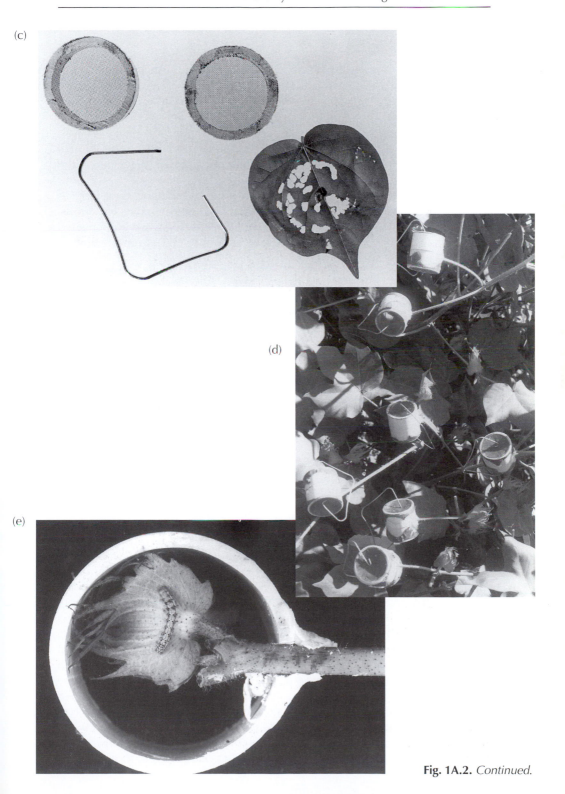

Fig. 1A.2. *Continued.*

Field plants bioassays

This type of bioassay is conducted in natural plant and environmental conditions. Spraying volume and formulation affect the effectiveness of pest management. The leaf and flower bud cages (Figs 1A.2b–e) are also used for the field bioassay. Second- and 3rd-instar larvae are used in the bioassay. Smaller screen meshes (100–200 µm) are used for field bioassays with neonate larvae. It should be noted that neonate mortality in the control insects brought about by natural entomopathogenic microorganisms on the phylloplane is often unacceptably high.

Portable meteorological stations are useful in the experimental plots to record the circadian temperature and humidity during the bioassay periods. Bioassays for recording the residual effect of *B. thuringiensis* are conducted by sequential caging of treated plants at 0, 2, 4 and 6 days from the time of the initial microbe application.

Types of bioassays

Dose–response

Dose–response is the most common type of bioassay among those used to determine biological effects of the bioinsecticide or the entomopathogen. In this bioassay, mortality is counted after a single period of time. For neonate larvae it is limited to 1–3 days. In the official bioassay with 3rd-instar larvae, the bioassay time is 7 or 8 days. The potency bioassay of *B. thuringiensis* is a special case of dose–response bioassay where the mortality of the experimental microbe is referenced against the international standard and expressed in international units per mg (IU mg^{-1}).

Time–response

Time–mortality bioassays are suitable mostly for 2nd- and 3rd-instar larvae, since the mortality of neonate larvae occurs too quickly to obtain time–mortality slopes. In order to obtain good slopes with 1st-instar larvae, it is recommended to make counts of dead larvae every 6 or 8 h. In comparison, the bioassays with older larvae can be prolonged to several days. This is explained in part by the fact that, unlike hatching larvae, 3rd instars survive for several days without feeding owing to the fat body that is used as an energy reserve when the larva cannot feed. Also, larval recovery from microbe intoxication increases survival (Dulmage *et al.*, 1978). In time–mortality bioassays with 3rd instars, most larval death occurs in the first 4–8 days, during which time–mortality is counted on successive days or every other day. The median lethal time (LT$_{50}$) calculation and analyses using computerized models are described in Chapter 6.

Feeding inhibition

These bioassays are conducted with 2nd or 3rd instars. Larvae are weighed on an analytical balance (0.1 mg accuracy). The mode of diet preservation should be considered carefully because of exposure to airborne contamination during larval handling. The concentration that cause a 50% reduction in larval weight as compared with that of the control larvae (EC_{50}) is the most common quantitative character in this type of bioassay. It is recommended as a criterion in comparing activities of *B. thuringiensis* preparations, since, for instance, weight reduction in young larvae caused by CryIA(c) proteins is a more sensitive parameter than the LC_{50} (mortality) in insect species susceptible to *B. thuringiensis* subsp. *kurstaki* (MacIntosh *et al.*, 1990).

Sublethal effect

Sublethal effects of *B. thuringiensis* are assessed in insects in two ways.

CONTINUOUS EXPOSURE
When larvae are exposed continuously to low levels of the pathogen, the follow up of the symptoms caused in the insect by the microbe preparation is extended to a full insect generation. Dulmage and Martinez (1973) demonstrated that sublethal doses of the microbe had marked effects on insect development. In bioassays for recording these effects, records of the following parameters are collected: larval period, percentage pupation, pupal weight, percentage adult emergence, adult fecundity and fertility.

SHORT-TERM EXPOSURE
In these assays, feeding of the larvae on dietary *B. thuringiensis* is discontinued after one of the exposure times, 24, 48 or 72 h, and the larvae are transferred to diets without the microbe. The control larvae are fed on untreated diets from the start of the bioassay. The microbe concentration, the length of exposure and combinations thereof determine the rate of reduction in the insect quality, including moth reproductive capacity (Salama *et al.*, 1981). These bioassays reflect pest management situations in the field when: (i) the larvae feed for a short time on the microbe and then escape feeding by penetrating into inner tissues of the plant; (ii) the microbial product covers the plant canopy only partially, so that defoliators feed on plant parts devoid of *B. thuringiensis*; and (iii) the product's activity at lethal concentrations is lost rapidly under direct sunlight and phylloplane effects.

Insect resistance to *B. thuringiensis* products

With the increasing uses of *B. thuringiensis* in the field, resistance of insects to the conventional *B. thuringiensis* products became a serious pest management problem. One of the most common examples of an insect becoming resistant to *B. thuringiensis* sprays in the field is the cruciferous pest, *Plutella xylostella*. The bioassay principles for evaluating resistance in genetically engineered *B. thuringiensis* plants are described in Subchapter 1B.

Inducement

Induction of resistance can be produced with Cry toxins of *B. thuringiensis* expressed either in *Escherichia coli* or in a recombinant strain with or without spores (see also Subchapter 1D). In the bioassay procedure used so far, the Cry toxins are either mixed in the diet or applied on the diet surface. First- or 2nd-instar larvae are reared on the diet and mortality is recorded after 5 or 7 days. Larval selection for resistance is conducted with one or two larvae per rearing cell exposed at subsequent generations to the same or two-fold higher concentrations, compared with the previous generation (Moar *et al.*, 1995). Another bioassay is based on the induction of resistance in large numbers (1000) of larvae reared collectively (Müller-Cohn *et al.*, 1996). The bioassays for selection for resistance are useful for testing: (i) cross-resistance, by exposing the resistant larvae to other Cry toxins and measuring the larval susceptibility to them; (ii) the stability of resistance, by using alternate treatments with or without the toxin; and (iii) reversion of resistance, by transferring the larvae to a diet without the microbe toxin on which they are reared for several generations.

In any bioassay protocol it is recommended to run preliminary bioassays with a dilution series to determine the LC_{50} and lower lethal concentrations that will be used for inducing resistance in the larvae. Also, it is suggested that a large larval population be used for the selection work to avoid sibling matings that will adversely affect moth fertility, and to represent better the initial susceptibility to the microbe.

Monitoring

In recent years, bioassays for tracing resistance in native insect populations have become an important tool for developing rational pest management with *B. thuringiensis*. One of the highest resistance build-ups in insects was monitored in *P. xylostella* as a result of foliar applications of spore–crystal commercial products. Bioassays for recording this resistance were based on the use of cabbage leaf discs (6-cm diameter). The discs were dipped in aqueous dilutions of wettable powder formulations of the HD-1 strain of *B. thuringiensis*, air dried and offered to about ten laboratory-reared 3rd-instar larvae. These larvae were the F_1–F_3 offspring of field-collected insects (Tabashnik *et al.*, 1990). A 2-day and a 5-day bioassay for recording the LC_{50} were effective in testing the stability of resistance. The resistance was significantly higher (about 20 times) for CryIC in native populations that had been treated with *B. thuringiensis* than in the susceptible laboratory colony. However, the resistance to spore–crystal formulations of *B. thuringiensis* subsp. *aizawai* was markedly lower than to the CryIC toxin (Tabashnik *et al.*, 1990).

In using bioassays for evaluating insect resistance to *B. thuringiensis*, it was shown that a shorter duration of bioassay (48 h) and the use of a single diagnostic concentration is a combination that requires less time and effort in evaluating insect resistance to *B. thuringiensis* in the field (Tabashnik *et al.*, 1993). The use of this combination for routine assays was suggested, since the loss of information in using this bioassay was not critical in comparison with the standard bioassay procedures requiring a dilution series of the microbe and a longer bioassay time (for 2nd and 3rd instars). These authors showed that the bioassay time was not significantly affected by the slope, SE of slope and 95% confidence limits for the $LC_{50,}$ whereas the control mortality increased as the bioassay time increased. This improvement in bioassay efficiency for monitoring *P. xylostella* resistance to *B. thuringiensis* was further developed by replacing the leaf-dip bioassays with a dietary bioassay and a single *B. thuringiensis* concentration which discriminates between susceptible and resistant populations of the insect in the field (Perez *et al.*, 1997). Microbe toxicity remained stable in the diets stored for 10 days at 26°C or 14 days at 5°C and, therefore, the dietary bioassay was suggested for use as a test kit in field work.

Types of preparations

Non-formulated

SPORE–CRYSTAL MIXTURES
Most commercial products are based on the spore–crystal complex. This mixture is the natural product obtained from the fermentation process. In products used in Japan, spores are inactivated to avoid detrimental effects to the silkworm, which is used commercially for the production of silk. The crystal of *B. thuringiensis* is responsible for the insecticidal effect in the majority of agricultural pests, but spore effects have been determined against several insects.

SPORES
Spores are plated on nutrient agar to determine viability. Dilution series of spores for the bioassay are prepared as for potency assays but without the *B. thuringiensis* standard.

In the pathogenicity classification of insect type III made by Heimpel and Angus (1959) for *B. thuringiensis*, both spores and crystals were necessary to kill *Ephestia kuehniella*. *Galleria mellonella* also belonged to this type (Burges *et al.*, 1976). Moreover, potentiation of the crystal protein by spores in moth larvae may explain some of the differences in insect host spectrum among *B. thuringiensis* subsp. *kurstaki* strains (Moar *et al.*, 1989). With the exception of *B. thuringiensis* products in Japan, where spores are inactivated, conventional preparations consist of spores and crystals at a ratio of 1:1. Therefore, bioassays of purified spores are useful to evaluate their role in developing moth control strategies with *B. thuringiensis.*

The insecticidal activity of *B. thuringiensis* spores has been investigated less in recent years, primarily because genetic manipulations for pest control are based solely on the crystal toxins. However, spore bioassays are useful in developing microbial pest management strategies with *B. thuringiensis* products containing both spores and crystals.

Crystals

Official potency bioassays are limited to the spore–crystal mixtures and therefore cannot be used for purified crystals. However, mortality, feeding inhibition and sublethal effects (see below) can be determined with dietary bioassays as was described for the spore–crystal mixtures and the commercial *B. thuringiensis* products. Methods of crystal and spore separation and purification are described by S. Braun (see Subchapter 1D). Spore irradiation with a low dose of γ-rays (3.0 Mrad) and addition of streptomycin to the diet were used to avoid effects of spores remaining as impurities in the crystal preparation (Li *et al.*, 1987). Adequate concentrations of surfactant for homogeneous dispersion have to be added to the purified crystal mixture. This dispersion has to be confirmed by microscope observations. Nanogram amounts of the purified crystals per gram of diet are needed for the neonate bioassays.

Cry proteins

The isolation and preparation of Cry proteins are described by S. Braun (see Subchapter 1D). Dilution series are used in the diets as described for the potency bioassays, but without using the *B. thuringiensis* standard. Both mortality (LC_{50}) and feeding inhibition (EC_{50}) parameters are useful to describe the protein activities in the insect larvae (MacIntosh *et al.*, 1990).

Formulated

CONVENTIONAL
Liquids. Liquid concentrates and aqueous mixtures of wettable powders have to be well shaken to disperse the active ingredients homogeneously in the aqueous phase before their use in laboratory assays or in the tank mix in the field.

Granules. The spore–crystal mixtures encapsulated in starch (Dunkle and Shasha, 1988) or embedded in wheat flour granules (Navon *et al.*, 1997) have to be released from the coating matrix so that the *B. thuringiensis* preparation will be dispersed homogeneously in the diet. In cornstarch preparations, digestion with amylase has been recommended to release the *B. thuringiensis* materials (McGuire *et al.*, 1997). In wheat flour-embedded products, the microbe materials are recovered from the granules by blending the product in 0.5% aqueous Tween 80 solution for 30 s (Navon, unpublished). The spore–crystal mixture is then separated from the starchy or granule material by centrifugation. This is necessary in order to avoid undesired feeding effects of those materials which can be separated from the spores and crystals by centrifugation (see Braun, Subchapter 1D).

GENETICALLY MANIPULATED

CellCap®. The δ-endotoxin is encapsulated in the dead walls of *Pseudomonas fluorescens* through transgenic manipulation. By this encapsulation the crystal toxins are protected from undesired environmental effects (Gelernter and Schwab, 1993). These transgenic commercial products are active against Lepidoptera by means of the CryIA(c) protein, and the Colorado potato beetle (CPB) by means of the CryIIIA protein from *B. thuringiensis* subsp. *san diego* (Herrnstadt *et al.*, 1986). The plant and field bioassays of these products against CPB are described by Zehnder and Gelernter (1989).

Transconjugational. These products are based on combinations of two δ-endotoxin proteins that together expand the activity against insects that are not affected by either of the toxic proteins when expressed alone. This combination was produced by transconjugation of a plasmid bearing a gene of one *B. thuringiensis* subspecies into the cells of another subspecies of the microbe (Carlton *et al.*, 1990). In addition, the activity of each of the transferred proteins was increased by genetic manipulation. Using this strategy, commercial products have been developed that are active against both lepidopterans and coleopterans or with a widened host range against lepidopterous pests (combining *B. thuringiensis* subsp. *kurstaki* and subsp. *aizawai*).

Bioassays of the transconjugational products are based on using two insect species in parallel, each susceptible to one of the two crystal proteins in the product. For example, *Helicoverpa armigera* larvae or other heliothine species can be used to assay the activity of the CryIA(c) toxin of *B. thuringiensis* subsp. *kurstaki*, whereas *S. littoralis* larvae or other *Spodoptera* species are used to determine the activity of CryIC toxin from *B. thuringiensis* subsp. *aizawai*. The transconjugant with both coleopteran and lepidopteran activity is used against the CPB and a *kurstaki*-sensitive lepidopterous species. Recombinant *Bt* products were highly active against a broad insect host range (Baum *et al.*, 1999).

Assessing tritrophic interactions

In these types of bioassays, interactions of the phytophagous insect and *B. thuringiensis* are expanded to more trophic levels such as plant allelochemicals and parasitoids or invertebrate predators (Navon, 1993).

Interactions among B. thuringiensis, *herbivorous insects and plant allelochemicals*

Two types of bioassays with *B. thuringiensis* are useful to evaluate effects of allelochemical(s) (Navon, 1993) on the microbe intoxication: neonate bioassays and last-instar bioassays (Navon et al., 1993).

NEONATE LARVAE
Bioassays of antagonistic allelochemicals such as cotton condensed tannin in interaction with CryIA(c) of the HD-73 δ-endotoxin showed that elevation in this tannin's concentration significantly increased the LC_{50} of *Heliothis virescens* neonates. In this neonate bioassay, prepare a dilution series of the microbe and determine the LC_{50}. Separately, prepare a dilution series of the allelochemical and weigh the larvae every day. Use the feeding inhibition records to determine the EC_{50}. Then use the microbial LC_{50} and the allelochemical EC_{50} and near concentrations in the same diet and record larval mortality. The shift in the microbial LC_{50} caused in the presence of the allelochemical concentration in the diet will indicate the type of interaction (antagonistic, compatible or synergistic).

MATURE LARVAE
Such bioassays are suitable for measuring growth and consumption of larvae (Navon *et al.*, 1992, 1993) fed artificial or natural food containing *B. thuringiensis*, with or without the allelochemical. With these bioassays sublethal feeding effects can be evaluated. The common nutritional indices used for this feeding bioassay are: relative growth rate (RGR) and relative consumption rate (RCR). RGR is defined as the larval weight gain per initial larval weight per 24 h, and RCR is defined as the amount of diet consumed per initial larval weight per 24 h.

Use newly ecdysed larvae as their midgut is essentially without food, and, unlike mid-instar larvae, the postecdysed larvae will start feeding upon their exposure to the diet. Prepare a regression line for differences between fresh and dry weights of diets for different weights of fresh diet. Use the regression equation to calculate dry weight of any fresh weight of 1–2 g diet cubes. To obtain an accurate larval weight, weigh the larvae after the 24 h bioassay, freeze them and dissect to remove the frozen gut content; then reweigh the larvae, lyophilize them and weigh again. Prepare a regression equation for conversion of fresh to dry larval weight without midgut content. Analyse the

RGR and RCR data by two-factor analyses of variance (*B. thuringiensis* preparation treatment × allelochemical treatment) (Navon *et al.*, 1993).

Interactions among B. thuringiensis, insects and natural enemies

Bioassays are useful for evaluating compatibilities between microbial and entomophagous pest management agents. Several bioassays are useful for the study of these tritrophic interactions. Some of the bioassays have been described in previous studies of interactions. Interaction bioassays among *H. armigera*, the larval endoparasitoid *Microplitis croceipes* (Braconidae) and *B. thuringiensis* (Blumberg *et al.*, 1997) are described below.

ADULT PARASITOID LONGEVITY
Female and male wasps are fed with *B. thuringiensis* commercial products, purified spores, or spore–crystal mixtures mixed with honey to determine their effects on longevity. Amaranth (C.I. No. 16185, S. No. 212, E. Merck, Darmstadt, Germany) is added (0.05% w/v) to the food to observe the colour in the adult digestive system as an indication of food ingestion. Adult survival is recorded daily and longevity is calculated.

ADULT PARASITOID OVIPOSITION
The host larvae at the stage that is suitable for parasitization are reared for 24 h on diets with a sublethal dilution series of *B. thuringiensis*. These larvae are subsequently exposed for 1 h to the parasitoid females for egg laying. Then the adult parasitoids are removed and the host larvae are dissected in a drop of saline solution under a binocular microscope to count the parasitoid eggs.

IMMATURE PARASITOIDS
The host larvae are fed for 24, 48 and 72 h on lethal and sublethal concentrations of dietary *B. thuringiensis*. In one set of bioassays the microbe-fed host larvae are exposed for oviposition by the parasitoid females. In another set, the host larvae are fed for 2, 4 and 6 days on the dietary microbe. Mortality and longevity of the host larvae are recorded daily, as is the number of parasitoid pupae obtained from the surviving larvae. Control host larvae with or without parasitoid oviposition are fed on diets without *B. thuringiensis*.

Experimental design and data preparation

Use 100 neonate larvae in two replicates for each concentration. Use no less than five concentrations per single *B. thuringiensis* strain. The number of cells in the rearing units depends on the source of supply, but a rearing unit cannot become a replicate if the number of cells is too small to fit into the probit model (see Chapter 7). For example, with a unit of 25 cells use two units as a single replicate of 50 larvae. Repeat the bioassays on 4 different days to allow for variations in insect quality, rearing conditions and experimental procedures. The mortality data of experiments conducted on different dates can be pooled to produce a single fitted dose–mortality curve. If this fit is poor, the causes should be identified (see Chapter 7). Note that using SAS computer packages for statistical analyses does not require a calculation of percentage mortality. Also, there is no need to correct the observed mortality according to Abbott's formula, since both control-adjusted and power models incorporate a control mortality parameter which is estimated from the data. Ignore control records showing more than 10% mortality. If the variance in mortality among experiments run on different days is high, the analyses for overdispersion (see Chapter 7) will show whether to pool the data together or to treat them as separate experiments. Observe the 95% confidence limits of the LC_{50}. If the 95% confidence interval is more than twice the value of the LC_{50}, the bioassay precision is questionable.

For potency bioassays, use the LC_{50} records in the potency formula (Dulmage *et al.*, 1971) as follows:

$$\frac{LC_{50}\ \text{standard}}{LC_{50}\ \text{sample}} \times \text{IU mg}^{-1}\ \text{standard} = \text{sample IU mg}^{-1}.$$

Bioassays of *B. thuringiensis* Against Coleopterous Insects

The *B. thuringiensis* subsp. *tenebrionis* (Krieg *et al.*, 1983) and subsp. *san diego* (Herrnstadt *et al.*, 1986) are active against coleopteran insects belonging mostly to Chrysomelidae species, of which the CPB, *Leptinotarsa decemlineata*, is the most important target insect for these microbes. Lower susceptibilities were recorded in other beetle species (Keller and Langenbruch, 1993). LC_{50} determinations in the CPB were made with leaf disc bioassays based on spore count of the microbial strain using a reference standard preparation (Keller and Langenbruch, 1993). Potency bioassays were developed for the CPB based on a standard *B. thuringiensis* subsp. *san diego* with an assigned potency of 50,000 CPB international units (CPB IU mg^{-1}) (Ferro and Gelernter, 1989). In these bioassays the full insecticidal power (i.e. the spore–crystal mixture) of this microbe was determined.

In plant bioassays of *B. thuringiensis* subsp. *san diego* against the CPB, terminal shoots of potato plants grown in the greenhouse are inserted into plastic reservoirs containing a nutrient solution. They are sprayed to runoff with a dilution series of commercial preparation of this microbe. Four 2nd-instar larvae of CPB per plant shoot are used. Cumulative mortality is recorded at 24-h intervals (Zehnder and Gelernter, 1989) and time–mortality calculated (see Chapter 7).

Concluding Remarks

The bioassays of *B. thuringiensis* have been steadily modified and new bioassays have been designed since the 1960s. The need to modify official bioassays and develop new ones derives from the discovery of new strains, the development of conventional and genetically manipulated *B. thuringiensis* formulations, the limited knowledge of environmental impact of the microbe and the necessity to study and monitor insect resistance to the microbe. For example, the mortality records (LC_{50}) used to determine the potency and activity of the microbe, the time–mortality (LT_{50}) and effective concentration (EC_{50}) became useful and accurate parameters in bioassay protocols. Standardized dietary bioassays and diverse leaf/plant bioassay procedures have been made more useful. In tritrophic interactions of *B. thuringiensis*, the microbe bioasssays are combined with compatibility assays of plant allelochemicals or interactions with the insect host and its larval parasitoid(s). The bioassays for monitoring insect resistance to the microbe have been made more efficient and practical. In the foreseeable future, it is expected that the need for new or improved bioassays as useful tools for developing rational control strategies with *B. thuringiensis* will continue.

References

Abbott Laboratories (1992) *B.t. Products Manual.* Abbott Publications.

Angus, T.A. (1954) A bacterial toxin paralyzing silkworm larvae. *Nature* 173, 545.

Baum, J.A., Timothy, B.J. and Carlton, B.C. (1999) *Bacillus thuringiensis.* Natural and recombinant bioinsecticide products. In: Hall, F.R. and Menn, J.J. (eds) *Biopesticide Use and Delivery.* Humana Press, New Jersey.

Beegle, C.C. (1989) Bioassay methods for quantification of *Bacillus thuringiensis* δ-endotoxin. In: Hickle, L.A. and Fitch, W.L. (eds) *Analytical Chemistry of* Bacillus thuringiensis. *ACS Symp. Series* 432, 14–21.

Blumberg, D., Navon, A., Keren, S., Goldenberg, S. and Ferkovich, S.M. (1997) Interactions among *Helicoverpa armigera* (Lepidoptera: Noctuidae), its larval endoparasitoid *Microplitis croceipes* (Hymenoptera: Braconidae), and *Bacillus thuringiensis. Journal of Economic Entomology* 90, 1181–1186.

Bonnefoi, A., Burgerjon, A. and Grison, P. (1958) Titrage biologique des preparations de spores de *Bacillus thuringiensis. Comptes Rendus de l'Académie des Sciences* 247, 1418–1420.

Burgerjon, A. (1959) Titrage et definition d'une unite biologique pour les preparations de *Bacillus thuringiensis* Berliner. *Entomophaga* 4, 201–206.

Burges, H.D. (1967) The standardization of products based on *Bacillus thuringiensis. Proceedings of the International Colloquium on Insect Pathology and Microbial Control,* Wageningen, The Netherlands, pp. 306–308.

Burges, H.D. and Jones, K.A. (1999) Formulation of bacteria, viruses and protozoa. In: Burges, H.D. (ed.) *Formulation of Microbial Biopesticides. Beneficial Microorganisms, Nematodes and Seed Treatments.* Kluwer Academic Publishers, The Netherlands.

Burges, H.D., Thomson, E.M. and Latchford, R.E. (1976) Importance of spores and δ-endotoxin protein crystals of *Bacillus thuringiensis* in *Galleria mellonella. Journal of Invertebrate Pathology* 27, 87–94.

Cannon, R.J.C. (1993) *Bacillus thuringiensis* use in agriculture: a molecular perspective. *Biological Reviews* 71, 561–636.

Carlton, B.C., Gawrone-Burke, C. and Johnson, T.B. (1990) Exploiting the genetic diversity of *Bacillus thuringiensis* for the creation of new bioinsecticides. In: *Proceedings of the Fifth International Colloquium on Invertebrate Pathology and Microbial Control*, Society for Invertebrate Pathology, Adelaide, Australia, pp. 18–22.

Dulmage, H.T. and Martinez, E. (1973) The effects of continuous exposure to low concentrations of endotoxin of *Bacillus thuringiensis* on the development of the tobacco budworm *Heliothis virescens. Journal of Invertebrate Pathology* 22, 14–22.

Dulmage, H.T., Boening, O.P., Rehnborg, C.S. and Hansen, G.D. (1971) A proposed standardized bioassay for formulations of *Bacillus thuringiensis* based on an international unit. *Journal of Invertebrate Pathology* 18, 240–245.

Dulmage, H.T., Graham, H.M. and Martinez, E. (1978) Interactions between the tobacco budworm, *Heliothis virescens,* and the δ-endotoxin produced by the HD-1 isolate of *Bacillus thuringiensis* var. *kurstaki:* Relationship between length of exposure to the toxin and survival. *Journal of Invertebrate Pathology* 32, 40–50.

Dulmage and Cooperators (1981) Insecticidal activity of isolates of *Bacillus thuringiensis* and their potential for pest control. In: Burges, H.D. (ed.) *Microbial Control of Pests and Plant Diseases 1970–1980.* Academic Press, London, pp. 193–222.

Dunkle, R.L. and Shasha, B.S. (1988) Starch-encapsulated *Bacillus thuringiensis:* a new method for increasing environmental stability of entomopathogens. *Environmental Entomology* 17, 120–126.

Ferro, D.N. and Gelernter, W.D. (1989) Toxicity of a new strain of *Bacillus thuringiensis* to Colorado potato beetle. *Journal of Economic Entomology* 82, 750–755.

Gelernter, W. and Schwab, G.E. (1993) Transgenic bacteria, viruses, algae and other microorganisms as *Bacillus thuringiensis* toxin delivery systems. In: Entwistle, P.F., Cory, J.S., Bailey, M.J. and Higgs, S. (eds) Bacillus thuringiensis, *an Environmental Biopesticide: Theory and Practice.* John Wiley & Sons, Chichester, pp. 89–124.

Hannay, C.L. (1953) Crystalline inclusion in aerobic spore-forming bacteria. *Nature* 172, 1004.

Heimpel, A.M. and Angus, T.A. (1959) The site of action of crystalliferous bacteria in Lepidoptera Larva. *Journal of Insect Pathology* 1, 152–170.

Herrnstadt, C., Soares, G.G., Wilcox, E.R. and Edwards, D.L. (1986) A new strain of *Bacillus thuringiensis* with activity against coleopteran insects. *Bio/Technology* 4, 305–308.

Keller, B. and Langenbruch, G.A. (1993) Control of coleopteran pests by *Bacillus thuringiensis.* In: Entwistle, P.F., Cory, J.S., Bailey, M.J. and Higgs, S. (eds)

Bacillus thuringiensis, *an Environmental Biopesticide: Theory and Practice.* John Wiley & Sons, Chichester, pp. 171–191.

Krieg, A., Huger, A.M., Langenbruch, G.A. and Schnetter, W. (1983) *Bacillus thuringiensis* var. *tenebrionis*: ein neuer gegenüber Larven von Coleopteren wirksamer Pathotyp. *Zeitschrift für Angewandte Entomologie* 96, 500–508.

Levinson, H.Z. and Navon, A. (1969) Ascorbic acid and unsaturated fatty acids in the nutrition of the Egyptian cotton leafworm, *Prodenia litura. Journal of Insect Physiology* 15, 591–595.

Li, R.S., Jarrett, P. and Burges, H.D. (1987) Importance of spores, crystals and δ-endo-toxins in the pathogenicity of different varieties of *Bacillus thuringiensis* in *Galleria mellonella* and *Pieris brassicae. Journal of Invertebrate Pathology* 50, 277–284.

MacIntosh, S.C., Stone, T.B., Sims, S.R., Hunst, P.L., Greenplate, J.T., Marrone, P.G., Perlak, F.J., Fischhoff, D.A. and Fuchs, R.L. (1990) Specificity and efficacy of purified *Bacillus thuringiensis* proteins against agronomically important insects. *Journal of Invertebrate Pathology* 56, 258–266.

McGuire, M.R., Galan-Wong, L.J. and Tamez-Guerra, P. (1997) Bacteria: Bioassay of *Bacillus thuringiensis* against lepidopteran larvae. In: Lacey, L.A. (ed.) *Manual of Techniques in Insect Pathology.* Academic Press, London, pp. 91–99.

Mechalas, B.J. and Anderson, N.B. (1964) Bioassay of *Bacillus thuringiensis* Berliner-based microbial insecticides II. Standardization. *Journal of Insect Pathology* 6, 218–224.

Menn, J.J. (1960) Bioassay of a microbial insecticide containing spores of *Bacillus thuringiensis. Journal of Insect Pathology* 2, 134–138.

Moar, W.J., Trumble, J.T. and Federici, B.A. (1989) Comparative toxicity of spores and crystals from the NRD-12 and HD-1 strains of *Bacillus thuringiensis* subsp. *kurstaki* to neonate beet armyworm (Lepidoptera: Noctuidae). *Journal of Economic Entomology* 82, 1593–1603.

Moar, W.J., Bosch, D., Frutos, C.R., Luo, K. and Adang, M.J. (1995) Laboratory devel-opment of resistance to *Bacillus thuringiensis* by beet armyworm, *Spodoptera exigua*: implications for the field. *Proceedings of the Beltwide Cotton Conference* Vol. II, San Antonio, Texas, pp. 881–885.

Müller-Cohn, J., Chaufaux, J., Buisson, C., Gilois, N., Sanchis, V. and Lereclus, D. (1996) *Spodoptera littoralis* (Lepidoptera: Noctuidae) resistance to CryIC and cross-resistance to other *Bacillus thuringiensis* crystal toxins. *Journal of Economic Entomology* 89, 791–797.

Navon, A. (1993) Control of lepidopteran pests with *Bacillus thuringiensis.* In: Entwistle, P.F., Cory, J.S., Bailey, M.J. and Higgs, S. (eds) Bacillus thuringiensis, *an Environmental Biopesticide: Theory and Practice.* John Wiley & Sons, Chichester, pp. 125–146.

Navon, A. and Gelernter, W. (1996) Proposals for addressing standardization issues. *Abstracts of the SIP 29th Annual Meeting and IIIrd International Symposium on* Bacillus thuringiensis, Cordoba, Spain, p. 29.

Navon, A., Wysoki, M. and Keren, S. (1983) Potency and effect of *Bacillus thuringiensis* preparations against larvae of *Spodoptera littoralis* and *Boarmia (Ascotis) selenaria. Phytoparasitica* 11, 3–11.

Navon, A., Meisner, J. and Ascher, K.R.S. (1987) Feeding stimulant mixtures for *Spodoptera littoralis* (Lepidoptera: Noctuidae). *Journal of Economic Entomology* 80, 990–993.

Navon, A., Zur, M. and Arcan, L. (1988) Effects of cotton leaf surface alkalinity on feeding of *Spodoptera littoralis* larvae. *Journal of Chemical Ecology* 14, 839–844.

Navon, A., Klein, M. and Braun, S. (1990) *Bacillus thuringiensis* potency bioassays against *Heliothis armigera, Earias insulana,* and *Spodoptera littoralis* larvae based on standardized diets. *Journal of Invertebrate Pathology* 55, 387–393.

Navon, A., Federici, B.A., Walsh, T.S. and Peiper, U.M. (1992) Mandibular adduction force of *Heliothis virescens* (Lepidoptera: Noctuidae) larvae fed the insecticidal crystals of *Bacillus thuringiensis. Journal of Economic Entomology* 85, 2138–2143.

Navon, A., Hare, J.D. and Federici, B.A. (1993) Interactions among *Heliothis virescens* larvae, cotton condensed tannin and the CryIA(c) endotoxin of *Bacillus thuringiensis. Journal of Chemical Ecology* 19, 2485–2499.

Navon, A., Keren, S., Levski, S., Grinstein, A. and Riven, J. (1997) Granular feeding baits based on *Bacillus thuringiensis* products for the control of lepidopterous pests. *Phytoparasitica* 25 (suppl.), 101S–110S.

Perez, C.J., Tang, J.D. and Shelton, A.M. (1997) Comparison of leaf-dip and diet bioassays for monitoring *Bacillus thuringiensis* resistance in field population of diamondback moth (Lepidoptera: Plutellidae). *Journal of Economic Entomology* 90, 94–101.

Plimmer, J.R. (1999) Analysis, monitoring and some regulatory implications. In: Hall, F.R. and Menn, J.J. (eds) *Biopesticide Use and Delivery.* Humana Press, New Jersey.

Salama, H.S., Foda, M.S., Sharaby, A., Matter, M. and Khalafallah, M. (1981) Development of some sublethal levels of endotoxins of *Bacillus thuringiensis* for different periods. *Journal of Invertebrate Pathology* 38, 220–229.

Sebesta, K., Farkas, J., Horska, K. and Vankova, J. (1981) Thuringiensin, the beta-exotoxin of *Bacillus thuringiensis.* In: Burges, H.D. (ed.) *Microbial Control of Pests and Plant Diseases 1970–1980.* Academic Press, London, pp. 249–282.

Singh, P. and Moore, R.F. (1985) *Handbook of Insect Rearing,* Vol. II. Elsevier, Amsterdam.

Tabashnik, B.E., Cushing, N.L., Finson, N. and Johnson, M.W. (1990) Field development of resistance to *Bacillus thuringiensis* in diamondback moth (Lepidoptera: Plutellidae). *Journal of Economic Entomology* 83, 1671–1676.

Tabashnik, B.E., Finson, N., Chilcutt, C.F., Cushing, N.L. and Johnson, M.W. (1993) Increasing efficiency of bioassays: evaluating resistance to *Bacillus thuringiensis* in diamondback moth (Lepidoptera: Plutellidae). *Journal of Economic Entomology* 86, 635–644.

Zehnder, G.H. and Gelernter, W.D. (1989) Activity of the M-ONE formulation of a new strain of *Bacillus thuringiensis* against the Colorado potato beetle (Coleoptera: Chrysomelidae): relationship between susceptibility and insect life stage. *Journal of Economic Entomology* 82, 756–761.

1B. Bioassays of Genetically Engineered *Bacillus thuringiensis* Plant Products

S.R. Sims

Whitmire Micro-Gen, St Louis, Missouri, USA

Introduction

Plant transformation has evolved from *Agrobacterium*-mediated transformation of *Nicotiana tabacum* (tobacco) protoplasts in the mid-1980s, to a revolutionary technology with essentially limitless potential for the improvement of global agriculture. Commercialized transgenic plants in the United States have quickly captured a large percentage of potential acreage. In 1998, plants containing genes coding for production of insecticidal proteins from *Bacillus thuringiensis* were planted on 2.6 M acres of the total 13.8 M cotton acres and on more than 10 M of the total 60 M maize acres. *B. thuringiensis* crops protect against insect damage and often provide significantly greater yields compared with non-transgenic crops cultivated using conventional pest management approaches. Assays using living insects play an important role in the research, development and registration of transgenic plant varieties. This section reviews some of the procedures that were instrumental in commercialization of the initial group of insect-resistant plant products (cotton, maize and potato) from the Monsanto Co., St Louis, Missouri. Although the assays described in this section were developed using the Lepidoptera-active CryIAb, CryIAc and CryIIA proteins and the Coleoptera-active CryIIIA protein, they have wide applicability to many other *B. thuringiensis* insecticidal proteins. Plant assays appropriate for evaluating the biological activity of transgenic plants containing insecticidal proteins from *B. thuringiensis* do not differ substantially from techniques used to study the insect resistance of non-transgenic crop plants (see Smith *et al.*, 1994). Bioassays of transgenic plants are most frequently used as a method to screen among many transformation 'events' to select for those which

© CAB *International* 2000. *Bioassays of Entomopathogenic Microbes and Nematodes* (eds A. Navon and K.R.S. Ascher)

produce the highest levels of protein and show the greatest potential for field evaluation. Bioassays sometimes serve as a back-up screening technique for enzyme-linked immunosorbent assay (ELISA) procedures which are capable of accurately determining low concentrations of transgenic proteins present in plant tissue matrices (Sims and Berberich, 1996). However, when ELISA techniques have not been refined, bioassay is of primary importance.

Purified insecticidal proteins

An adequate supply of purified protein is essential for conducting the safety assessment studies required for transgenic plant product registration. Because it is difficult to purify large amounts of protein from plant tissues, production of insecticidal protein is typically done using recombinant *Escherichia coli* and microbial fermentation techniques. Protein is purified from crude refractile bodies of *E. coli* using non-chromatographic methods and lyophilized. Purity of the final lyophilized preparation is determined by BCA (bicinchoninic acid) assay and ELISA and the protein is tested for equivalence to the protein produced in transgenic plants. If similar in length (amino acids), size (determined by SDS–PAGE) and amino acid sequence to the plant protein, the *E. coli* purified protein is considered to be 'substantially equivalent' and qualifies for use in registration studies and other tests.

Production of transgenic plants and *B. thuringiensis* proteins in plant tissues

The major crop plant species (maize, rice, cotton, potato) that produce *B. thuringiensis* proteins have been transformed by one of two techniques. One method uses 'disarmed' *Agrobacterium tumefaciens* bacteria to transfer foreign DNA into host cells. The other technique, especially useful with monocotyledonous plants, is free DNA transfer to plant protoplasts using microprojectile bombardment (see Hinchee *et al.*, 1994). The level of expression, and plant tissue location, of the transgenic proteins is dependent on a component of the transgenic gene construct known as the promoter sequence. If the promoter is 'constitutive', such as the commonly used 35S promoter from cauliflower mosaic virus (CaMV), then transgenic protein production occurs in most plant tissues. If the promoter is 'tissue-specific', then transgenic protein expression can be limited to areas such as reproductive tissues or leaves and stems. The rate of transgenic protein production over the developmental stages of a plant is usually related to overall protein production. Thus, tissues from vegetative stage plants typically contain the highest protein concentrations, whereas transgenic protein levels decline in reproductive and later stages of plant development.

Types of Bioassays

Leaf assays

Most initial screening assays of maize, potatoes and cotton are no-choice tests that use leaves, or leaf sections, from young plants which typically have the highest concentration of *B. thuringiensis* proteins in leaf tissues. Leaves are detached, placed on top of moist filter paper in Petri dishes and infested with one or more neonate larvae of the appropriate test species. An alternative approach is to cut out circular leaf discs with a cork borer and place the discs on top of agar, containing antimicrobials, within individual wells of microtitre assay plates. Neonate larvae are added and the wells are covered with plastic film (Mylar®, E.I. DuPont de Nemours and Co., Wilmington, Delaware), heat sealed using a tacking iron and ventilated using pin holes. In leaf assays, larval growth (weight, instar) and survivorship is measured after 2–4 days; damage to the leaf tissue is most easily scored using an index comparing transgenic with control leaf tissue. If transgenic events cannot be ranked using these assays (for example when all larvae die and/or there is little damage to leaves), other approaches become necessary. Later instar larvae of the test species are relatively less susceptible to transgenic proteins and can be used instead of neonates. Also, species with different susceptibility can be used in the assay. For example, neonate larvae of *Ostrinia nubilalis* (Hübner), *Helicoverpa zea* (Boddie) and *Spodoptera frugiperda* (J.E. Smith) can be used to study the insecticidal protein levels in transgenic maize events while *Pectinophora gossypiella* (Saunders), *H. zea* and *Heliothis virescens* (F.) are useful for evaluating cotton.

Pollen assays

Evaluating expression of *B. thuringiensis* protein in pollen is important for ecological risk assessment because many non-target beneficial insects use pollen as a food source. Also, pollen is critical for larval growth of some pest species. The presence of *B. thuringiensis* protein in maize pollen, for example, is an important aspect of plant resistance to second-generation *O. nubilalis* since neonate larvae often develop exclusively on pollen prior to attaining a size large enough to penetrate sheath collar tissue. The procedure described below for *B. thuringiensis* protein expression in maize pollen illustrates one bioassay approach. Pollen was collected from maize plants during the first 7 days of pollen shed. To prevent protein degradation, pollen was stored in tightly capped vials at $-80°C$ until tested. Microtitre plates (24-well) with agar + mould inhibitor (Diet 1 in Table 1B.1) were used. Each well received 50–100 µg of pollen and two neonate *O. nubilalis* larvae. Completed wells were sealed with Mylar film and ventilated with pin holes. Test duration was 4 days at 26°C. Larval survivorship and stunting were scored. At 4 days, there was ≥ 85% survival on the control pollen and ≥ 85% of surviving larvae were second instar (Sims, unpublished data).

Root assays

Considerable research effort is being made to engineer crops with resistance to root-feeding insects and nematodes. The 'Mount Everest' of transgenic maize research is to develop plants resistant to corn rootworm larvae (*Diabrotica* sp.). Ortman and Branson (1976) described a hydroponic growth pouch method for studying maize resistance to larvae of *Diabrotica virgifera virgifera* LeConte and *Diabrotica undecimpunctata howardi* Barber. The growth pouches they used were clear plastic with an absorbent paper wick inside that is folded near the top to form a perforated trough. The wick provides support for one pregerminated maize seedling that is infested with rootworm larvae after about 10 cm of root growth. Another method for studying root effects on *Diabrotica* larvae is to use a plastic plant tissue culture tray (Phytatray®, Sigma Chemical Co., St Louis, Missouri) with a layer of 2.5% agar (~ 6 mm deep) on the bottom covered with sterile filter paper. Maize is grown in 7.6-cm diameter pots for 7–8 days until foliage height is approximately 7.6 cm. Then the soil is gently washed off the roots using running water and individual plants are placed on top of the filter paper/agar base. Ten neonate *D. undecimpunctata howardi* larvae (~ 24 h old) are added and the top is replaced. Test duration is 7 days at L:D (light : dark) 14:10, 26°C, and 70% RH (relative humidity). Under these conditions, control mortality is low (≤ 10%) and larvae feed only on the roots (S.R. Sims, unpublished data). A soil assay useful for root resistance evaluation uses 7.6-cm diameter plastic pots with drainage holes covered with polyurethane foam and commercial potting soil such as MetroMix® (The Scotts Company, Marysville, Ohio). Plants are grown as for the root assay described above. At 7.6–10.1 cm of leaf growth, the soil at the base of the maize is lightly 'cultivated' using a metal spatula and the soil is inoculated with 10 neonate larvae (24–48 h old) of *D. undecimpunctata howardi* pre-fed on the roots of non-transgenic germinated maize seeds. The test is scored after 7 days by washing the soil off the roots into two stacked US Standard sieves, No. 40 (425 μm opening) on the bottom and No. 20 (850 μm opening) on top. After washing most soil through the sieves, living rootworm larvae are floated out of the remaining soil and roots by immersing the sieves into a tray of saturated magnesium sulphate solution (Sims, unpublished data).

Callus assays

Refined plant transformation techniques can produce thousands of transformed cell lines requiring considerable effort for callus plate maintenance and plant regeneration. Bioassay can be used to screen for, and eliminate, events with low levels of insecticidal protein. This is extremely useful since in most crops, such as maize, callus events producing low levels of insecticidal protein rarely develop into plants with high protein expression (S.R. Sims, unpublished data). The precedent for callus bioassay comes from work on non-transgenic plants. Williams *et al.* (1983, 1985) demonstrated that callus derived from insect-resistant maize genotypes was also resistant. Callus can be evaluated by either direct infestation with insects or diet incorporation (described below). Callus (0.2–0.6 g) can be tested within wells of 24-well microtitre plates. Because callus is largely (> 80%) water, preblotting callus on sterile filter paper and placement of the callus on top of a sterile filter paper disc, such as a 1/2″ Difco Concentration Blank, reduces excess free moisture. Transfer of callus to wells using sterile technique and covering the bottom of wells with antimicrobial agar (propionic + phosphoric acid at 3.0 ml l^{-1}) minimizes microbial contamination. One or more neonate larvae are added to each well, which is then sealed with Mylar and ventilated with pin holes. Test duration is 4–7 days at 23–28°C after which larval survivorship and size/weight relative to controls are recorded. Because of its extreme sensitivity to CryIAb protein, *Manduca sexta* (L.) is useful for assay of transgenic maize callus to verify transformation and protein expression. However, *M. sexta* sensitivity is a drawback when an attempt is made to prioritize or rank the relative potency of different transgenic constructs (i.e. different genes, promoters, coding sequences, etc.) or prioritize the relative potency of different transgenic events sharing the same construct. Although *O. nubilalis* is less sensitive to CryIAb protein than *M. sexta* (MacIntosh *et al.*, 1990), many *B. thuringiensis* gene constructs in maize callus produce 100% mortality of neonate *O. nubilalis* larvae. Such constructs are difficult to rank regarding relative insecticidal activity. As with leaf assays, one solution is to use other species of Lepidoptera that will consume maize callus but are less sensitive to the insecticidal proteins used in the constructs. *H. zea, Diatraea grandiosella* (Dyar) and *S. frugiperda* are extremely useful in this regard since all survive and develop normally on maize callus and moult to second instars within 5 days at 26°C on non-transgenic callus (S.R. Sims, unpublished data). Potent *B. thuringiensis* maize constructs can be evaluated with later instar larvae of *O. nubilalis* since they are less sensitive to insecticidal proteins than neonates. A growth inhibition assay uses third-instar *O. nubilalis* larvae (4–8 mg) fed for 7 days on rootworm diet (Marrone *et al.*, 1985). A sample of at least 12 larvae is weighed, added to callus from one line (one larva per well), allowed to feed for 4 days, and then reweighed. The percentage growth inhibition (= growth reduction) of treatment groups compared with non-transgenic controls can be calculated using the following equation:

$$\% \text{ growth inhibition} = 1.0 - [(F_t \div I_t) \div (F_c \div I_c)]$$

where F = final mean weight, I = initial mean weight, t = treatment, c = control.

Diet Incorporation of Insecticidal Proteins, Transgenic Callus and Plant Material

Diet incorporation procedures are essential for quantifying the biological activity of transgenic proteins and plant materials. These procedures include surface overlay assays in which a thin layer of protein or plant material in aqueous solution or gel is applied to a diet surface and allowed to dry before being infested with test insects. Transgenic proteins can be success-fully added to virtually any insect diet. For dry diets, protein and diet can be mixed with water to form a slurry, then quickly frozen on dry ice and lyophilized (Sims and Martin, 1997). The agar-based diets described below are useful for testing many insect species that are targets for control using transgenic plants.

Assay diets

The special considerations that need to be addressed for successful diet incor-poration of transgenic material centre around the heat and chemical lability of insecticidal proteins. Agar with a low gelling point (preferably lower than 45°C) is required to avoid denaturation of heat-sensitive proteins. Serva agar (Boehringer Ingelheim Bioproducts, Heidelberg, Germany) is an excellent choice for protein incorporation although it is considerably more expensive than standard agar products. Antimicrobials commonly used for insect diets should be checked for their effect(s) on the proteins studied. Both formalde-hyde and a phosphoric/propionic acid blend (see diet recipes) can be used with CryIAb, CryIAc, CryIIA and CryIIIA proteins with no obvious effect on bioactivity. It may be expedient to substitute commercially available insect diets for the diets listed here. Commercially available diets, such as the tobacco budworm diet from Bio-Serv, Inc. (Frenchtown, New Jersey), should be prepared according to manufacturer's instructions, with low-temperature gelling agar being substituted for regular agar. It may also be necessary to increase the total water added to achieve optimal diet viscosity.

Basic insect diets used for bioassay work

Three diets are described (diets 1, 2 and 3 in Tables 1B.1–1B.3) which have proved to be especially useful for research on transgenic plants. These diets are prepared as follows:

- Add agar to approximately 60% of the total water (540 ml water for the Lepidoptera and Colorado potato beetle diets, 600 ml for southern corn rootworm diet) in a Fernbach culture flask and agitate until mixed. Reserve remaining water in a separate container as cooling water. A

Table 1B.1. Diet 1: soybean flour–wheatgerm diet.

Ingredient	Amount (g)
Soybean flour (Nutrisoy Flour no. 40)	48.5
Wheatgerm	41.4
Wesson salt	11.8
Sugar	48.5
Vitamin mix	11.2
Methyl paraben	1.18
Aureomycin (6% soluble powder)	1.18
Sorbic acid	1.18
Mould inhibitor blend*	3 ml
Agar (Serva agar, Feinbiochemica GmbH and Co.)	14
Water, distilled	930 ml

*420 ml propionic acid + 45 ml phosphoric acid (85%) + 535 ml distilled water.
Source: King and Hartley, 1992.

Table 1B.2. Diet 2: southern corn rootworm diet.

Ingredient	Amount (g)
Wheatgerm (Bio-Serv, Inc., Cat. no. 1661)	41.0
Casein (Bio-Serv, Inc., Cat. no. 1100)	48.0
Sucrose (Bio-Serv, Inc., Cat. no. 3900)	57.5
Wesson salt mix (U.S. Bio-Chem., Cleveland, Ohio, no. 21420)	14.0
Alphacel (ICN Biomedicals, Costa Mesa, California, no. 900453)	20.5
Cholesterol (Bio-Serv, Inc., Cat. no. 5180)	0.10
Methyl parasept (Kalama Chemical, Piscataway, New Jersey)	1.50
Sorbic acid (Bio-Serv, Inc., Cat. no. 6967)	1.0
Streptomycin sulphate (ICN Biomedicals, Inc., Cat. no. 100556)	0.2
Chlortetracycline (ICN Biomedicals, Inc., Cat. no. 190327)	0.2
Vanderzant vitamins (ICN Biomedicals, Inc., Cat. no. 903244)	13.5
Linseed oil (Bio-Serv, Inc., Cat. no. 5680)	1.5 ml
Potassium hydroxide (10%) (Fisher, no. P250)	12.5 ml
Mould inhibitor blend*	1.5 ml
Water, distilled	1000 ml
Agar (Serva agar, Feinbiochemica GmbH and Co.)	21.5

*420 ml propionic acid + 45 ml phosphoric acid (85%) + 535 ml distilled water.
Source: Marrone *et al.*, 1985.

Fernbach flask with large volume helps to prevent 'boil overs' when agar is heated. (Note: the amount of agar used may be increased or decreased by at least 20% without significantly affecting the utility of the final mixture as a test diet.)

- Microwave water–agar mixture for 10 min on high setting or until agar is completely dissolved and mix has come to a boil. Pour hot agar into a Waring blender or equivalent.

Table 1B.3. Diet 3: Colorado potato beetle assay diet.

Ingredient	Amount (g)
Colorado Potato Beetle Diet (Bio-Serv, Inc., Frenchtown, New Jersey, no. F9380)	147.5
Potato flakes (any brand, powdered)	10.0
Methyl parasept (Kalama Chemical, Piscataway, New Jersey)	1.0
Sorbic acid (Bio-Serv, Inc., Cat. no. 6967)	1.0
Mould inhibitor blend*	2.5 ml
Water, distilled	930 ml
Agar (Serva agar, Feinbiochemica GmbH and Co.)	13

*420 ml propionic acid + 45 ml phosphoric acid (85%) + 535 ml distilled water.

- Add all the cooling water and blend for approximately 15 s. If a variable transformer is used to regulate voltage, set blender on high and transformer at approximately 30% of maximum.
- Add all dry and liquid diet ingredients and blend for at least 2 min at 40–75% transformer output voltage.
- Pour blended diet into 500-ml polyethylene squeeze bottles and cap with lids having spouts trimmed to 5–10 mm long.
- Set bottles of warm liquid diet into water bath preheated to 48–54°C.

Addition of insect diet to test samples

- Introduce test sample (protein or plant material) into an appropriate mixing container such as a 50-ml disposable centrifuge tube.
- Add warm liquid insect diet to sample in a ratio of 4:1 (example: 24 ml insect diet is added to a 6 ml solution or suspension of the test sample or 32 ml diet to an 8 ml test sample).
- Cap centrifuge tube containing insect diet + sample and blend for at least 15–20 s using a vortex mixer (battery-powered hand-held mixers can also be used).
- Pour small subsamples (0.75–2.0 ml) of treated diet into individual wells of a multiwell assay tray.
- To prepare larger (≥ 40 ml) samples, add diet to sample (4:1 ratio) in blender, blend, then pour treated diet directly into 32-, 50- or 96-well trays. Smaller capacity (< 1 litre) blenders work better with small test samples. Wash blender thoroughly with hot soapy water and rinse between treatments.
- Allow at least 10–20 min drying/hardening time for treated diet under sterile airflow hood.
- Cover all wells containing treated diet with Mylar film using a tacking iron.

Addition of insects to test sample

- Peel Mylar covering off sample wells to be inoculated with test insects.
- Transfer one neonate larva, of appropriate species, to each well using a moistened camel-hair brush or equivalent. Wells can also be infested by pipetting ova suspended in agar or xanthan gum solution. After inoculation of ova, the liquid carrier solution is evaporated under a sterile hood.
- Reseal Mylar with tacking iron. Avoid touching iron to the Mylar directly over the treated diet because static electricity often causes larvae to cling to the Mylar where they are killed by contact with the heated iron.
- Use hypodermic needle or insect pin to poke at least one ventilation hole in the Mylar above each well.
- Incubate completed assay tray at 20–30°C under an L:D 14:10 photoperiod.
- Score number of surviving larvae, at each dose, after 4–8 days. If wells have been inoculated with ova, then score the number of wells containing one or more living larva.
- For estimation of EC_{50} (treatment concentration reducing larval weight to 50% that of control), weigh individuals or replicate groups of surviving larvae and calculate mean weight of the larvae in each treatment.

Growth inhibition assays

Mortality assays are commonly used to calculate the LC_{50} or LD_{50} values for insecticidal proteins. However, it is often necessary to quantify concentrations of purified or transgenic insecticidal proteins that are too low to study using dose–mortality response. A larval growth inhibition assay is useful for evaluation of these samples. For example, extremely low concentrations (< 1 ng protein ml^{-1} of diet) of CryIAb and CryIAc proteins, purified and in a plant tissue matrix, were quantified using a *H. virescens* larval growth inhibition assay (Sims and Berberich, 1996). In this study, bioassay and ELISA estimates of protein concentration were similar. In other situations, insect bioassay can be superior to an ELISA because bioassay is less affected by suboptimal protein extraction efficiency or interference effects from other chemicals in plant tissues (Sims and Berberich, 1996). A growth inhibition bioassay can also be used to measure quantitatively levels of active insecticidal proteins in plant–soil matrices for soil degradation studies (Sims and Holden, 1996; Sims and Ream, 1997). Bioassay requires that all doses have an equal amount of plant tissue (Sims and Berberich, 1996). The addition of appropriate negative control tissue, such as tissue from non-transgenic isolines, controls for stunting effects from plant metabolites and secondary plant compounds such as the terpenoid aldehydes in cotton (Lukefahr and Martin, 1966). The possibility of interaction effects (synergy, antagonism) between insecticidal proteins and the control plant matrix should always be investigated by appropriate 'spiking' experiments using purified protein. For plants and tissues lacking a validated, optimized ELISA, bioassay is a powerful tool for detecting minute amounts of active CryIA proteins. *H. virescens* is a useful species for growth assays to test many insecticidal proteins affecting Lepidoptera because of the large difference between LC_{50} and EC_{50} values. However, larvae of many, if not most, other susceptible species of Lepidoptera show similar growth inhibition in response to selected insecticidal proteins (Sims, 1997). Larval growth inhibition studies can be designed so that larvae are either weighed individually or in replicated groups. The group design does not provide an estimate of weight variation among individual larvae but it minimizes work and is easier when the test larvae are small. A starting cohort of ≥ 24 larvae on test diet usually provides adequate survivors (≥ 20) for final weighing. At least three groups of larvae should be tested per day and the assay should be replicated two or three times on separate days. Optimum duration of the growth assay depends on the test species and temperature. For *H. virescens*, assays are typically scored after 6 or 7 days at 28°C, whereas *P. gossypiella* may require 10–14 days at a similar temperature. Larval growth response to sublethal doses of insecticidal proteins is most properly analysed using non-linear regression analysis. A logistic model that has proved to be useful for analysis is the following:

$$\text{Weight} = W_0/[(1 + (\text{conc.}/EC_{50})^B]$$

In this model: W_0 = the expected control weight, conc. = the amount of transgenic tissue (e.g. mg 30 ml^{-1} diet) or purified protein (e.g. ng ml^{-1} diet), EC_{50} = the EC_{50} expected for the transgenic tissue or pure protein, and B = the logistic function 'slope' parameter. This model establishes a single 'best fit' curve for each series of concentrations (within a treatment) and corresponding larval weight responses. The SAS non-linear regression procedure NLIN is used to fit the model to the mean total larval weight data. The output of interest includes estimates, standard errors, and confidence intervals for W_0, EC_{50}, and B.

Surface overlay assay

- Prepare diet as for diet incorporation assay except use additional 20% water.
- Pour diet into wells of multiwell rearing/assay trays. Microtitre assay trays (24- or 96-well) are especially useful.
- Use a repeating Eppendorf (or equivalent) pipette to dispense 30–50 μl (96-well tray) or 200 μl (24-well tray) of test sample on to the surface of the diet.
- Allow the test samples to dry completely before adding insects. This takes approximately 2–4 h at 20–30% RH but longer at higher RH.
- Cover all wells containing treated diet with Mylar film using a tacking iron.
- Add insects to test samples and incubate as described for diet incorporation.

- Score number of living larvae at each dose after 4–7 days. For EC_{50} determination, weigh surviving larvae from treatment replications and calculate mean larval weight per treatment.

Protoplast diet overlay

Protoplasts are single plant cells lacking cell walls. Leaf protoplast electroporation of plants such as maize is a simple, rapid and sensitive procedure for detecting gene activity and active protein production of novel gene constructs. The relatively small amount of *B. thuringiensis* protein present in cell lysate samples, however, necessitates the use of both a sensitive test insect and a sensitive test system. Bioevaluation of protoplasts involves a surface overlay procedure in which the protoplasts are layered on to insect diet in multiwell assay trays followed by careful drying under a sterile airflow hood. The sensitivity of *M. sexta* makes it useful for testing many *B. thuringiensis* proteins active against Lepidoptera species. The surface overlay technique increases the concentration of the sample eaten by *M. sexta* larvae and further enhances assay sensitivity. Assay with *M. sexta* is best done using the southern corn rootworm diet (Diet 2). A 96-well microtitre plate can be prepared by filling each well with approximately 200 µl of diet and allowing the diet to harden. Using sterile technique under a biohood, 50 µl (200 µl per well in 24-well trays) of test sample is pipetted on to the diet surface of each well. Excess sample liquid is allowed to dry under the biohood over 3–5 h. When drying is complete, one neonate *M. sexta* larva is added to each well and the wells are sealed with Mylar and ventilated with a pin hole. Test duration is 72 h at ~28°C. Survivorship and stunting of survivors are scored. To facilitate comparison among samples, the cell lysate should be titrated and the dilutions used to generate dose–response data. Alternatively, as stated previously, a less sensitive lepidopteran species can be used. The level of insecticidal protein in the sample can be determined using immunoassay (ELISA) for comparison with larval survivorship data.

Callus diet incorporation

ELISA-determined levels of *B. thuringiensis* protein in callus and insect development/survivorship results are not always consistent. Chimeric expression of protein in the callus may increase the variability of ELISA results; this effect is related to the specific *B. thuringiensis* protein evaluated and the efficiency of the ELISA (S.R. Sims, unpublished data). Older callus, while retaining significant bioactivity, can contain higher levels of phenolic and other compounds that interfere with ELISA, resulting in underestimation of active *B. thuringiensis* protein. In these situations, better results can often be obtained if the callus tissue is homogenized prior to sampling for insect bioactivity or ELISA analysis. Callus is homogenized in sterile distilled water or phosphate buffer using a small blender or a tissue homogenizer. The homogenized callus is then added to the appropriate insect diet using the procedures described above. The water content of the callus being processed should be omitted from the diet since this water is added back with sample incorporation. Diet incorporation of callus is set up, scored and analysed using larval growth inhibition procedures.

Dried plant material diet incorporation

Several key points should be observed for effective incorporation of dried transgenic plant material into insect diets. Plant material should be freeze-dried rather than air- or oven-dried to minimize denaturation of sensitive proteins. The resulting dried plant material should be finely ground so that insects cannot avoid exposure when feeding on the amended diet. Initial tissue grinding can be done in a heavy-duty Waring blender using dry ice to avoid excessive heat and prevent plant material from clumping. A mortar and pestle can be used for the final grinding process. Ground material should be stored frozen (preferably at $< -80°C$) prior to use. The maximum amount of plant material that can be incorporated is different for each insect species and plant system but generally equals the amount of non-transgenic material that does not excessively stunt the growth (or cause mortality) of the test insects. Antibiosis effects of cotton and potato plant tissues, for example, will severely stunt test insect growth. One must control for this stunting by maintaining an equal amount of plant tissue in each sample. For example, if test samples have 6, 2, 0.75, 0.25 and 0.10 g of transgenic tissue then the same tubes, respectively, must also contain 0, 4, 5.25, 5.75 and 5.90 g of non-transgenic tissue. The non-transgenic tissue should come from an isogenic line that is essentially identical to the transgenic line except for the expression of transgenic proteins. It is also important to evaluate a dilution series of pure protein 'spiked' into diet containing a standard amount of control tissue (6.0 g in the example above). The LD_{50} or EC_{50} values of these data can then be used to estimate the concentration of bioactive protein in the transgenic plant tissue. For example, if the EC_{50} of the transgenic plant sample is 300 µg ml^{-1} and the EC_{50} of the pure protein spiked into diet containing control tissue is 0.003 µg ml^{-1}, then the concentration of protein in the tissue is estimated as the ratio (0.003 µg ml^{-1}) ÷ (300 µg ml^{-1}) = 10 µg g^{-1}. Within the limits of assay sensitivity, this technique can be used to evaluate the concentration of protein in any transgenic plant part.

Bioassays for monitoring resistance of Lepidoptera to *B. thuringiensis* proteins in transgenic plants

Arguably the most controversial issue associated with transgenic crops involves the potential for insect resistance and the sufficiency of proposed pre-emptive resistance management procedures (McGaughey and Whalon, 1992; Tabashnik, 1994). A critical component of all resistance management strategies is an efficient resistance monitoring programme. Data from monitoring programmes serve as a check on the effectiveness of resistance management strategies and, more importantly, permit early detection of resistant phenotypes. Under ideal circumstances, discovery of resistance in its early stages allows remedial measures to be implemented prior to control failures (ffrench-Constant and Roush, 1990). Historically, resistance to conventional chemical and microbial insecticides has been monitored and compared using log-dose probit mortality responses of insect strains. This approach involves calculation of a resistance ratio using the LD_{50} or LC_{50} of the field test strain divided by the LD_{50} or LC_{50} of a reference susceptible strain and statistical comparison of the $LD_{50}s/LC_{50}s$ and slopes of the probit regression lines (Staetz, 1985; Robertson and Preisler, 1992). Similarly, 'baseline' susceptibility studies on insects targeted for control by transgenic plants have generated $LC_{50}s$ and slope estimates for different populations exposed to *B. thuringiensis* protein incorporated into insect diet (Stone and Sims, 1993; Siegfried *et al.*, 1995). $LC_{50}s$ and slope estimates are suitable for distinguishing resistant phenotypes at a high frequency but they are not adequately sensitive for detecting resistance when the incidence of resistance is low, for example $10^{-3}-10^{-4}$ (Roush and Miller, 1986). 'Diagnostic' doses that unambiguously discriminate between resistant and susceptible phenotypes are a more efficient means of finding resistant phenotypes because all individuals tested provide useful data (Roush and Miller, 1986; ffrench-Constant and Roush, 1990).

Growth inhibition response evaluation

Larval growth inhibition assays using diagnostic doses can be used to monitor for changes in insect susceptibility to transgenic proteins (Sims *et al.*, 1996). Since the growth response of individual larvae is important in resistance monitoring, a modification of the non-linear regression equation given previously was used to calculate EC_{99} (the concentration required to reduce larval weight to 1% that of the mean control weight) and 95% confidence interval (CI) values for *H. virescens* and *H. zea*. The equation is:

$$\text{Weight} = W_0/[(1 + (100 - 1)(\text{concentration}/EC_{99})^B]$$

where EC_{99} = the EC_{99} expected for the transgenic protein and the other parameters are as previously described for the EC_{50} calculation. First-instar larvae were tested against concentrations of CryIAc protein using the previously described growth inhibition assay. The estimated EC_{99} values and 95% CI for *H. virescens*, 0.058 µg ml^{-1} (0.030–0.086), and *H. zea*, 28.8 µg ml^{-1} (−7.4–65.1), were more than 100 times lower than the corresponding LC_{99} estimates for CryIAc protein. Although the EC_{99} calculation provides a reasonable diagnostic dose estimate, in practice the lowest concentration providing the requisite degree of larval growth inhibition is the most efficient choice. For example, the EC_{98} (6.6 µg ml^{-1}, 0.1–13.0) of *H. zea* is more practical because it provides discrimination (stunting) of all susceptible larvae at a much reduced concentration (Sims *et al.*, 1996). All healthy larvae of *H. zea* and *H. virescens* tested on control diet were 3rd–5th instars and weighed >10 mg after 7 days. Therefore, the diagnostic dose in this situation should prevent susceptible larvae from reaching 3rd instar. Due to variability in larval growth rates, this criterion would involve doses producing a mean larval weight of ≤1.0 mg. Above this weight, a significant percentage of susceptible larvae might still reach 3rd instar and the incidence of false positives would increase (Sims *et al.*, 1996). Diagnostic EC_{99} concentrations for CryIAc protein will result in some larval mortality for both *H. virescens* and *H. zea* but this does not reduce the efficiency of the growth assay because both dead and stunted larvae are classified as susceptible. If a single diagnostic dose is used for resistance monitoring, samples of susceptible populations from across the geographic range of that species should be tested to validate the dose. A multipopulation approach was used to establish discriminating doses of microbial *B. thuringiensis* products against Australian *Helicoverpa armigera* and *H. punctigera* (Forrester and Forsell, 1995).

Concluding Remarks

While mortality assays have traditionally been used to study the potency of microbial *B. thuringiensis* preparations, the relatively low concentrations of *B. thuringiensis* insecticidal protein in transgenic plants often requires the use of larval growth inhibition assays for protein detection and quantification. If an appropriate assay insect is selected, the lower limit of insecticidal protein detection can approach the sensitivity of validated immunoassays such as ELISA. Insect assays are relatively easy to set up, score and analyse, and they provide an essential tool for prioritizing transgenic plant lines for further development. Diet incorporation assays evaluating transgenic proteins are not seriously compromised by the presence of moderate levels of plant material or other contaminants such as soil, and the assays provide important tools for regulatory and environmental risk assessment studies. Perhaps of greatest importance, bioassays are essential in resistance monitoring efforts aimed at early detection of reduced insect susceptibility to the transgenic proteins in plants.

References

ffrench-Constant, R.H. and Roush, R.T. (1990) Resistance detection and documentation: the relative roles of pesticidal and biochemical assays. In: Roush, R.T. and Tabashnik, B.E. (eds) *Pesticide Resistance in Arthropods.* Chapman & Hall, New York, pp. 4–38.

Forrester, N.W. and Forsell, L. (1995) Development of discriminating dose assays for *Bacillus thuringiensis* subspecies *kurstaki* in Australian *Helicoverpa* spp. In: Feng, T-Y. *et al.* (eds) Bacillus thuringiensis *Biotechnology and Environmental Benefits*, Vol. I. Hua Shiang Yuan Publ. Co., Taipei, Taiwan, pp. 251–257.

Hinchee, M.A.W., Corbin, D.R., Armstrong, C.L., Fry, J.E., Sato, S.S., DeBoer, D.L., Petersen, W.L., Armstrong, T.A., Conner-Ward, D.V., Layton, J.G. and Horsch, R.B. (1994) Plant transformation. In: Văsil, I.K. and Thorpe, T.A. (eds) *Plant Cell and Tissue Culture.* Kluwer Academic Publishers, Dordrecht, The Netherlands, pp. 231–270.

King, E.G., Jr. and Hartley, G.G. (1992) Multiple-species insect rearing in support of research. In: Anderson, T.E. and Leppla, N.C. (eds) *Advances in Insect Rearing for Research and Pest Management.* Westview Press, Boulder, Colorado, pp. 159–172.

Lukefahr, M.J. and Martin, D.F. (1966) Cotton plant pigments as a source of resistance to the bollworm and tobacco budworm. *Journal of Economic Entomology* 59, 176–179.

MacIntosh, S.C., Stone, T.B., Sims, S.R., Hunst, P.L., Greenplate, J.T., Marrone, P.G., Perlak, F.J., Fischhoff, D.A., and Fuchs, R.L. (1990) Specificity and efficacy of purified *Bacillus thuringiensis* proteins against agronomically important insects. *Journal of Invertebrate Pathology* 56, 258–266.

Marrone, P.G., Ferri, F.D., Mosley, T.R. and Meinke, L.J. (1985) Improvements in laboratory rearing of the southern corn rootworm, *Diabrotica undecimpunctata howardi* Barber (Coleoptera: Chrysomelidae), on an artificial diet and corn. *Journal of Economic Entomology* 78, 290–293.

McGaughey, W.H. and Whalon, M.E. (1992) Managing resistance to *Bacillus thuringiensis* toxins. *Science* 258, 1451–1455.

Ortman, E.E. and Branson, T.F. (1976) Growth pouches for studies of host plant resistance to larvae of corn rootworms. *Journal of Economic Entomology* 69, 380–382.

Robertson, J.L. and Preisler, H.K. (1992) *Pesticide Bioassays With Arthropods.* CRC Press, Boca Raton, Florida.

Roush, R.T. and Miller, G.L. (1986) Considerations for design of insecticide resistance monitoring programs. *Journal of Economic Entomology* 79, 293–298.

Siegfried, B.D., Marcon, P.C., Witkowski, J.F., Wright, R.J. and Warren, G.W. (1995) Susceptibility of field populations of the European corn borer, *Ostrinia nubilalis* (Hübner) (Lepidoptera: Pyralidae), to the microbial insecticide, *Bacillus thuringiensis* Berliner. *Journal of Agricultural Entomology* 12, 257–263.

Sims, S.R. (1997) Host activity spectrum of the CryIIa *Bacillus thuringiensis* subsp. *kurstaki* protein: Effects on Lepidoptera, Diptera, and non-target arthropods. *Southwestern Entomologist* 22, 395–404.

Sims, S.R. and Berberich, S.A. (1996) *Bacillus thuringiensis* CryIA protein levels in raw and processed seed of transgenic cotton: determination using insect bioassay and ELISA. *Journal of Economic Entomology* 89, 247–251.

Sims, S.R. and Holden, L.R. (1996) Insect bioassay for determining soil degradation of *Bacillus thuringiensis* var. *kurstaki* [CryIA(b)] protein in corn tissue. *Environmental Entomology* 25, 659–664.

Sims, S.R. and Martin, J.W. (1997) Effect of the *Bacillus thuringiensis* insecticidal proteins CryIA(b), CryIA(c), CryIIA, and CryIIIA on *Folsomia candida* and *Xenylla grisea* (Insecta: Collembola). *Pedobiologia* 41, 412–416.

Sims, S.R. and Ream, J.E. (1997) Soil inactivation of the *Bacillus thuringiensis* subsp. *kurstaki* CryIIA insecticidal protein within transgenic cotton tissue: laboratory microcosm and field studies. *Journal of Agriculture and Food Chemistry* 45, 1502–1505.

Sims, S.R., Greenplate, J.T., Stone, T.B., Caprio, M.A. and Gould, F.L. (1996) Monitoring strategies for early detection of Lepidoptera resistance to *Bacillus thuringiensis* insecticidal proteins. In: Brown, T.M. (ed.) *Molecular Genetics and Evolution of Pesticide Resistance.* American Chemical Society Symposium Series Number 645, pp. 229–242.

Smith, C.M., Khan, Z.R. and Pathak, M.D. (1994) *Techniques for Evaluating Insect Resistance in Crop Plants.* CRC Press, Boca Raton, Florida.

Staetz, C.A. (1985) Susceptibility of *Heliothis virescens* (F.) (Lepidoptera: Noctuidae) to permethrin from across the cotton belt; a five year study. *Journal of Economic Entomology* 78, 505–510.

Stone, T.B. and Sims, S.R. (1993) Geographic susceptibility of *Heliothis virescens* and *Helicoverpa zea* (Lepidoptera: Noctuidae) to *Bacillus thuringiensis. Journal of Economic Entomology* 86, 989–994.

Tabashnik, B.E. (1994) Evolution of resistance to *Bacillus thuringiensis. Annual Review of Entomology* 39, 47–79.

Williams, W.P., Buckley, P.M. and Taylor, V.N. (1983) Southwestern cornborer growth on callus initiated from corn genotypes with different levels of resistance to plant damage. *Crop Science* 23, 1210–1212.

Williams, W.P., Buckley, P.M. and Davis, F.M. (1985) Larval growth and behavior of the fall armyworm on callus initiated from susceptible and resistant corn hybrids. *Journal of Economic Entomology* 78, 951–954.

1C. Bioassays of *Bacillus thuringiensis* subsp. *israelensis*

O. Skovmand[1] and N. Becker[2]

[1]*Intelligent Insect Control, Montpellier, France;* [2]*German Mosquito Control Association, Waldsee, Germany*

Introduction

Bacillus thuringiensis subsp. *israelensis* is a spore-forming bacterium producing at least four major protein toxins that are confined to a crystalliferous inclusion body. These toxins have been shown to be highly effective against a wide variety of mosquito and blackfly species in different climatic zones and proved to be environmentally safe (de Barjac, 1990; Becker and Margalit, 1993; Becker, 1997). Since 1981, *B. thuringiensis* subsp. *israelensis* has been used worldwide in routine control programmes against mosquitoes and blackflies.

The efficacy of microbial control agents is affected by diverse biotic and abiotic factors: susceptibility of the target larvae, temperature, quality of water, intensity of sunlight, density of larval populations, and presence of filter-feeding non-target organisms (Mulla *et al.*, 1990; Becker *et al.*, 1992a,b). Before using *B. thuringiensis* subsp. *israelensis* products in microbial control programmes, it is of high priority to understand the impact of these factors. This is especially true with regard to the calculation of dosage, the use of the right formulation under given environmental conditions, and the optimal timing for treatment.

It is important to evaluate the activity of the formulations used in control programmes by means of laboratory bioassays in order to assess the potency of the microbial product and the efficacy of the product against the indigenous mosquito species in the laboratory.

Standardized bioassay protocols of *B. thuringiensis* subsp. *israelensis* have been developed by the World Health Organization (WHO) in cooperation with the Institut Pasteur, Paris (WHO, 1979; de Barjac and Thiery-Larget, 1984; de Barjac, 1985) and the USDA (McLaughlin *et al.*, 1984). In the following part of this chapter, protocols for bioassays of *B. thuringiensis* subsp. *israelensis* will be discussed.

Bioassay Method for the Titration of *B. thuringiensis* subsp. *israelensis* Preparations with the IPS 82 Standard

(According to the Institute Pasteur, The WHO Collaboration Center for Entomopathogenic Bacilli, Paris, France)

In this bioassay, potency of *B. thuringiensis* subsp. *israelensis* is determined in reference to the standard IPS 82. Vials with the IPS 82 standard of *B. thuringiensis* subsp. *israelensis* can be provided by Dr Isabelle Thiery, Institut Pasteur, 28, rue Dr Roux, F-75724, Paris Cedex 15, France.

Weigh 50 mg of the standard powder and mix with 10 ml deionized water in a 20-ml penicillin flask containing 15 glass beads (6-mm diameter). Homogenize this suspension vigorously with a crushing vibration machine for 10 min at 700 strokes min^{-1}. Add 0.2 ml of this suspension to 19.8 ml of deionized water. Homogenize this suspension in a 22-mm diameter test tube for a few seconds on a vortex-type shaker at maximum speed. Take aliquots of 15, 30, 60, 90 and 120 µl from the basic suspension, and pipette each into a plastic cup containing 150 ml of deionized water. The final concentration of the dilutions will be 0.005, 0.01, 0.02, 0.03 and 0.04 mg l^{-1} of the standard IPS 82. Use four cups per concentration. The controls contain 150 ml water per cup without the microbe.

In a similar manner, a dilution series from an initial suspension is made with the test microbe preparation. Twenty-five fourth-instar larvae of *Aedes aegypti* are placed in each cup with a Pasteur pipette. The selection of fourth instars should be strict; the larvae are 4–5 mm long, with the head capsule being about double the width of the body. This size can be obtained within 4–7 days after eclosion depending on the larval nutrition. Older larvae are not suitable for tests, since they do not feed on the day preceding pupation. The range of dilutions should exceed that of the standard to be certain that a reliable regression line is obtained. Labour can be saved by first conducting a range-finding bioassay, with a few widely spaced concentrations of the test material. The results are used to decide on the concentrations to be used in the accurate assay and, in part, as replication of the bioassay.

Each series of bioassays consists of at least 400 larvae exposed to the standard *B. thuringiensis* preparation, 500–1000 larvae for the test preparations, and 100 larvae for the controls. The tests should be conducted at 27 ± 1°C. Mortality is recorded after 24 and 48 h by counting live and dead larvae. The 48-h observation is useful in routine work to confirm previous data and to check that only *B. thuringiensis* components are involved in the bioassay. If some pupae are observed, they have to be removed and deducted from the mortality count. Since dead larvae can be consumed by the living larvae, mortality is determined by deducting the live larvae from the original number of larvae used at the start of the bioassay.

Control mortalities exceeding 5% should be corrected by Abbott's formula. Bioassays with a control mortality higher than 10% should be discarded. Probit analysis is used for the regression line of mortality–concentration from which the LC_{50}s of the standard and the test microbe are estimated. The LC_{50}s are used to calculate the potency of the test preparation in International Toxic Units (ITU) according to the formula:

$$\frac{LC_{50} \text{ of the standard}}{LC_{50} \text{ of the test preparation}} \times 15{,}000 \text{ ITU mg}^{-1}$$

$$= \text{ITU mg}^{-1} \text{ of test preparation}$$

To enhance the precision of the bioassays, they should be repeated on at least three different days, and the standard deviation calculated.

More homogeneous results can be obtained by feeding the larvae for the first 24 h only. For the dispersion of the standard and test materials in water, a Waring blender, glassmill, rotating rod homogenizer or sonicator (bath or rod sonicator) is recommended. The homogenization of the product in water should be at low intensity, to simulate the mixing conditions in outdoor control programmes. Conversely, mixing at high speed will improve the bioassay precision as described in 'Sample preparation' below. In order to improve the assessment of susceptibility of indigenous species in the natural habitat, water from the breeding sites is used instead of distilled water. The LC_{99} evaluated for field-collected larvae could be defined as the minimum effective dose and serve as a guideline for the assessment of the field tests (Becker and Rettich, 1994).

Factors Affecting the Bioassay and Potency Determination for *B. thuringiensis* subsp. *israelensis* Products

The efficacy of microbial products can be affected by a variety of factors (Becker *et al.*, 1992a,b; Lacey, 1997) (see Introduction). Some of the factors are listed below.

Dilution

Statistical literature on probit analysis often recommends the use of two or three concentrations and a large number of insects per concentration, provided that the confidence limits for the mortality data are narrow (Finney, 1971). This experimental design does not fit the microbe bioassays where the variance in mortality within concentrations and tests is very high. Therefore it is recommended that 6–8 concentrations be used.

The dilution ratio between concentrations varies with the insect species, the bacterium and the product. With the 1:0.5 ratio recommended in the WHO protocols for *B. sphaericus* against *Culex quinquefasciatus*, six sets of mortality data with linear dose–response between 0 and 100% can be obtained. However, in the bioassays of *B. thuringiensis* subsp. *israelensis* against *Aedes aegypti*, it was shown in an interlaboratory study comparing methods and results from five laboratories (Skovmand *et al.*, 1998) that at least one or two out of the six concentrations were out of line. This in turn will render the bioassay statistically non-valid. Therefore, it is suggested that the dilution ratio of 1:0.7 to 1:0.8 be used in the standard protocols for *B. thuringiensis* subsp. *israelensis*. The dilution series according to the WHO protocols ranges from 1.0 to 0.01 ml pipetted from a stock mixture.

Sample preparation

Most products, including the standard IPS 82, consist of multicellular particles. Intensive homogenization of the particles of the original product will turn them into a finely divided form. This procedure will steepen the slope of the dose–mortality curve and decrease the LC_{50}. The mortality rates for the new microbe standard, concentrated liquids and some new wettable powders of *B. thuringiensis* subsp. *israelensis* are obtained with finely divided particles (i.e. single cells) that will not be affected by the rate of homogenization. So, it is expected that the mortality lines (Finney, 1971) of the products will be high and parallel (Skovmand *et al.*, 1997). This means that the laboratory bioassay dose–responses with the new products are now closer to those obtained in mosquito control programmes.

Factors affecting the bioassay

Larval instar

Larval sensitivity to bacterial toxins is reduced as the larvae develop (Becker *et al.*, 1992a,b); 2nd-instar *Ae. vexans* were about 10 times more sensitive than 4th instars at a water temperature of 25°C. It is therefore essential that larvae of more or less the same age are always used for the bioassay.

Insect species

Large differences in sensitivity among mosquito species can be found due to differences in their feeding habit, ability to activate the protoxin, and toxin binding to midgut cell receptors (Lacey and Singre, 1982; Davidson, 1989). For example, larvae of *Culex pipiens* were 2–4 times less susceptible to *B. thuringiensis* subsp. *israelensis* than *Aedes* species of the same instar.

Temperature

The feeding rate of *Aedes vexans* decreases with temperature, and this effect leads to a reduction in consumption of bacterial toxins. With a rise in water temperature from 5 to 25°C, sensitivity of 2nd instars of this mosquito species to *B. thuringiensis* subsp. *israelensis* increases by more than 10 times. The influence of water temperature on the efficacy of the bacterium against the insects differs also with larval instar. Second-instar *Ae. vexans* are twice as sensitive to the microbe as 4th-instar larvae at a water temperature of 15°C. This difference in sensitivity was 10 times higher at 25°C (Becker *et al.*, 1992b).

Larval density

At a larval density of ten *Ae. vexans* 4th instars per cup with 150 ml water the LC_{50} was 0.0162 ± 0.004 mg l^{-1}; with 75 larvae per cup the LC_{50} increased by about 7 times (0.1107 ± 0.02 mg l^{-1}). This indicates that when the number of larvae increases, larger amounts of *B. thuringiensis* subsp. *israelensis* have to be used to reach mortality levels obtained at lower mosquito populations (Becker *et al.*, 1992b).

State of nutrition

Feeding the larvae before starting the bioassay caused an increase in the LC_{50}. In contrast, the LC_{50} of larvae that were fed sparsely throughout their growth prior to the bioassay, and not at all on the bioassay day itself, was very low. These results were obtained in a comparative study conducted among several laboratories (Skovmand *et al.*, 1998). When water polluted with bacteria was used in the test cups, 2–3 times more *B. thuringiensis* had to be applied in order to achieve the same effect as was obtained when clean water was used (Mulla *et al.*, 1990).

Types of larval feeding

Aedes aegypti larvae prefer to feed from the bottom of the container, whereas *Cx. pipiens* and even more so *Anopheles* species are surface feeders (Dahl *et al.*, 1993), although all three mosquito species will also feed on microbe particles in suspension. Wettable powders commonly consist of large particles and will precipitate within minutes to the bottom of the test cups (Skovmand *et al.*, 1997). In contrast, fluid formulations that are based on minute particles will disperse homogeneously in water. Thus, the availability to and the efficacy of the spore–crystal mixtures of the product against the larvae will depend on the feeding behaviour of the mosquito species. Also, the type of container may affect the bioassay results, as protoxins will stick to vial surfaces that are electrically charged.

Sunlight

Ultraviolet sunlight destroys the bacterial toxins and kills the vegetative cells and spores. Therefore, bioassay cups with water should not be exposed to the sun or to irradiation by UV lamps. In small-scale experiments, a severe sunlight inactivation of the microbial product in clean water has been reported (Becker *et al.*, 1992a).

Heat

Exposing the vegetative cells, live spores and toxins for 10 min to 80°C will inactivate them. Therefore, microbial products should not be stored close to heat sources. Furthermore, to prolong viability, fluid products should be stored at 5°C and the standard powders at −18°C. Liquid products should not be stored below the freezing point, as that will cause the proteins to aggregate and thus will increase particle size. To overcome this problem, a thorough homogenization of the product is made before use in the bioassay (Skovmand *et al.*, 1997).

Concluding Remarks

Potency bioassays were designed to determine the insecticidal activity of the spore–crystal mixture of *B. thuringiensis* subsp. *israelensis* referenced against an international standard. To achieve precision in this bioassay, several factors have to be optimized or considered: (i) insect quality, feeding behaviour, mosquito species and density; (ii) environmental conditions of water quality, temperature and light; and (iii) type of formulation. Therefore, strict protocol guidelines are needed to avoid undesirable changes in these factors. However, mortality levels of potency bioassays differ from those obtained in outdoor control programmes due to differences in particle sizes of the different formulations. Thus, for the same formulations, mortality records from both field tests and laboratory bioassays have to be compared to define more accurately the dosing of *B. thuringiensis* subsp. *israelensis* in mosquito management programmes.

References

Becker, N. (1997) Microbial control of mosquitoes: management of the upper Rhine mosquito population as a model programme. *Parasitology Today* 13, 485–487.

Becker, N. and Margalit, J. (1993) Use of *Bacillus thuringiensis israelensis* against mosquitoes and black flies. In: Entwistle, P.E., Cory, J.S., Bailey, M.J. and Higgs, S. (eds) Bacillus thuringiensis, *an Environmental Biopesticide: Theory and Practice*. John Wiley & Sons, Chichester, pp. 147–170.

Becker, N. and Rettich, F. (1994) Protocol for the introduction of new *Bacillus thuringiensis israelensis* products into the routine mosquito control program in Germany. *Journal of the American Mosquito Control Association* 10, 527–533.

Becker, N., Ludwig, M., Beck, M. and Zgomba, M. (1992a) The impact of environmental factors on the efficacy of *Bacillus thuringiensis* against *Culex pipiens*. *Bulletin of the Society of Vector Ecology* 18, 61–66.

Becker, N., Zgomba, M., Ludwig, M., Petric, D. and Rettich, F. (1992b) Factors influencing the activity of *Bacillus thuringiensis* var. *israelensis* treatments. *Journal of the American Mosquito Control Association* 8, 285–289.

Dahl, C., Sahlèn, G., Grawe, J., Johanison, A. and Amneus, H. (1993) Differential particle uptake by larvae of three mosquito species (Diptera: Culicidae). *Journal of Medical Entomology* 30, 537–543.

Davidson, E.W. (1989) Variations in binding of *Bacillus thuringiensis* toxin and wheat germ agglutinin to larval midgut cells of six species of mosquitoes. *Journal of Invertebrate Pathology* 53, 251–259.

de Barjac, H. (1985) Standardized test for the determination of the susceptibility of mosquito larvae to bacterial toxins. *WHO Report* VBC/ECV/85.25.

de Barjac, H. (1990) Characterization and prospective view of *Bacillus thuringiensis* subsp. *israelensis*. In: de Barjac, H. and Sutherland, D.J. (eds) *Bacterial Control of Mosquitoes and Black Flies*. Rutgers University Press, New Jersey, pp. 10–15.

de Barjac, H. and Thiery-Larget, I. (1984) Characterization of IPS 82 as standard for biological assay of *Bacillus thuringiensis* H-14 preparations. *WHO Mimeograph Document* VBC/84.892. Geneva, Switzerland.

Finney, D.J. (1971) *Probit Analysis*. Cambridge University Press.

Lacey, L.A. (1997) Bacteria: laboratory bioassay of bacteria against aquatic insects with emphasis on larvae of mosquitoes and black flies. In: Lacey, L.A. (ed.) *Manual of Techniques in Insect Pathology*. Academic Press, New York, pp. 79–90.

Lacey, L.A. and Singre, S. (1982) Larvicidal activity of new isolates of *Bacillus sphaericus* and *Bacillus thuringiensis* (H-14) against Anopheline and Culicine mosquitoes. *Mosquito News* 42, 537–542.

McLaughlin, R.E., Dulmage, H.T., Alls, R., Couch, T.L., Dame, D.A., Hall, I.M., Rose, R.I. and Versoi, P.L. (1984) US standard bioassay for the potency assessment of *Bacillus thuringiensis* serotype H-14 against mosquito larvae. *Bulletin of the Entomological Society of America* 30, 26–29.

Mulla, M.S., Darwazeh, H.A. and Zgomba, M. (1990) Effect of some environmental factors on the efficacy of *Bacillus sphaericus* 2362 and *Bacillus thuringiensis* (H-14) against mosquitoes. *Bulletin of the Society of Vector Ecology* 15, 166–175.

Skovmand, O., Hoegh, D., Rederson, H.S. and Rasmussen, T. (1997) Parameters influencing potency of *Bacillus thuringiensis* var. *israelensis* products. *Journal of Economic Entomology* 90, 361–368.

Skovmand, O., Thiery, I., Benzon, G., Sinegre, G., Monteny, N. and Becker, N. (1998) Potency of products based on *Bacillus thuringiensis* var. *israelensis*: Interlaboratory variations. *Journal of the American Mosquito Control Association* 14, 298–304.

WHO (1979) An interim standardized bioassay method for the titration of experimental and commercial primary powders and formulations of *Bacillus thuringiensis* serotype H-14. *Third Meeting of the Scientific Working Group on Biological Control of Insect Vector of Diseases*. Document TDR/BCV-SWG/3/79.3. Annex, pp. 28–29.

1D. Production of *Bacillus thuringiensis* Insecticides for Experimental Uses

S. Braun

The Biotechnology Laboratory, Institute of Life Sciences, The Hebrew University of Jerusalem, Jerusalem, Israel

The number of investigators participating in the research of microbial insecticides is large and their backgrounds are diverse. Some are skilled microbiologists, whereas others lack microbiological experience. This contribution addresses both groups. Hence, in addition to topics directly related to its title, the text contains a general description of *Bacillus thuringiensis* and its δ-endotoxin; it also deals with basic microbiological concerns such as safety issues. A considerable part of this chapter comprises detailed protocols of useful techniques, most of them tested in the author's laboratory, where they performed satisfactorily. It reviews current techniques of producing *B. thuringiensis* samples ranging from several grams to several hundred grams of the insecticide powder, the latter quantity being sufficient for small field trials. However, production of larger amounts of *B. thuringiensis* as well as formulation and storage of insecticides, that require specialization and expensive equipment, are beyond the scope of this review. Recently, formulations of *B. thuringiensis* were described in depth in a book by Burges (1998).

Isolation of New Strains of *B. thuringiensis*

B. thuringiensis is a ubiquitous soil organism (Martin and Travers, 1989). One may search for *B. thuringiensis* in dead or diseased target insects, although this association is not exclusive. However, some insect-infested niches are especially rich in crystal-forming insecticidal *B. thuringiensis* strains (Chaufaux *et al.*, 1997). Thus, in a random sample of residue from an animal feed mill rich in Lepidoptera, 65% of colonies having *B. thuringiensis* morphology formed parasporal crystals (Meadows *et al.*, 1992), while only 10% did in soil-derived samples (Travers *et al.*, 1987). High incidence of *B. thuringiensis* was also reported in the phylloplane (Smith and Couche, 1991).

B. thuringiensis is abundant in rich topsoil. It is rarely found in subterranean samples or in the desert. The usual range of *B. thuringiensis* in the top soil is 10^2–10^4 colony-forming units per gram (CFU g^{-1}) of soil (Martin, 1994). Since fertile soil is rich in many microorganisms, including various spore-forming bacilli, screening for new isolates is rendered less tedious by strict selection as described below.

An effective technique has been established by Travers *et al.* (1987). Germination of *B. thuringiensis* spores is strongly inhibited by acetate. Incubation of a soil sample in a rich fermentation medium containing 0.25 M sodium acetate results in germination, and, hence, loss of heat stability of most *Bacillus* spores. Germinated spores and non-spore-forming organisms are then destroyed by heat-shock, while non-germinated *B. thuringiensis* spores survive. Heat-shocked samples containing viable *B. thuringiensis* spores are seeded on agar medium and allowed to sporulate. This method also selects for *B. sphaericus, B. megaterium, B. cereus,* and, occasionally, for some other *Bacillus* species (Martin and Travers, 1989). Dense, round *B. sphaericus* colonies are easily recognizable. Colonies of the 'pancake' morphology typical for *B. thuringiensis* are examined by light microscopy for the presence of inclusion bodies (crystals).

Although there is a tendency to distinguish between *B. thuringiensis* and *B. cereus* species solely by the inability of the latter to produce δ-endotoxin (González *et al.*, 1982; Meadows *et al.*, 1992), these closely related species differ, as demonstrated by the successful design of specific DNA probes based on 16S rRNA genes (te Giffel *et al.*, 1997). In our limited test samples, *B. thuringiensis* strains were invariably selected by the acetate selection procedure, while only 2–5% of spores of several *B. cereus* species survived the procedure. Others have also reported the specificity of the acetate selection protocol (Brownbridge, 1989). This specificity may be related to the defective germination mechanism in spores of strains forming parasporal inclusions. Such strains have thinner spore coats and there is deposition of protoxin on the spore surface (Aronson *et al.*, 1986).

Addition of selective antibiotics such as polymixin B and penicillin to the isolation media for vegetative *B. thuringiensis* cells has been used by some investigators (Saleh *et al.*, 1970; Akiba and Katoh, 1986).

Procedure 1 for isolation of *B. thuringiensis* is adapted from Travers *et al.* (1987) and Martin and Travers (1989).

Identification and Characterization of *B. thuringiensis* Isolates

The traditional taxonomic key for identification of various species in the genus *Bacillus* is based upon biochemical tests (Gordon *et al.*, 1973). The

Procedure 1. Isolation of *B. thuringiensis*

Soil sample preparation

B. thuringiensis spores are readily inactivated by exposure to UV radiation. Therefore, soil samples should be taken preferably from 2–5 cm below the surface or from permanently shadowed places. Dry soil samples can be collected conveniently in sterile plastic bags. Sealed bags can be stored at ambient temperature. Moist soil (about 20 g) can be collected in sterile glass tubes (50 ml) stoppered with a loosely fitting sterile cotton-wool plug, and allowed to dry at ambient temperature.

Media preparation

L-broth medium contains (g): bacto-tryptone 10, yeast extract 5, NaCl 5 in 1 l of distilled water at pH 6.8. For isolation of *B. thuringiensis*, L-broth is supplemented with sodium acetate (0.25 M, pH 6.8). T3 medium contains (g): bacto-tryptone 3, bacto-tryptose 2, yeast extract 1.5, MnCl$_2$ 0.005 in 1 l of 50 mM phosphate buffer (pH 6.8). Agar (20 g l^{-1}) is added to a hot (about 80°C) medium for the preparation of L- or T3-agar media, and stirred until dissolution. All media are autoclaved for 15 min at 121°C. The sterile agar medium is poured into Petri plates at 50–60°C. The medium should be swirled before pouring to ensure that it is evenly mixed. If the agar surface is wet at the time of inoculation, it can be dried for about one hour in a sterile hood, while the base of the Petri dish is inverted and placed on the edge of the lid for support.

   Bacto-tryptone (pancreatic digest of casein) and bacto-tryptose (hydrolysate of meat and vegetable proteins) are trade names of products supplied by Difco Laboratories Inc. Similar products may be obtained from other suppliers of microbiological media such as Biolife Italiana, Oxoid Ltd, Sheffield Products, E. Merck or others.

Acetate selection

Soil sample (1 g) is added to L-broth (20 ml) supplemented with sodium acetate in a baffled Erlenmeyer flask (125 ml). The mixture is shaken at 30°C and 250 rpm for 4 h. Samples of about 0.5 ml in 10 ml-test tubes are heat-treated for 3 min in a water bath at 80°C, and used to inoculate L-agar plates (these plates are made without acetate). Colonies that are formed after overnight growth at 30°C are transferred on to T3-agar plates, allowed to sporulate for 40 h at 30°C and then examined under the microscope for the presence of crystals.

same principle has been extended into classification of *B. thuringiensis* subspecies (Heimpel, 1967; de Barjac, 1981). Currently, there exists a definitive classification of *B. thuringiensis* strains into 45 flagellar (H) serovars (de Barjac and Bonnefoi, 1968; de Barjac, 1981; de Barjac and Franchon, 1990).

The full list of *B. thuringiensis* serovars according to H-serotype is supplied by the International Entomopathogenic *Bacillus* Center of the Pasteur Institute, Paris, in the catalogue of its *B. thuringiensis* and *B. sphaericus* collection.

No correlation exists between the H-serotype, parasporal crystal serotype and activity spectrum. The activity spectrum can be verified only in biological assays or predicted by the type of insecticidal crystal proteins (Aronson *et al.*, 1986). Crystal proteins encoded by *cry* genes have been classified as CryI to CryVI (Table 1D.1) depending on host specificity and amino acid homology (Höfte and Whiteley, 1989; Tailor *et al.*, 1992). Cytolysins (Cyt) are specifically toxic to dipteran larvae, while they have a broad cytolytic activity *in vitro* (Ishii and Oba, 1994).

In 1995, a revised nomenclature of *cry* genes was submitted to *Microbiological and Molecular Biology Reviews* (Crickmore *et al.*, 1995). This nomenclature, while retaining the main feature of the previous, is based upon amino acid homology rather than on host specificity. The detailed description of the new nomenclature can be found at WWW site: http://www.biols.susx.ac.uk/Home/Neil_Crickmore/Bt/index.html.

Screening for novel *cry* genes and their sequencing is becoming an important tool in the research of *B. thuringiensis* insecticides, and in the development of recombinant strains with broader host specificity. Numerous probes have been developed for hybridization analysis or for fingerprinting *cry* genes using PCR (polymerase chain reaction) techniques (Visser, 1989; Chak *et al.*, 1994; Ceron *et al.*, 1995; Shin *et al.*, 1995; Feitelson and Narva, 1997).

Table 1D.1. Host specificity of Cry protoxins.[1]

Homology group	Host specificity	Variations in host specificity
CryI	Lepidoptera	CryIA(b)[2] and CryIC[3] may confer toxicity against Diptera, CryIB[4] may confer toxicity against Coleoptera
CryII	Lepidoptera and Coleoptera	CryIIA – Lepidoptera and Diptera CryIIB and IIC – only Lepidoptera
CryIII	Coleoptera	
CryIV	Diptera	
CryV	Lepidoptera and Coleoptera	
CryVI	Nematodes	

[1]Lereclus *et al.*, 1993; [2]Haider *et al.*, 1987; [3]Smith *et al.*, 1996; [4]Bradley *et al.*, 1995.

Management and Preservation of *B. thuringiensis* Strains

Numerous *B. thuringiensis* strains are preserved because they carry larval toxin genes of potential industrial significance. It is, therefore, crucial that these genes are not lost or modified during short- and, especially, long-term storage, as might happen through involuntary or voluntary selection processes or through contamination. It is seemingly easy to preserve spore formers because of the excellent stability of spores, especially in the genus *Bacillus*. To the despair of an industrial microbiologist, the biological activity of *B. thuringiensis* depends upon the expression of insecticidal crystal protein genes (*cry*) that are normally associated with several large plasmids (González and Carlton, 1980, 1984) and determine the host specificity of *B. thuringiensis* isolates. Many *B. thuringiensis* strains also contain bacteriophages (de Barjac *et al.*, 1974; Inal *et al.*, 1990). Lysis frequently ensues following sequential transfer of dense *B. thuringiensis* mats (D. Klein and S. Braun, unpublished). Methods of preservation should ensure the unchanged character of these inherently unstable elements. Despite the excellent stability of some natural plasmids of *B. thuringiensis*, habitual growth in rich medium could favour fast-dividing populations devoid of large plasmids. High cell densities, which a rich medium is able to support, increase the probability of plasmid exchange and of propagation of phages.

Although an ideal industrial microorganism should, preferably, not have transposable elements leading to variability within the population, this property of *B. thuringiensis* allows the tailoring of insecticidal properties of second-generation *B. thuringiensis* products, such as AGREE, CONDOR or CUTLASS, to specific pest complexes in the crop by a simple transfer of plasmids without recourse to recombinant DNA technology (González *et al.*, 1982).

Laboratories involved in screening soil or other samples may maintain a large number of cultures both in short-term and in long-term storage. Thus, the cost, the amount of time needed for maintenance, and the necessity to avoid cross-contamination should be carefully considered in the selection of the preservation method.

A distinction should be made between the preservation of strains available from collections (low value strains) and your own important strains that have not yet been deposited (high value strains) with a collection. Low value strains are readily available, and, pending routine biological assays to assure their correct preservation, do not require sophisticated preservation methods. If in use for more than a few months, these strains may be preserved in frozen suspension, which allows for repeated use of the same storage unit.

For your own unique strains, whose loss is irreplaceable, the surest method of long-term preservation is in freeze-dried stocks. This method, however, is time-consuming and requires professionalism and specialized equipment. A general discussion of various preservation methods is found

Procedure 2. Preservation of *B. thuringiensis* strains on nutrient agar slants.

B. thuringiensis can be preserved for at least 3 months on nutrient agar (NA) slants. Commercial nutrient agar, such as Difco NA, contains 1.5% agar, while 2% agar is used for *B. thuringiensis* cultures. Agar may be added either to the premixed formulation or to the nutrient broth (g l^{-1}, bacteriological peptone 3, beef extract 5). Most *B. thuringiensis* strains grow and sporulate well on NA. Agar slants are prepared in sterile screw-cap glass test tubes (20 or 40 ml). As in all preservation methods, it is important to inoculate heavily to ensure the formation of a uniform mat. Inoculated slants are incubated for 40 h at 28–30°C. The caps are then tightened and the test tubes can be stored at ambient temperature or, preferably, in a refrigerator. When needed, the slants are washed with water containing 0.01% sodium lauryl sulphate to suspend the spores. This suspension in sterile tubes can be kept for 2–3 months in the refrigerator, and used for routine inoculations.

Procedure 3. Long-term storage of *B. thuringiensis* spores in frozen suspensions.

A convenient method for long-term storage of a large number of strains in frequent use is to store spore suspensions in water containing sodium lauryl sulphate (0.01%) and glycerol (15%) frozen at −70°C. Spores are harvested from slants in the storage medium as described above, made up to a concentration of about 10^{10} spores ml^{-1} and distributed in sterile 2-ml plastic tubes. The tubes are frozen in liquid nitrogen and stored in a freezer. Inoculation from the stored samples can be made repeatedly by passing an inoculation loop over the frozen surface and streaking it on NA plates. It is advisable never to open more than one tube at a time to prevent cross-contamination. This storage is stable for at least 5 years and, probably, much longer. Even vegetative cells can be maintained for at least several years suspended in 30% glycerol at −70°C (Bernhard *et al.*, 1997).

Procedure 4. Preservation of freeze-dried spores in vacuum-sealed ampoules.

The spores are harvested from the agar slants as described above, in a solution of spray-dried skimmed milk (10% w/v). The milk solution is earlier made sterile by filtration through a membrane with a pore cut-off below 0.2 μm. The spore suspension (10^{10} spores ml^{-1}) is dispensed into sterile 1-ml glass ampoules in aliquots of 0.2 ml. The ampoules are then stoppered with cotton plugs and placed in a freezer for 1 h. Subsequently, the frozen suspension is freeze-dried for 8 h at a pressure below 0.2 mbar. The ampoule is then sealed by flame under vacuum.

in Snell (1991). Detailed technical information concerning methods described in Procedures 2–4 is presented by Dulmage (1983), Malik (1991) and Bernhard *et al.* (1997).

Fermentation of *B. thuringiensis* for Insecticide Preparation

Preparation of *B. thuringiensis* insecticide usually involves batch growth of the organism. Modern industry and laboratories produce *B. thuringiensis* in submerged fermentation.

Semi-solid fermentation, although seemingly simple, requires highly skilled personnel. Moreover, it is haunted by poor reproducibility and the danger of contamination (Dulmage, 1983). However, in several countries (China, Brazil), the concept of decentralized local insecticide production by farmers has been attempted (Bernhard and Utz, 1993; Salama and Morris, 1993; Capalbo, 1995). Media for semi-solid fermentation comprise cheap agricultural surplus or waste products. Large surface area of the medium, the prerequisite for good oxygen transfer, is attained by the addition of fibrous material such as bran. Humidity of the medium is maintained at about 50–60%. The process continues for 5–7 days. The spore content in the final product is in excess of 10^{10} per gram. Contamination may result in inactive or even dangerous material; therefore, precautions must be taken to minimize this risk by the use of centrally distributed clean inocula.

The transcription of most δ-endotoxin genes is regulated by sporulation σ-factors: σ^{28} and σ^{35} (Aronson *et al.*, 1986). Thus, high cell density has to be reached in the logarithmic culture as well as high sporulation rate in the stationary phase. The exception to this rule is the *cry*IIIA gene whose expression is regulated by the σ^A promoter, although it continues to accumulate during sporulation (Sekar *et al.*, 1987; Baum and Malvar, 1995).

Optimal fermentation parameters leading to cell densities of 10^{10} cells ml^{-1} require significant research and development for any specific strain, and, thus, are the domain of commercial producers. However, moderately good batches resulting in 10^8–10^9 spores ml^{-1} can be achieved routinely without too much effort for almost any strain. The formula for success is the use of metabolically balanced media under conditions of good aeration. Under poorly aerobic or anaerobic conditions, as well as in the presence of excess sugars or amino acids, the sporulation and, therefore, δ-endotoxin production is inhibited (Bernhard and Utz, 1993). Correlation has been found between the volume coefficient of oxygen transport, k_La, and spore productivity in submerged fermentation (Flores *et al.*, 1997) for *B. thuringiensis kurstaki* HD-1; k_La was suggested as the scale-up parameter for *B. thuringiensis* fermentation processes.

Growth of *B. thuringiensis* is optimal at 24–30°C. Within this wide optimum range, lower temperatures are preferable, since upper optimum temperatures involve increased oxygen demand that may reduce the sporulation rate. Cultivation of *B. thuringiensis* at 42°C induces plasmid loss (González and Carlton, 1984), and, thus, may be detrimental to the toxicity.

Bacilli are typical carrion dwellers. They are well adapted to digest varied and complex substrates. They are excellent producers of proteases, amylases, glucanases and other lytic enzymes (Aronson and Geiser, 1992). Typical media contain glucose or a complex carbohydrate and protein or protein hydrolysates. Cheap media are made, usually, of molasses, corn steep liquor, $CaCO_3$, meals of soybean, cottonseed, maize and fish, starch, wheat bran, rice husk, etc. (Salama *et al.*, 1983; Mummigati and Raghunathan, 1990; Morris *et al.*, 1997). Sporulation of *B. thuringiensis* is strongly influenced by trace element concentration, particularly Mn^{2+} ions.

Use of cheap media, however, leads to great variability in fermentation results as well as to the presence of ballast particles in the final product. Batch-to-batch variability of commercial agricultural products has to be determined. This activity may become extensive and it is justified only if a mass-production method is tested. Otherwise, standard media components that are quality controlled and allow for reproducible results are preferable.

For the production of δ-endotoxin from *B. thuringiensis* Berliner, the optimal concentrations of carbohydrate and protein were 20 g l^{-1} and 12.5 g l^{-1}, respectively. Spore yield using this ratio with industrial media components in a pilot-scale fermenter (600 l) reached 10^9 spores ml^{-1} (Vecht-Lifshitz *et al.*, 1989). Very similar ratios of protein to carbohydrate produced excellent results with several other *B. thuringiensis* serovars (Salama *et al.*, 1983; Morris *et al.*, 1997).

Sugars are catabolysed with accumulation of acids, acetate, lactate and hydroxybutyrate. Sugar excess in the early stationary phase is, sometimes, accompanied by accumulation of polyhydroxybutyrate granules in the cells. These inclusions, which may be mistaken for the parasporal bodies, are an important energy reserve source linked to the formation of spores and crystal proteins during the sporulation phase (Liu *et al.*, 1994). Although *B. thuringiensis* growth is not influenced significantly by pH between 5.5 and 8.5, neutral or slightly alkaline conditions are preferable for sporulation. Therefore, in shake-flask cultures, where control of pH by titration is inconvenient, $CaCO_3$ is sometimes included in the medium. The $CaCO_3$ residue in the spent medium is harvested with the crude insecticide, and thus enters as ballast in bioassay. It is possible to remove it by lowering the pH to 5 by careful addition of HCl to the culture before harvesting. With most strains, there is no need for the pH control in the media recommended here. In these media, sugar is exhausted before the pH drops below 5. The remaining protein or amino acids are then metabolized resulting in increased pH, reaching pH 8.5 at sporulation.

In well-formulated media, the sporulation is completed and lysis occurs within 18–24 h, even at relatively low inoculation rates. The culture may be harvested within 4–5 h after the appearance of cells containing refractile forespores, even if lysis is not complete. The yield of fermentation may be assessed by counting vegetative and sporulating cells under a phase micro-scope. To obtain reliable results, all counting must be done before the spor-angia have lysed. Microscopic examination is a quick and reliable method of quality control, especially in monitoring repeated batches of the same organism. Poor growth and/or sporulation indicate a failure. It is important to stress that although cell density, sporulation rate and δ-endotoxin yield are generally interdependent, no strict correlation exists. Thus, bioassay is the only real measure of the success of fermentation.

Media for Preparation of *B. thuringiensis* Insecticide

Many investigators have used insecticides prepared from cultures grown in media without carbohydrates, such as L-broth or T3, for bioassays. We have found that balanced protein–carbohydrate media give consistently better yields (S. Braun and D. Klein, unpublished).

The NZB medium described in Procedure 5 was tested with more than 30 established strains of *B. thuringiensis*. It usually supports production of 10^8–10^9 spores ml^{-1} medium (Table 1D.2) with excellent toxicity to target insects (Navon *et al.*, 1990, 1994).

The mixture of high-grade (glucose) and low-grade (glycerol) carbon sources ensures moderate pH shift during fermentation. In most cases, NZB medium may be used without CaCO$_3$. It seems that both glycerol and lac-tate (the latter is present at a concentration of about 1 M in corn steep liquor) are metabolized only after exhaustion of glucose and amino acids. Unlike rich substrates, these relatively poor substrates not only do not inhibit sporulation, but provide an energy source for successful sporulation. NZB medium was used for shake-flask production as well as for small (10–50 l) fermenter batches.

Table 1D.2. Yield of *B. thuringiensis* spores on NZB medium.

Strain	Serotype	Serovar	LC_{50} (µg g^{-1})[1] *Spodoptera littoralis*	*Helicoverpa armigera*	Spore yield (ml^{-1})
HD 1	H 3a, 3b, 3c	*kurstaki*	–	1.40	4×10^8
HD 73	H 3a, 3b, 3c	*kurstaki*	–	0.03	1×10^8
HD 248	H 7	*aizawai*	5.40	–	2×10^8
HD 263	H 3a, 3b, 3c	*kurstaki*	–	0.53	5×10^8
HD 266	H 3a, 3b, 3c	*kurstaki*	–	0.87	1×10^9
HD 269	H 3a, 3b, 3c	*kurstaki*	–	0.82	3×10^8
HD 273	H 5a, 5b	*galleriae*	–	7.00	2×10^8
HD 277	H 4a, 4c	*keniae*	–	1.20	2×10^8
HD 283	H 7	*aizawai*	6.60	–	1×10^8
HD 307	H 1	*thuringiensis*		8.40	1×10^8
HD 309	H 1	*thuringiensis*		1.60	2×10^9

[1]Dietary bioassays; 1st instar; 96 h.

Procedure 5. Preparation of NZB medium (adapted from Navon *et al.*, 1990).

The NZB medium contains (g l^{-1}): glucose 10, glycerol 10, corn steep liquor 10 (for standard corn steep liquor containing 30% solid residue after evaporation, otherwise recalculate for 3 g l^{-1} corn steep liquor solids in the medium), yeast extract 5, NZ-amine B (casein digest, Sheffield Products, Norwich, Connecticut) 10, MgCl$_2$ 2, CaCO$_3$ 2. The pH is adjusted to 7.0 by the addition of NaOH prior to the addition of CaCO$_3$.

Very good results have been obtained with BM medium (Salama *et al.*, 1983). This medium, supplemented with protein and complex carbohydrate sources such as cotton seed meal, various leguminous seeds or fodder yeast, resulted in production yields of about 10^9 spores ml^{-1} culture for *B. thuringiensis kurstaki* (HD 1 and HD 73) and *entomocidus*. Particulate material that is present in this medium makes it poorly suitable for shake-flask experiments. However, it is excellent for semi-industrial pilot production of *B. thuringiensis* insecticides.

Procedure 6. Preparation of BM medium with cotton seed meal (adapted from Salama *et al.*, 1983).

BM medium contains (g l^{-1}): glucose 6, yeast extract 2, K$_2$HPO$_4$ 4.3, CaCO$_3$ 2. Add finely ground (particle size below 0.1 mm) cotton seed meal (20 g l^{-1}). Sterilize for at least 45 min at 121°C.

Fermentation Laboratory and Equipment

Even the smallest microbiological operation requires a considerable amount of specialized equipment. The equipment list includes: a standard laminar flow hood for sterile work, a bench-top autoclave, a dry-air oven, a microwave oven and a culture incubator.

Small samples of microbial insecticides (up to 50 g) are usually produced in shake-flasks. Shake-flasks are also required for growing inocula. To allow growth at different temperatures, at least two rotatory shaker-incubators per laboratory are recommended. Practical shake-flasks (Erlenmeyer flasks) for the purposes of fermentation may vary in size between 0.25 and 2 l. They can be smooth or baffled. Baffled shake-flasks prevent vortex formation and ensure 2–3 times higher oxygen transfer rates than the standard flasks, and, thus, support high sporulation rates. Flasks with baffles on the conical part tend to splash medium on the plug, leading to possible contamination and decreasing the oxygen flow. The particulate materials from the media as well as microorganisms tend to accumulate on the baffles. In our laboratory, we prefer shake-flasks with baffles on the bottom of the flask.

Oxygen transfer rates decline with increase in the volume of medium. Low volumes of medium should be used for the preparation of *B. thuringiensis*, usually no more than 20% of the flask volume.

Bench-top fermenters can attain better oxygen transfer rates than shake-flasks; three- to fourfold that of the baffled flasks, thus allowing more efficient use of the medium and better sporulation conditions for *B. thuringiensis*. They are indispensable when larger samples of the insecticide are needed.

Procedure 7. Shake-flask culture of *B. thuringiensis* (adapted from Navon *et al.*, 1990).

NZB medium (100–200 ml per flask) is dispensed into baffled 1-l Erlenmeyer flasks. The flasks are sterilized at 121°C for 20 min. Growth of some *B. thuringiensis* strains may be inhibited by the products of reaction that occur between glucose and amino acids at the temperature of sterilization. It is preferable, therefore, to sterilize the aqueous glucose solution (40% w/w), which is made slightly acidic by the addition of a drop of HCl, and to add this solution to the sterilized medium. The cultures are inoculated from spore suspension prepared by washing agar slants containing well-sporulated *B. thuringiensis* mats with sterile 0.01% aqueous sodium lauryl sulphate. The inoculation rate is, usually, about 10^5–10^6 spores ml^{-1} medium. The inoculated media are incubated at 30°C in a rotatory shaker (rotating at 250 rpm in a 25.4 mm circular orbit) for 20–44 h, depending upon the completion of sporulation and lysis. Spores, crystals and debris are collected by centrifugation, and washed twice with distilled water. The resulting paste is freeze-dried for better preservation and stored in a refrigerator. This procedure yields 8–11 g l^{-1} of dry powder.

In contrast to the shake-flask cultures that demand little more than the equipment of a standard microbiological laboratory, production of biomass even in bench-top fermenters requires considerably more specialized equipment, skilled personnel and safety measures. Spray formed during aeration of fermenters and spill of biomass during handling may pose considerable safety problems regarding contamination by pathogens. All fermenters should be located in a place where spillage could be easily contained and cleaned. A drainage channel in the floor under the fermenter is strongly recommended. Supplies of coolant, deionized water, steam, electricity, compressed air and heating gas are essential.

Air supply

A simple diaphragm pump is required that is able to deliver up to 2 vvm (volumes of air per volume of fermenter per minute) under the counterpressure of about 0.5 bar. Diaphragm pumps are oil-free and do not foul the lines with oil deposits, which can lead to contamination. The best material for air supply and other lines is steel piping with a smooth interior surface. A water trap and a crude particle filter with a stainless steel sieve of opening size 74 μm should be installed on the air line. Bench-top fermenters are usually supplied with polycarbonate or stainless steel cartridge filters with polytetrafluoroethylene (PTFE) membrane filter elements. These filters, though expensive, are effective in wet conditions and durable. If only a few batches of insecticide are produced each year, disposable plastic filter units are recommended.

Steam

Depending on the type and manufacturer, modern fermentation vessels are sterilized either by placing the entire fermentation vessel into an autoclave or by sterilization *in situ*. Bench-top fermenters which can be sterilized *in situ* are considerably more expensive but better suited for *B. thuringiensis* fermentations using media containing insoluble material, since they can be agitated during sterilization. Agitation prevents sedimentation of particles; thus more reliable sterilization is achieved.

Some bench-top and laboratory fermenters up to a volume of 20 l are heated electrically; for others a simple steam generator may be purchased. Sterile samples can be taken using a steam-sterilized sample valve or a sterile syringe through a rubber septum. Pressure reducers should be installed in both air and steam lines. All steam lines should be well isolated by a suitable lagging able to withstand the sterilization temperatures. This will prevent the forming of condensate and protect personnel from burns.

Water

The mains water usually satisfies the requirement of all bench-top fermenters for coolant in terms of inlet pressure and supply requirements. Hard water, however, may cause the blockage of small-diameter pipes in the heat exchanger in prolonged use. For media preparation, high-quality water is preferable, since microelement concentration in mains water is subject to seasonal and other changes. In some places drinking water is routinely fluorinated. All these elements may influence the reproducibility of fermentation. Therefore, deionized water should be used routinely.

Fermenter

Most demands for the small-scale testing of *B. thuringiensis* can be satisfied with bench-top fermenters (working volume 10–15 l) supplied with baffles and a standard Rushton turbine rotor. Numerous suppliers offer similar products of essentially the same design and quality. The main consideration for selecting a certain model should be the quality of service the supplier may offer.

Fermentation parameter control

Only temperature measurement and control are essential to small-scale pro-duction of *B. thuringiensis*. Yet, it is advisable to have the option to mea-sure, register and control pH and to measure dissolved oxygen. The fermenter may thus be operated more successfully, and data for future scale-up and process optimization may be collected. With *B. thuringiensis* the shear forces are not critical, and the oxygen transfer rate is the most cru-cial factor in scaling-up. Industrial-scale fermenters may be limited in impeller speed and air supply. We have been successful in forecasting process parameters of the 30 m^3 fermenter using data collected on the 15 l bench-top. Thus, the impeller speed of the production fermenter can be scaled-up from that of a small experimental vessel using the following sim-ple formula for constant power input per volume:

$$(D_1/D_2)^2 = (n_1/n_2)^3$$

where D_1 and D_2 are impeller diameters, and n_1 and n_2 the respective impeller speeds. This equation is applicable for any geometrically similar

Procedure 8. Production of *B. thuringiensis* insecticide in a 15 l fermenter (adapted from Cohen *et al.*, 1991).

NZB medium (10 l) lacking glucose but containing polypropylene glycol antifoam (MW 1200, 2 g l^{-1}) is sterilized *in situ* at 121°C for 30 min with stirring (250 rpm). A glucose solution in water (40% w/w) is autoclaved separately and added to the rest of the medium just before inoculation. The temperature of the medium is brought to 28°C, and is controlled with no more than 1°C deviation. The impeller speed is maintained at 350 rpm, the air flow at 6 l min^{-1} and the pressure at 0.5 bar. The medium is then inoculated from a 15–20 h shake-flask culture (10 ml). The culture is allowed to grow for 20–44 h. A mixture of spores, crystals and debris is harvested by centrifugation, washed twice with distilled water and freeze-dried. This procedure yields 70–80 g of dry insecticide.

fermentation vessel. Data are then collected within the acceptable impeller speed range and transferred to the production process using the same scale-up criterion.

Transgenic Expression of *B. thuringiensis* Insecticidal Proteins

The precise biochemical and toxicological characterization of *B. thuringiensis* insecticidal proteins has become much less tedious with the advent of genetic engineering. Numerous genes encoding these proteins have been cloned and expressed in a variety of non-toxic bacteria such as *Escherichia coli* (Schnepf and Whiteley, 1981; McPherson *et al.*, 1988; Moar *et al.*, 1994), *B. subtilis* (Klier *et al.*, 1982) and *Pseudomonas fluorescens* (Obukowicz *et al.*, 1986; McPherson *et al.*, 1988) as well as in several acrystalloferous (Cry⁻) strains of *B. thuringiensis* (Moar *et al.*, 1994; Delécluse *et al.*, 1995). Transgenically expressed *cry* genes allow assessment of the toxicity of every crystal protein separately from other components, and establishment of the strict structure–activity relationships of various protein domains and of specific amino acids within the domain.

Domain-specific studies of *B. thuringiensis* insecticidal proteins have led to the understanding of their mode of action and to the development of industrially important crystal protein mutants and hybrids (Visser *et al.*, 1993). Insecticide delivery was improved by expressing δ-endotoxin proteins in microorganisms colonizing roots such as *P. fluorescens* (Obukowicz *et al.*, 1986) or leaves such as *B. cereus* (Moar *et al.*, 1994) in other bacteria, viruses, algae (Gelernter and Schwab, 1993) and, finally, in plants (Ely, 1993).

Wide dissemination of transformation procedures and the specialized nature of plasmid construction tools makes this a topic of special review outside the scope of this publication. A wealth of suitable organisms and vectors can be obtained from individual researchers, enterprises and institutions as well as from the official collections of bacilli (see Appendix 1D.1).

Convenient vehicles of production of transgenic δ-endotoxins are either (Cry⁻) strains of *B. thuringiensis* or *E. coli* transfected with plasmids harbouring suitable crystal protein genes. *E. coli* cells frequently 'fail to produce a significant amount of toxin protein when they are expressed from their native promoters' (Donovan *et al.*, 1988). For expression sufficient to produce inclusion bodies of the crystal protein in *E. coli*, the native promoter has to be replaced with a strong inducible promoter such as *lac*, or with a weak constitutive promoter. Thus, the cryIIA operon was expressed constitutively under the T7 promoter (Moar *et al.*, 1994); inclusion bodies were observed in *E. coli* grown in LB medium (Luria or Lenox Broth, g l⁻¹: pancreatic digest of protein 10.0, NaCl 5.0, yeast extract 5.0 at pH 7.0) after 24 h.

Transfecting the cloned gene with its upstream flanking sequences to a *Bacillus* species maximizes the expression from native promoters, and enables sufficient yield of insecticidal crystal proteins (Chambers *et al.*,

1991). The cloning is normally carried out in one of many *E. coli* to *B. thuringiensis* shuttle vectors, such as pHT 3101 (Lereclus *et al.*, 1989).

Small (several mg) samples of either *E. coli* or *B. thuringiensis* may be produced in shake-flasks on the LB medium. For better sporulation of *B. thuringiensis* or other Bacilli, $MgCl_2 \cdot 2H_2O$ (2 g l^{-1}) and $MnCl_2 \cdot 4H_2O$ (0.02 g l^{-1}) have to be added to the LB medium.

Optimization of production of recombinant proteins is a tedious procedure, and not worthwhile to undertake for non-commercial purposes. However, if the experimental set-up necessitates production of several grams of recombinant protein, it is recommended to use a small bench-top fermenter in fed-batch mode. Various techniques of high cell density recombinant protein expression are well described (for review, see Yee and Blanch, 1992).

Microbiological Safety

The genus *Bacillus* contains only one obligatory human pathogen, *B. anthracis*, which is responsible for anthrax in man and animals. *B. anthracis* is closely related to *B. cereus* and *B. thuringiensis*. *B. cereus* is commonly misidentified as *B. anthracis* (Collins and Lyne, 1984). Unlike the former, *B. anthracis* is non-motile, sensitive to penicillin and does not grow at 45°C.

Strains of *B. thuringiensis* that are used in the manufacture of commercially available biological insecticides have undergone extensive safety testing as a part of registration requirements, and have established an excellent safety record.

However, even the most harmless *B. thuringiensis* strains could cause allergies in some people, or even cause infections in people with a damaged immune system. Isolation of new *B. thuringiensis* strains requires caution. Soil and other microbiologically undefined samples should be treated as if they harbour human, animal or plant pathogens. Some strains of *B. thuringiensis* contain enterotoxin (Mikami *et al.*, 1995). Therefore, although *B. thuringiensis* is a minimal hazard organism, good microbiological practices should always be followed. Besides minimizing danger, these practices reduce the possibility of contamination and, thus, experimental errors.

Preparation of Spores and Crystals of *B. thuringiensis*

In some bioassays spores and crystals may be tested separately. For potency, parasporal crystals can be physically separated from spores, residual vegetative cells and cell debris by a variety of different methods (Goodman *et al.*, 1967; Delafield *et al.*, 1968). They are based upon various principles such as differences in buoyancy, hydrophobicity or surface charge. The degree by which such differences appear between parasporal crystals on the one hand, and spores, vegetative cells and cellular debris on the other hand are strain specific. Therefore, none of the many methods published over the years give satisfactory results with all strains. The use of different methods may be necessary to isolate parasporal crystals from a variety of strains. Some of the methods that we have found particularly useful in our work are described in Procedure 9.

Procedure 9. Separation of spores and crystals of *B. thuringiensis* (adapted from Delafield *et al.*, 1968).

Stock solutions: (i) buffer A: 20 mM phosphate buffer pH 7.0 (KH_2PO_4 1.07 g l^{-1}, K_2HPO_4 2.12 g l^{-1}) with 0.01% Triton X-100; (ii) buffer B: 3 M phosphate buffer pH 7.0 (KH_2PO_4 160 g l^{-1}, K_2HPO_4 318 g l^{-1}); (iii) System Y (Sacks and Alderton, 1961): Polyethylene glycol (PEG) 4000 11.2 g l^{-1}, buffer B 34 ml l^{-1}. The mixture is prepared by adding buffer B to the suspension of spores and crystals (in the vessel used for homogenization) followed by the solid PEG and water to the desired volume.

An aliquot of well-autolysed culture of *B. thuringiensis* containing about 4×10^{12} spores is precipitated by centrifugation. The residue, except for the dense lower part consisting of some unlysed cells and debris, is suspended in buffer A containing NaCl (1 M) and centrifuged. This operation is repeated at least four times. The residue is washed in buffer A containing NaCl (0.2 M) and then suspended in the minimal amount of the same buffer without salt. Residual unlysed cells are then removed by extracting it five times with System Y as follows. The spore and crystal suspension in buffer A is transferred into a 4-l Waring blender. System Y (1.5 l) is prepared in the bowl of the blender as described above, and homogenized for 2 min at top speed. The suspension is then centrifuged at 1500 *g* for several minutes in a swinging bucket rotor. The duration of the centrifugation depends upon the size and the geometry of the centrifuge tubes. It should be sufficient to achieve phase separation without precipitating spores and crystals. It is important to carry out the centrifugation at ambient temperature without cooling. About three-quarters of the upper phase containing spores and crystals is carefully removed by suction without disturbing the interface. Fresh blank upper phase is added to replace the removed volume, and the procedure of homogenization and centrifugation is repeated. The combined upper phases are diluted with one volume of distilled water and centrifuged.

    The residue is washed three times with buffer A and resuspended in the same buffer (180 ml). The suspension is mixed with an aqueous solution (20 g + 80 ml water) of sodium dextrane sulphate 500. Solid PEG 6000 (13 g), NaCl (7.5 g), and buffer B (3.3 ml) are added. After dissolution of solids, the volume of the mixture is brought to 600 ml by adding a well-stirred mixture of the same composition but with buffer A without bacterial particles. The mixture is transferred to a separating funnel, shaken vigorously, and kept for 30 min at 5°C. The mixture separates into two phases: the upper phase, rich in PEG, contains mostly spores; the lower phase, rich in dextrane sulphate, is enriched in crystals. The upper phase is removed, cleared of particles by centrifugation (25,000 *g*, 30 min) and returned to the separating funnel for repeated extraction.

Spore residues obtained from the upper phase in the first two rounds of extraction are collected. Crystals in the lower phase are practically free of spores (only 0.1% of all particles in the lower phase are spores). The lower phase is diluted with an equal volume of distilled water. The residue of crystals (about 60% of the initial count) is collected by centrifugation and washed several times with water.

Crystals dissolve completely in NaOH (0.1 N) or in 8 M urea solution (pH 8.5) containing β-mercaptoethanol (10%). The small amount of remaining insoluble material contains mostly spores. It is cleared by centrifugation. For further use in toxicity measurements, alkaline or urea solutions have to be dialysed cold against several changes of Tris-HCl buffer (10 mM, pH 8.4) (Somerville *et al.*, 1968).

Preliminary separation of spores and crystals improves the efficiency of this procedure. This may be achieved by the extraction of spores in the foam produced by shaking an aqueous suspension of spores and crystals (Gingrich, 1968; Sharpe *et al.*, 1979; Li *et al.*, 1987).

A relatively simple method of separating debris, spores and protein crystals uses ion exchange chromatography (Procedure 10; Murty *et al.*, 1994).

Procedure 10. Isolation of protein crystals of *B. thuringiensis* (adapted from Murty *et al.*, 1994).

Materials and solutions: (i) CM-cellulose, H$^+$ form, 0.7 meq g^{-1}; (ii) buffer A: sodium acetate buffer (50 mM, pH 5); (iii) buffer A + Ca^{2+}: buffer A containing CaCl$_2$ (15 mM); (iv) buffer B: sodium phosphate buffer 50 mM, pH 7.0; (v) buffer C: Tris-HCl buffer (10 mM, pH 8) containing EDTA (1 mM).

A well-autolysed culture of *B. thuringiensis* (0.5 ml, *c.* 10^9 spores ml^{-1}) is precipitated by centrifugation. The residue, except for the dense lower part consisting of some unlysed cells and debris, is suspended in buffer A + Ca^{2+}, centrifuged, resuspended in the same buffer (0.1 ml) containing phenylmethanesulphonyl fluoride (PMSF) (1 mM) and sonicated using, for instance, a Braun Labsonic L sonicator (100 W) for 30 s.

The resulting suspension is loaded on to the CM-cellulose column (15 ml, 1.4 × 10 cm) pre-equilibrated with buffer A. The column is then eluted with buffer A (150 ml) and buffer B (150 ml), which remove the cell debris and spores. Finally, the crystals are specifically eluted with buffer C. It is advisable to collect fractions (5 ml) and assess the crystal-containing fractions by light microscopy and SDS–PAGE.

Residual spores may be removed on a continuous sodium bromide (30–70% w/v) density gradient (Li *et al.*, 1987). Different types of parasporal inclusion bodies may be separated on discontinuous density gradients. Thus, bipyramidal and cuboidal crystals from *B. thuringiensis kurstaki* HD-1 were separated by the centrifugation (52,000 *g*, 1 h) of the crystals' suspension layered on a discontinuous gradient containing 30, 32, 33, 34, 35 and 36% (w/v) sodium bromide (Moar *et al.*, 1989).

Another method by which the toxin can be isolated from the spore–crystal mixture without separation (Luo and Adang, 1994) is shown in Procedure 11.

Procedure 11. Solubilization of crystal protein with KOH and reconstitution of protoxin (adapted from Luo and Adang, 1994).

The crystal–spore–debris mixture obtained after sporulation and lysis of *B. thuringiensis* culture (1 l of about 10^8 spores ml^{-1}) is centrifuged at 10,000 *g* for 15 min and washed several times with distilled water to remove all material that adsorbs at 260 nm. The pellet is then gently stirred for 30 min at ambient temperature with KOH solution (30 ml, 50 mM) containing β-mercaptoethanol (1%). The insoluble material is removed by centrifugation for 30 min at 27,000 *g*. The protoxin is precipitated from the alkaline supernatant by lowering the pH to 5.0 with HCl and recovered by centrifugation for 15 min at 27,000 *g*.

It is important to remember that solubilization and reconstitution of crystal proteins may lead to significant changes in toxicity caused by partial hydrolysis, proteolysis and denaturation. Decreases and increases in toxicity of reconstituted crystal proteins due to modification of solubility in the alimentary tract of the target insect have been reported (Du *et al.*, 1994).

References

Akiba, Y. and Katoh, K. (1986) Microbial ecology of *Bacillus thuringiensis*. V. Selective medium for *Bacillus thuringiensis* vegetative cells. *Applied Entomology and Zoology* 21, 210–215.

Aronson, A.I. and Geiser, M. (1992) Properties of *Bacillus thuringiensis* and its intracellular crystal proteins. In: Doi, R.H. and McGloughlin, M. (eds) *Biology of Bacilli: Applications to Industry.* Butterworth-Heinemann, New York, pp. 219–249.

Aronson, A.I., Beckman, W. and Dunn, P. (1986) *Bacillus thuringiensis* and related insect pathogens. *Microbiology Reviews* 50, 1–24.

Baum, J.A. and Malvar, T. (1995) Regulation of insecticidal crystal protein production in *Bacillus thuringiensis*. *Molecular Microbiology* 18, 1–12.

Bernhard, K. and Utz, R. (1993) Production of *Bacillus thuringiensis* insecticide for experimental and commercial uses. In: Entwistle, P.F., Cory, J.S., Bailey, M.J. and Higgs, S. (eds) Bacillus thuringiensis, *an Environmental Biopesticide: Theory and Practice.* John Wiley & Sons, New York, pp. 255–257.

Bernhard, K., Jarrett, P., Meadows, M., Butt, J., Ellis, D.J., Roberts, G.M., Pauli, S., Rodgers, P. and Burges, H.D. (1997) Natural isolates of *Bacillus thuringiensis*: Worldwide distribution, characterization and activity against insect pests. *Journal of Invertebrate Pathology* 70, 59–68.

Bradley, D., Harkey, M.A., Kim, M.K., Biever, K.D. and Bauer, L.S. (1995) The insecticidal CryIB crystal protein of *Bacillus thuringiensis* ssp. *thuringiensis* has dual specificity to coleopteran and lepidopteran larvae. *Journal of Invertebrate Pathology* 65, 162–73.

Brownbridge, M. (1989) Isolation of new entomopathogenic strains of *Bacillus thuringiensis* and *B. sphaericus*. *Israel Journal of Entomology* 23, 109–113.

Burges, H.D. (1998) *Formulation of Microbial Biopesticides, Beneficial Microorganisms, Nematodes and Seed Treatment.* Kluwer Academic Publishers, Dordrecht, The Netherlands.

Capalbo, D.M.F. (1995) *Bacillus thuringiensis*: Fermentation process and risk assessment, a short review. *Memórias do Instituto Oswaldo Cruz* 90, 135–138.

Ceron, J., Ortiz, A., Quintero, R., Guereca, L. and Bravo, A. (1995) Specific PCR primers directed to identify *cry*I and *cry*III genes within a *Bacillus thuringiensis* strain collection. *Applied and Environmental Microbiology* 61, 3826–3831.

Chak, K.-F., Chao, D.-C., Tseng, M.-Y., Kao, S.-S., Tuan, S.-J. and Feng, T.-Y. (1994) Determination and distribution of *cry*-type genes of *Bacillus thuringiensis* isolates from Taiwan. *Applied and Environmental Microbiology* 60, 2415–2420.

Chambers, J.A., Jelen, A., Gilbert, M.P., Jany, C.S., Johnson, T.B. and Gawron-Burke, C. (1991) Isolation and characterization of a novel insecticidal gene from *Bacillus thuringiensis* subsp. *aizawai*. *Journal of Bacteriology* 173, 3966–3976.

Chaufaux, J., Marchal, M., Gilois, N., Jehanno, I. and Buisson, C. (1997) Research on natural strains of *Bacillus thuringiensis* in different biotopes throughout the world. *Canadian Journal of Microbiology* 43, 337–342.

Cohen, E., Rozen, H., Josef, T., Braun, S. and Margulies, L. (1991) Photoprotection of *Bacillus thuringiensis kurstaki* from ultraviolet irradiation. *Journal of Invertebrate Pathology* 57, 343–351.

Collins, C.H. and Lyne, P.M. (1984) *Microbiological Methods.* Butterworth, London, pp. 356–360.

Crickmore, N., Zeigler, D.R., Feitelson, J., Schnepf, E., Lambert, B., Lereclus, D., Baum, J. and Dean, D.H. (1995) Revision of the nomenclature for the *Bacillus thuringiensis* pesticidal *cry* genes. In: *Program and Abstracts of the 28th Annual Meeting of the Society for Invertebrate Pathology*, Bethesda, Maryland.

de Barjac, H. (1981) Identification of H-serotypes of *Bacillus thuringiensis*. In: Burges, H.D. (ed.) *Microbial Control of Pests and Plant Diseases, 1970–1980.* Academic Press, London.

de Barjac, H. and Bonnefoi, A. (1968) A classification of strains of *Bacillus thuringiensis* Berliner with a key to their differentiation. *Journal of Invertebrate Pathology* 11, 335–347.

de Barjac, H. and Franchon, E. (1990) Classification of *Bacillus thuringiensis* strains. *Entomophaga* 35, 233–240.

de Barjac, H., Sisman, J. and Cosmao-Dumavoir, V. (1974) Description de 12 bacteriophages isolés a partir de *Bacillus thuringiensis*. *Comptes Rendus de l'Academie des Sciences Hebdominares* 279, 1939–1942.

Delafield, F.P., Somerville, H.J. and Rittenberg, S.C. (1968) Immunological homology between crystal and spore protein of *Bacillus thuringiensis*. *Journal of Bacteriology* 96, 713–720.

Delécluse, A., Rosso, M.-L. and Ragni, A. (1995) Cloning and expression of a novel toxin gene from *Bacillus thuringiensis* subsp. *jegathesan* encoding a highly mosquitocidal protein. *Applied and Environmental Microbiology* 61, 4230–4235.

Donovan, W.P., González, J.M., Jr, Gilbert, M.P. and Dankocsik, C.C. (1988) Isolation and characterization of EG2158, a new strain of *Bacillus thuringiensis* toxic to coleopteran larvae and nucleotide sequence of the toxin gene. *Molecular and General Genetics* 214, 365–372.

Du, C., Martin, P.A.W. and Nickerson, K.W. (1994) Comparison of disulfide contents and solubility at alkaline pH of insecticidal and noninsecticidal *Bacillus thuringiensis* protein crystals. *Applied and Environmental Microbiology* 60, 3847–3853.

Dulmage, H.T. (1983) Guidelines on local production for operational use of *Bacillus thuringiensis*, especially serotype H-14. In: Vandekar, M. and Dulmage, H.T. (eds) *Guidelines for production of Bacillus thuringiensis H-14: Proceedings of a Consultation held in Geneva, Switzerland, 25–28 October, 1982.* UNDP/World Bank/WHO, pp. 51–60.

Ely, S. (1993) The engineering of plants to express *Bacillus thuringiensis* δ-endotoxins. In: Entwistle, P.F., Cory, J.S., Bailey, M.J. and Higgs, S. (eds) *Bacillus thuringiensis, an Environmental Biopesticide: Theory and Practice.* John Wiley & Sons, New York, pp. 105–124.

Feitelson, J.S. and Narva, K.E. (1997) Primers and probes for the identification of *Bacillus thuringiensis* genes and isolates. US Patent 5,667,993 (Mycogen Corporation).

Flores, E.R., Perez, F. and Delatorre, M. (1997) *Bacillus thuringiensis* scale up; oxygen transfer, spore productivity. *Journal of Fermentation and Bioengineering* 83, 561–564.

Gelernter, W. and Schwab, G.E. (1993) Transgenic bacteria, viruses, algae and other organisms as *Bacillus thuringiensis* delivery systems. In: Entwistle, P.F., Cory, J.S., Bailey, M.J. and Higgs, S. (eds) *Bacillus thuringiensis, an Environmental Biopesticide: Theory and Practice.* John Wiley & Sons, New York, pp. 89–104.

Gingrich, R.E. (1968) A flotation procedure for producing spore-free crystals from commercial formulations of *Bacillus thuringiensis* var. *thuringiensis. Journal of Invertebrate Pathology* 10, 180–184.

González, J.M. and Carlton, B.C. (1980) Patterns of plasmid DNA in crystalliferous and acrystalliferous strains of *Bacillus thuringiensis. Plasmid* 3, 92–98.

González, J.M. and Carlton, B.C. (1984) A large transmissible plasmid is required for crystal toxin production in *Bacillus thuringiensis* variety *israelensis. Plasmid* 11, 28–38.

González, J.M., Brown, B.J. and Carlton, B.C. (1982) Transfer of *Bacillus thuringiensis* plasmids coding for δ-endotoxin among strains of *Bacillus thuringiensis* and *B. cereus. Proceedings of the National Academy of Sciences of the USA* 79, 6951–6955.

Goodman, N.S., Gotfried, R.J. and Rogoff, M.H. (1967) Biphasic system for separation of spores and crystals of *Bacillus thuringiensis. Journal of Bacteriology* 94, 485.

Gordon, R.E., Haynes, W.C. and Pang, C.H.N. (1973) The genus *Bacillus*. USDA Agriculture Research Service, Agriculture Handbook No. 427. Washington, DC.

Haider, M.Z., Ward, E.S. and Ellar, D.J. (1987) Cloning and heterologous expression of an insecticidal δ-endotoxin gene from *Bacillus thuringiensis* var. *aizawai* IC1 toxic to both lepidoptera and diptera. *Gene* 52, 285–290.

Heimpel, A.M. (1967) A taxonomic key proposed for the species of 'crystalliferous' bacteria. *Journal of Invertebrate Pathology* 9, 364–375.

Höfte, H. and Whiteley, H.R. (1989) Insecticidal crystal proteins of *Bacillus thuringiensis. Microbiology Reviews* 53, 242–255.

Inal, J.R.M., Karunkaran, V. and Burges, H.D. (1990) Isolation and propagation of phages naturally associated with the *aizawai* variety of *Bacillus thuringiensis*. *Journal of Applied Bacteriology* 68, 17–22.

Ishii, T. and Oba, M. (1994) The 23-kilodalton CytB protein is solely responsible for mosquito larvicidal activity of *Bacillus thuringiensis kyushiensis*. *Current Microbiology* 29, 91–94.

Klier, A., Fargette, F., Ribier, J. and Rapaport, G. (1982) Cloning and expression of the crystal protein genes from *Bacillus thuringiensis* strain Berliner. *EMBO Journal* 1, 791–799.

Lereclus, D., Arant, S.O., Chaufbauz, J. and Lecadet, M.M. (1989) Transformation and expression of a cloned δ-endotoxin gene in *Bacillus thuringiensis*. *FEMS Microbiology Letters* 60, 211–217.

Lereclus, D., Delécluse, A. and Lecadet, M.-M. (1993) Diversity of *Bacillus thuringiensis* toxins and genes. In: Entwistle, P.F., Cory, J.S., Bailey, M.J. and Higgs, S. (eds) *Bacillus thuringiensis, an Environmental Biopesticide: Theory and Practice*. John Wiley & Sons, New York, pp. 37–70.

Li, R.S., Jarret, P. and Burges, H.D. (1987) Importance of spores, crystals and δ-endotoxins in the pathogenicity of different varieties of *Bacillus thuringiensis* in *Galleria mellonella* and *Pieris brassicae*. *Journal of Invertebrate Pathology* 50, 277–284.

Liu, W.-M., Bajpai, R. and Bihari, V. (1994) High-density cultivation of spore-formers. *Annals of the New York Academy of Sciences* 721, 310–325.

Luo, K. and Adang, M.J. (1994) Removal of adsorbed toxin fragments that modify *Bacillus thuringiensis* CryIC δ-endotoxin iodination and binding by sodium dodecyl sulfate treatment and renaturation. *Applied and Environmental Microbiology* 60, 2905–2910.

Malik, K.A. (1991) Maintenance of microorganisms by simple methods. In: Kirsop, B.E. and Doyle, A. (eds) *Maintenance of Microorganisms and Cultured Cells: a Manual of Laboratory Methods*, 2nd edn. Academic Press, London, pp. 121–132.

Martin, P.A.W. (1994) An iconoclastic view of *Bacillus thuringiensis* ecology. *American Entomologist* 40, 85–90.

Martin, P.A.W. and Travers, R.S. (1989) Worldwide abundance and distribution of *Bacillus thuringiensis* isolates. *Applied and Environmental Microbiology* 55, 2437–2442.

McPherson, S.A., Perlak, F.J., Fuchs, R.L., Marrone, P.G., Lavrik, P.B and Fischhoff, D.A. (1988) Characterization of the coleopteran-specific protein gene of *Bacillus thuringiensis* var. *tenebrionis*. *Bio/Technology* 6, 61–66.

Meadows, M.P., Ellis, D.J., Butt, J., Jarrett, P. and Burges, H.D. (1992) Distribution, frequency, and diversity of *Bacillus thuringiensis* in an animal feed mill. *Applied and Environmental Microbiology* 58, 1344–1350.

Mikami, T., Horikawa, T., Murakami, T., Sato, N., Ono, Y., Matsumoto, T., Yamakawa, A., Murayama, S., Katagiri, S. and Suzuki, M. (1995) Examination of toxin production from environmental *B. cereus* and *Bacillus thuringiensis*. *Journal of the Pharmaceutical Society Japan* 115, 743–748.

Moar, W.J., Trumble, J.T. and Federici, B.A. (1989) Comparative toxicity of spores and crystals from the NRD-12 and HD-1 strains of *Bacillus thuringiensis* subsp. *kurstaki* to neonate beet armyworm. *Journal of Economic Entomology* 82, 1593–1603.

Moar, W.J., Trumble, J.T., Hice, R.H. and Backman, P.A. (1994) Insecticidal activity of the CryIIA protein from the NRD-12 isolate of *Bacillus thuringiensis* subsp. *kurstaki* expressed in *E. coli* and *Bacillus thuringiensis* and in a leaf-colonizing strain of *B. cereus*. *Applied and Environmental Microbiology* 60, 896–902.

Morris, O.N., Kanagaratnam, P. and Converse, V. (1997) Suitability of 30 agricultural products and byproducts as nutrient sources for laboratory production of *Bacillus thuringiensis* subsp. *aizawai* (HD 133). *Journal of Invertebrate Pathology* 70, 113–120.

Mummigati, S.G. and Raghunathan, A.M. (1990) Influence of media composition on the production of δ-endotoxin by *Bacillus thuringiensis* var. *thuringiensis*. *Journal of Invertebrate Pathology* 55, 147–151.

Murty, M.G., Srinivas, G., Roop, S.B. and Sekar, V. (1994) A simple method for separation of protein crystals from *Bacillus thuringiensis* using carboxymethyl cellulose column chromatography. *Journal of Microbiological Methods* 19, 103–110.

Navon, A., Klein, M. and Braun, S. (1990) *Bacillus thuringiensis* potency bioassays against *Heliothis armigera, Earias insulana,* and *Spodoptera littoralis* larvae based on standard diets. *Journal of Invertebrate Pathology* 55, 387–393.

Navon, A., Ishaaya, I., Baum, D., Ben-Ari, Z., Braun, S., Yablonski, S. and Keren, S. (1994) Activity of *Bacillus thuringiensis* preparations for control of the vine moth *Lobesia botrana*: laboratory and field experiments 1991–1993. *Alon Hanotea* 48, 360–366 (in Hebrew).

Obukowicz, M.G., Perlak, F.J., Kusano-Kretzmer, K., Mayer, E.J. and Watrud, L.S. (1986) Integration of the δ-endotoxin gene of *Bacillus thuringiensis* into the chromosome of root-colonizing strains of pseudomonads using Tn5. *Gene* 45, 327–331.

Sacks, L.E. and Alderton, G. (1961) Behavior of bacterial spores in aqueous polymer two-phase systems. *Journal of Bacteriology* 82, 331–341.

Salama, H.S. and Morris, O.N. (1993) History and usage of *Bacillus thuringiensis* in developing nations. In: Entwistle, P.F., Cory, J.S., Bailey, M.J. and Higgs, S. (eds) *Bacillus thuringiensis, an Environmental Biopesticide: Theory and Practice.* John Wiley & Sons, New York, pp. 255–267.

Salama, H.S., Foda, M.S., Dulmage, H.T. and El-Sharaby, A. (1983) Novel fermentation media for production of δ-endotoxin from *Bacillus thuringiensis*. *Journal of Invertebrate Pathology* 41, 8–19.

Saleh, S.M., Harris, R.F. and Allen, O.N. (1970) Method of determining *Bacillus thuringiensis* var. *thuringiensis* Berliner in soil. *Canadian Journal of Microbiology* 15, 1101–1104.

Schnepf, H.E. and Whiteley, H.R. (1981) Cloning and expression of the *Bacillus thuringiensis* crystal protein gene in *E. coli*. *Proceedings of the National Academy of Sciences of the USA* 78, 2893–2897.

Sekar, V., Thompson, D., Maroney, M., Bookland, R. and Adang, M. (1987) Molecular cloning and characterization of the insecticidal crystal protein gene of *Bacillus thuringiensis* var. *tenebrionis*. *Proceedings of the National Academy of Sciences USA* 84, 7036–7040.

Sharpe, E.S., Herman, A.I. and Toolan S.C. (1979) Separation of spores and parasporal crystals of *Bacillus thuringiensis* by flotation. *Journal of Invertebrate Pathology* 34, 315–316.

Shin, B.S., Park, S.H., Choi, S.K., Koo, B.T., Lee, S.T. and Kim, J.I. (1995) Distribution of *cry*V-type insecticidal protein genes in *Bacillus thuringiensis kurstaki* and *Bacillus thuringiensis entomocidus*. *Applied and Environmental Microbiology* 61, 2402–2407.

Smith, G.P., Merrick, J.D. and Ellar, D.J. (1996) Mosquitocidal activity of the CryIC δ-endotoxin from *Bacillus thuringiensis* subsp. *aizawai. Applied and Environmental Microbiology* 62, 680–684.

Smith, R.A. and Couche, G.A. (1991) The phylloplane as a source of *Bacillus thuringiensis* variants. *Applied and Environmental Microbiology* 57, 311–331.

Snell, J.J.S. (1991) General introduction to maintenance methods. In: Kirsop, B.E. and Doyle, A. (eds) *Maintenance of Microorganisms and Cultured Cells: a Manual of Laboratory Methods*, 2nd edn. Academic Press, London, pp. 21–30.

Somerville, H.J., Delafield, F.P. and Rittenberg, S.C. (1968) Biochemical homology between crystal and spore protein of *Bacillus thuringiensis. Journal of Bacteriology* 96, 721–726.

Tailor, R., Tippett, J., Gibb, G., Pells, S., Pike, D., Jordan, L. and Ely, S. (1992) Identification and characterization of a novel *Bacillus thuringiensis* δ-endotoxin entomocidal to coleopteran and lepidopteran larvae. *Molecular Microbiology* 6, 1211–1217.

te Giffel, M.C., Beumer, R.R., Klijn, N., Wagendorp, A. and Rombouts, F.M. (1997) Discrimination between *B. cereus* and *Bacillus thuringiensis* using specific DNA probes based on variable regions of 16S rRNA. *FEMS Microbiology Letters* 146, 47–51.

Travers, R.S., Martin, P.A.W. and Reichelderfer, C.F. (1987) Selective process for efficient isolation of soil *Bacillus* spp. *Applied and Environmental Microbiology* 53, 1263–1266.

Vecht-Lifshitz, S.E., Gandman, M. and Zomer, E. (1989) Process optimization and scale-up of the *Bacillus thuringiensis* fermentation. *Israel Journal of Entomology* 23, 239–246.

Visser, B. (1989) A screening for the presence of four different crystal protein gene types in 25 *Bacillus thuringiensis* strains. *FEMS Microbiology Letters* 58, 6783–6788.

Visser, B., Bosch, D. and Honeé, G. (1993) Domain-specific studies of *Bacillus thuringiensis* crystal proteins: a genetic approach. In: Entwistle, P.F., Cory, J.S., Bailey, M.J. and Higgs, S. (eds) *Bacillus thuringiensis, an Environmental Biopesticide: Theory and Practice*. John Wiley & Sons, New York, pp. 71–88.

Yee, L. and Blanch, L.W. (1992) Recombinant protein expression in high cell density fed-batch cultures of *E. coli. Bio/Technology* 10, 1550–1556.

Appendix 1D.1: Main Collections of *B. thuringiensis* Strains

1. IEBC International Entomopathogenic *Bacillus* Center
   (WHO Collaborating Center)
   Unite des Bacteries Entomopathogenes
   Institut Pasteur
   25, rue du Dr Roux
   F-75724 Paris Cedex 15
   France

2. H. Dulmage (HD) Collection
   Dr L.K. Nakamura
   Microbial Properties Research
   Northern Regional Research Center
   1815 N. University Street
   Peoria, IL 61604
   USA

3. *Bacillus* Genetic Stock Center
   Prof. Donald H. Dean, Director
   Departments of Biochemistry,
   Molecular Genetics and Entomology
   The Ohio State University
   484 W. 12th Avenue
   Columbus, OH 43210–1292
   USA

Bioassays of Replicating Bacteria Against Soil-dwelling Insect Pests

2

T.A. Jackson and D.J. Saville

AgResearch, Lincoln, New Zealand

Introduction

Soil-dwelling insects pose some of the most intractable of insect pest problems. As these insects are hidden within the soil, their damage is hard to anticipate, and once damage is obvious remediation is difficult. Many persistent chemicals, the mainstay of control in the past, are no longer available owing to environmental pollution or high costs of production. For these reasons, attention is focused on insect pathogens as possible biocontrol agents for this important group of pests.

The most economically important groups of soil-dwelling pests include root-feeding beetles (Scarabaeidae and Curculionidae), termites (Isoptera) and a variety of dipteran and lepidopteran species. Most research on biological control of soil-dwelling pests has been focused on the Scarabaeidae (Jackson and Glare, 1992) where a number of successful programmes have demonstrated the potential of microbial control. Microbial control agents are often highly specific to a particular pest species or group. To meet the challenge posed by the wide range of soil-dwelling pests, new species and strains of microorganisms are required and the best possible strains must be selected for field testing and development as biocontrol agents.

A number of significant bacterial pathogens have been isolated from soil-dwelling pests. Non-spore-forming bacteria are frequently isolated from the cadavers of dead insects and a number of species have been directly implicated in pathogenicity. Septicaemia caused by the bacterial species from the genera *Xenorhabdus* and *Photorhabdus* is the result of vectoring by entomopathogenic nematodes (Kaya and Gaugler, 1993). Some species, for example *Serratia marcescens*, are able to cross the protective membrane

© CAB *International* 2000. *Bioassays of Entomopathogenic Microbes and Nematodes* (eds A. Navon and K.R.S. Ascher)

barriers and invade the haemocoel unassisted. However, these pathogens are highly variable in their pathogenicity, suggesting that specific conditions are required for invasion. Amber disease of the New Zealand grass grub (*Costelytra zealandica*) is caused by specific, gut-colonizing strains of *Serratia entomophila* and *S. proteamaculans* which carry a specific 200-kb plasmid (Glare *et al.*, 1996). This unusual chronic disease results in cessation of feeding, leading to death (Jackson *et al.*, 1993). Some varieties of the spore-forming *Bacillus thuringiensis* (*Bt*) have shown activity against soil-dwelling scarabaeid and dipteran pests (Suzuki *et al.*, 1992; Smits *et al.*, 1993), but most common *Bt* toxins show little effect against pests that have evolved in the microbial-rich soil. *Bacillus popilliae* are obligate pathogens of Scarabaeidae with a range of strains associated with different scarab species. Thus, while bacteria constitute a resource that has the potential to be used for control of soil-dwelling insects, the best strains for development as biological control agents can only be differentiated with effective bioassay methods.

Bioassays of soil-dwelling insects with replicating bacteria pose a number of specific challenges. First, the insects themselves are often long-lived and difficult to rear in the laboratory. Second, bacterial pathogens of insects must be ingested prior to activity. Once in the gut they may release toxins, replicate or invade the host's tissues. Non-spore-forming replicating bacteria are inherently unstable, subject to death in storage and loss of mobile genetic elements. Replicating bacteria will, by definition, multiply within the host tissue and cause difficulty in interpretation of concepts such as infective or lethal dose. Koch's postulates must also be treated with caution, as once treated with bacteria the insect will become contaminated with the microorganisms, which generally ensures reisolation. Lastly, spore-forming bacteria often require specific conditions for germination and growth and their effects can be difficult to assess.

Overcoming these challenges is only possible with well designed bioassays and the purpose of any bioassay (the experimental hypothesis) must be clearly defined at the outset. Bioassays can be used for pathogen screening, differentiation between pathogens or selection of strains with specific attributes.

Obtaining Larvae for Bioassay

A regular supply of test insects is a prerequisite for carrying out bioassays with bacteria. Laboratory-reared insects are usually used for screening activity of new chemicals or general activity of microbial toxins and these can be standardized for age and quality. As specificity of the bacterial/insect interaction is often a key factor in the isolation of pathogens for soil-dwelling pests (Jackson, 1996), the target insect must be used in bioassays. However, long life cycles and diapause requirements have made scarab rearing difficult.

Species that can be reared in the laboratory include *Maladera matrida* (Gol'berg *et al.*, 1991), *Papuana* spp. (Theunis and Aloali'i, 1998) and some Cetoninae. Where the full life cycle cannot be achieved in the laboratory, larvae for testing can be reared from eggs. Adult beetles can often be collected in large numbers and confined in cages with food and an appropriate substrate for oviposition. Larvae can then be reared for testing. This process is, however, laborious and not always successful. Where laboratory rearing is not possible it will be necessary to collect larvae from the field, although the quality and uniformity of field-collected insects is more questionable.

As test bacteria must be ingested, it is important that the test insects are feeding, preferably at a similar rate. Feeding is usually greater at the start of each larval instar. For scarab larvae, the early 3rd instar is usually the optimum stage for testing as consumption by earlier instars is very low. Late-instar stages can become quiescent prior to moulting, which invalidates their use in the test assay. Insects may also vary in susceptibility to a microorganism according to life stage. In all cases, the stage of insect development should be noted; this can be indicated by head capsule width and body size.

The quality of the insects collected is also a major concern in bioassays. Survival of field-collected insects in the laboratory will vary with the amount of care in handling as well as conditions in the field at the time of sampling. Larvae collected from the field can suffer from direct damage due to handling or internal bruising which will lead to 'blue disease' and internal breakdown of tissues and septicaemia within a few days. Many insect larvae are highly aggressive if placed together. Cuticular lesions produced by the mandibles will lead to septicaemia and death. Larvae of the grass grub (*C. zealandica*) and the European cockchafer (*Melolontha melolontha*) appear to be highly aggressive and cannot be maintained in crowded conditions, while some other species are amenable to higher density storage. The effects of combat can be minimized by storage at low temperature immediately after collection. For the aggressive species, it is best to collect insects carefully from the soil and place in individual compartments in trays, which can be quickly placed in storage bins at low temperature. Intrinsic factors such as disease can also have an effect on the survival of any cohort. Acute disease can lead to high levels of mortality in both treated and control groups of larvae. Disease can also synergize the effect of other organisms. For example, *Cyclocephala hirta* larvae infected by *B. popilliae* showed increased susceptibility to entomopathogenic nematodes (Thurston *et al.*, 1993), while exposure of 2nd instar grass grub larvae to *S. entomophila* increased susceptibility to fungal disease (Glare, 1994).

Thus stage of larvae, handling and collection conditions, and the state of health of the insects can lead to high levels of mortality among the insects over the duration of any bioassay and can lead to problems in differentiation between treatments and interpretation of assay results.

This is reflected in an inflation in the sample size required for a statistically significant result as the mortality in the control group increases. For example, suppose the researcher wants to be 90% sure of obtaining a 5% significant result if the true situation is that 60% of the larvae die because of the experimental treatment. For a true control mortality of 5%, this means a true treated group mortality of 62% (= 5 + (60 × 95/100)) and a required sample size of n = 12 per group (interpolated from Table 2.1). For a control mortality of 20% the corresponding treated group mortality is 68%, and n = 20 (Table 2.1). For a control mortality of 40% the corresponding treated group mortality is 76%, and $n \cong 39$ (Table 2.1). That is, the sample size required for a statistically significant result increases substantially as the mortality in the control group increases. As a result the objective in any bioassay should be to keep the non-treatment deaths to a minimum.

Table 2.1. Numbers of test insects required to be 90% sure (power) of establishing a difference in percentage mortality at the 5% level of significance using a one-sided Fisher–Irwin test for varying theoretical levels of percentage mortality in the control and treated groups.

True % mortality in control group	True % mortality in treated group				
	95	90	80	70	60
	Number of insects required per group				
5	5	6	8	10	13
10	6	8	10	12	17
20	8	10	12	18	30
30	10	12	18	31	53
40	13	17	30	53	116

Sample sizes are reproduced from Haseman (1978).

Preliminary Assays to Isolate Pathogens

Bioassays can be carried out with samples ranging from macerates containing mixtures of unknown organisms to pure cultures of named isolates from a culture collection. Where it is suspected that morbidity or mortality in an insect population is caused by a pathogenic organism, exposure of test insects to simple macerates of the diseased insects can reveal the presence of a pathogen. In the most simple form of assay, insects can be macerated, applied to soil or fed to test insects and the treated cohort observed for symptoms of disease or death. This approach was used during the original investigations of amber disease of grass grub (Trought *et al.*, 1982) and is highly appropriate for bacterial diseases with simple life cycles and direct oral routes of infection. Once a mixed culture has been identified as containing pathogenic microbes, clonal populations can be isolated and each pure culture tested for pathogenicity. The test should provide a clear end point, such as production of disease symptoms or death, to enable differentiation between disease-producing isolates and others. Purified, disease-producing isolates can then be characterized by microscopic, biochemical and molecular methods. Care should always be taken when isolating and culturing bacteria with unknown properties. Procedures should be designed according to safe laboratory practice (e.g. Barkley and Richardson, 1994) to minimize the risks from exposure to unknown microorganisms.

Isolation and Culture of Bacteria

Bacteria are highly variable in their culture requirements. Some bacteria, such as *Serratia*, *Pseudomonas* and *Bacillus* spp., can be cultured *in vitro* on standardized growth media. Other insect pathogens are fastidious in their growth requirements and can be produced only *in vivo*. These include the Rickettsia and *B. popilliae*. When a bacterial strain showing definite insecticidal activity has been isolated, the clone should be placed in a culture collection and stored under ultra-deep-freeze or lyophilized. Bacteria for testing should not be subcultured too many times from the original characterized clones, as there is always the chance of loss of mobile genetic elements (e.g. plasmids), mutation or contamination. All of these factors can produce inconsistent results in bioassays.

Bacteria from culture collection should be plated out on to agar plates to ensure purity and cultured in a broth culture medium. Culture methods and details on bacteriological media are provided in Gerhardt *et al.* (1994) or other standard microbiological texts. Non-culturable bacteria must be isolated from insect hosts and processed to standardized, purified preparations.

In order to make comparisons between the effects of strains and species of bacteria, equivalent doses of microbe must be delivered to each test group. Small size and large numbers make estimation of bacterial dose difficult and this is an area where variation and errors can often occur.

Calculating dose with culturable bacteria

The standard technique for calculating the concentration of bacteria in a suspension is by dilution plate count (Gerhardt *et al.*, 1994). Using this method the original suspension is processed through a series of stepwise dilutions and mixings until a measurable density of cells is contained in suspension (Fig. 2.1). An aliquot (0.1 ml) of suspension is then dropped on to an agar plate and spread across the surface using a glass or wire 'hockey stick'. General nutrient agar supporting a wide range of species may be used or special media may be necessary. As it is often difficult to anticipate the cell density, two or three dilutions may be plated, usually in duplicate. With experience the number of dilutions necessary can be reduced as there is greater confidence in the expected yield of bacteria. For example, *Serratia* spp. grown in overnight shake-flask culture of nutrient broth yield $1-5 \times 10^9$ cells ml^{-1}. Fermenter broths can yield ten times this number. The plates are then incubated for 24 h or more, according to the growth rate of the bacterium, and the resultant colonies counted. It is recommended that the number of colonies on the plate should be between 30 and 300 for the best estimate of cell density in the original culture, which is calculated proportionally according to the dilutions (Fig. 2.1). Duplicate plates may give an accurate estimate of the concentration in the dilution tube but this may not be truly representative of the concentration in the original sample. A better estimate of the original concentration will be obtained by preparing more than one series of dilutions. While tedious, this will avoid problems in later interpretation of results. Stability of the original culture is also of importance,

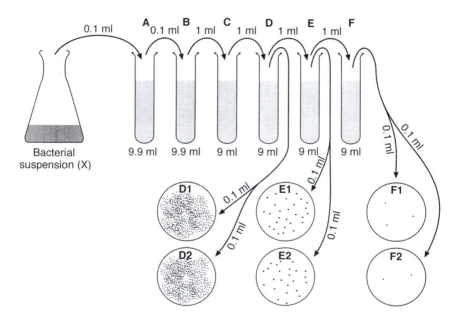

Fig. 2.1. Estimating the density of a suspension of culturable bacteria by dilution plate counting. (Density of cells ml^{-1} in X = number of cells (*n*) × tube dilution factor/volume of suspension plated.)

For the example above:

Bacteria ml^{-1} in X = $[((D1 + D2)/2) \times 10^6 + ((E1 + E2)/2) \times 10^7] / (2 \times 0.1)$
                    = $[((270 + 210)/2) \times 10^6 + ((29 + 23)/2) \times 10^7] / (2 \times 0.1)$
                    = $2.4 \times 10^9 \ ml^{-1}$

as the calculations of cell density cannot be made until growth has occurred on the plates, which will take one or more days. Stock cultures can then be adjusted by dilution to the required cell densities, but loss of viability in storage can render fine adjustment of cell density invalid. The plate count dilution procedure can be modified to meet particular needs and increase the number of samples that can be processed, for example using multiple microdots of 2–10 µl on a single agar plate, but the basic principles apply to all methods.

Calculation of dose with non-culturable bacteria

Dilution plating cannot be used for bacteria that do not grow on agar media, therefore alternative methods are required. If a pure culture can be obtained or the microbes have a distinctive morphology, estimates may be made with a bacterial counting chamber (e.g. Weber (Weber Scientific International Ltd, Teddington, UK) or Petroff–Hausser® (Arthur H. Thomas Co., Philadelphia, Pennsylvania)). Bacterial counting chambers differ from standard haemocytometers in that the depth of the chamber is 20 µm in comparison to the standard chamber depth of 100 µm. The counting chamber (Fig. 2.2) is defined by large, triple-lined grids (0.2 × 0.2 mm) divided into 16 smaller squares (0.05 × 0.05 mm). To estimate bacterial densities the original suspension must first be diluted (or concentrated) to an approximate cell density of $1–3 × 10^8$. This can be determined by trial and error. A drop of suspension (5–10 µl) should then be added to the counting chamber and the slide scanned at low magnification to ensure even spread has occurred with minimum aggregation of cells. If spread is uneven or cells are excessively clumped, the test sample should be remixed and diluted again. Bacterial dispersal can be aided by the addition of a surfactant. Cells should be counted, at ×400–1000 magnification, if they are within the square or touching the upper or right side of the square. Those touching the left or lower sides should not be counted. Small squares should be counted following a preset pattern to avoid double counting of some squares. Koch (1994) suggests that best estimates of density are obtained with bacterial numbers of between 5 and 15 per square and recommends counting at least 600 bacteria from a single slide. Our own variability analysis with *B. popilliae* spores indicates that a more cost-effective method of estimation is to make four to six slide preparations counting about 50 cells per slide, since we found most variation occurred between squares within grids and between slides within suspensions, and relatively little variation between grids within slides (Box 2.1).

Direct counting of bacterial cells works best for 'large' bacteria such as *B. popilliae* (sporangium approximately 6 × 1.5 µm) which will sink to the base of the chamber in 1–2 min. For smaller bacteria (Enterobacteriaceae and Micrococcidae), it is necessary to focus down through each square to

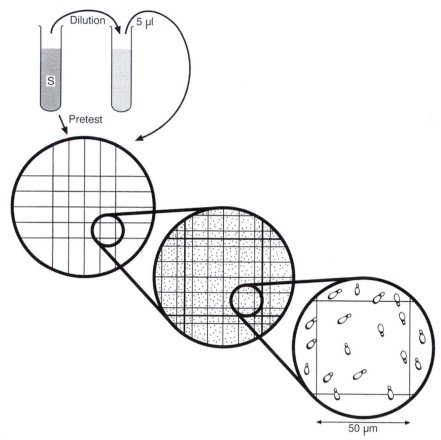

Fig. 2.2. Direct microscopic count for cell density estimates using a Weber bacterial counting chamber. (Density of cells ml^{-1} in S = n(average/square) × (2×10^7) × dilution factor.)

For the example above:

Bacteria ml^{-1} in S = $((10 + 13 + 14 + 7 + 9) / 5) \times (2 \times 10^7) \times (1/0.2)$
$$= 10.6 \times (2 \times 10^7) \times 5$$
$$= 1.06 \times 10^9$$

count all bacteria within the chamber volume. Information on improving counts for these bacteria is contained in Koch (1994).

Experimental Arenas

In any bioassay, insects can be tested as individuals or in groups. The advantage of treating groups of insects is that more insects can be handled

Box 2.1. Counting more squares, more grids or more slides? (i) Variance components for enumeration of suspensions of *B. popilliae* using a Weber bacterial counting chamber. (ii) The most cost-effective usage of time in enumerating suspensions of *B. popilliae*.

(i) Analysis of variance table for square root transformed counts of five squares (sq) in each of 3 grids (g) for each of three slide preparations (sl) for each of ten *B. popilliae* suspensions.

Source of variation	Degrees of freedom	Sum of squares	Mean square	Expected mean square	Estimated component
Suspension	9	617.3	68.587	–	–
Slides within suspensions	20	46.6	2.332	$15\sigma_{sl}^2 + 5\sigma_g^2 + \sigma_{sq}^2$	$s_{sl}^2 = 0.134$
Grids within slides	60	19.0	0.316	$5\sigma_g^2 + \sigma_{sq}^2$	$s_g^2 = 0.019$
Squares within grids	360	80.2	0.223	σ_{sq}^2	$s_{sq}^2 = 0.223$
Total	449	763.1	–	–	–

(ii) Cost-effective usage of time.
Given the above estimated variance components, the standard error of the mean (SEM) for any one suspension, given '*s*' squares counted in each of '*g*' grids on each of '*n*' slides, is:

$$SEM = \sqrt{\frac{1}{n}\left(0.134 + \frac{0.019}{g} + \frac{0.223}{gs}\right)}$$

Assuming that it takes $c = 8$ seconds (s) to count a square, no time to change squares, 8 s ($= c$) to change grids and 105 s ($= 13.125c$) to prepare and set up a new slide, the associated total cost (C) is:

$$C = ngsc + 13.125nc + ngc$$
$$= nc(gs + g + 13.125)$$

Given an estimate of the total time available, such as $C = 800$ s $= 100c$, we can express n in terms of g and s:

$$n = \frac{100}{gs + g + 13.125}$$

Continued

Box 2.1. *Continued*

This value can be substituted into SEM, or (SEM)^2 for convenience, yielding:

$$(SEM)^2 = \frac{gs + g + 13.125}{100}\left(0.134 + \frac{0.019}{g} + \frac{0.223}{gs}\right)$$

The objective now is to find the values of *g* and *s* which minimize (SEM)^2. We used a grid search in Excel. The values of *g* and *s* which yielded the minimum (SEM)^2 of 0.038, together with the associated values of *n*, were:

Squares (*s*)	Grids (*g*)	Slides (*n*)
2	2	5.2
3	2	4.7
4	1	5.5
5	1	5.2

This minimum SEM corresponds to a 95% confidence interval of about ± 20% on the backtransformed scale.

Note: If 95% confidence intervals are required for each suspension enumerated, the d.f. for error needs also to be considered. This consideration acts in favour of more slides rather than fewer slides.

in a set period of time. The disadvantage is that density-dependent competition can lead to high levels of mortality. For some social insects, such as termites, grooming removes pathogen propagules and the sensitivity of grouped insects is less than that of individuals held alone (Boucias *et al.*, 1996). Grass grub larvae and many other scarabs are prone to combat mortality and thus difficult to maintain in groups for long periods of time. The level of combat-related mortality can be reduced by lowering the temperature of the assay, but this may conflict with the temperature requirements of the microbe for infection. In general, short-term assays can be carried out successfully with individual larvae in soil-free compartments in trays. Longer assays (> 2 weeks) should be carried out in soil. Combat mortality can be avoided if the larvae are held in individual tubes.

Dosing

Application of microorganisms to the soil

To obtain a standardized dose of bacteria in the soil, microbes can be mixed through a bulk of soil prior to placement in pots or other experimental containers. Alternatively, a measured dose can be applied directly to the

experimental container. In our experiments with *Serratia* spp. we have pre-
ferred the latter approach where a measured dose of bacteria is applied in
water to a semi-dry soil (1 ml to 20 g soil, increasing soil moisture by
approximately 5% w/v). In this volume of water, bacteria enter the soil structure and are mixed through the soil by the feeding larvae.

Application to food

Insect larvae will often feed on artificial diet or vegetable material. Many
artificial diets contain antibiotics and, therefore, are not suitable for assessing replicating bacteria. Some scarab larvae will feed on small cubes of
carrot or other vegetables which can be impregnated with bacteria. We
have found that high density *Serratia* populations on fresh carrot are reasonably stable for 2–3 days in humid conditions but that low numbers of
bacteria may grow on nutrients provided by the carrot. In assays for *Bt*
toxins, Chilcott and Wigley (1993) incorporated bacteria and toxins into
artificial diet. Development of an artificial diet for scarabs was described
by Wigley (1992).

Forced feeding

Placement of bacteria into the insect mouthparts can provide a direct dose
of known cell density. Using a syringe with a blunted 30-gauge needle
attached to a Burkhardt applicator or similar precision dose delivery equipment, 1–2 µl can be applied directly into the oral cavity of the test larva. For
grass grub larvae the insect is held with the dorsal surface uppermost, the
blunt needle is inserted under the labrum and can be gently manipulated
between the basal sections of the mandibles allowing direct placement of
the measured dose into the oesophagus (Fig. 2.3). Sometimes regurgitation
can lead to loss of inoculum. By gently pushing the test insect further on to
the blunt needle the test dose can be placed directly into the midgut with no
apparent damage to the treated larva. This procedure was used by Dingman
(1996) for inoculation of Japanese beetle with *B. popilliae.*

Injection

Direct injection can provide an indication of the ability of bacteria to grow
within the haemocoel or can be used to determine pathogenicity for bacteria that show variable rates of infection due to difficulties in crossing the
cuticular membranes. Bacteria of the Enterobacteriaceae usually show limited and highly variable pathogenicity *per os*, but will provide a reproducible high degree of pathogenicity after injection. Bacteria of the genera

Xenorhabdus and *Photorhabdus* are highly pathogenic once injected and are usually vectored into insects by their symbiotic nematode carriers, *Steinernema* and *Heterorhabditis* spp. In assays of Japanese beetle with *B. popilliae*, bacteria can be injected either as spores or vegetative cells using a microinjector loaded with a syringe and a 27-gauge needle (Klein 1992, 1997). Third-instar larvae are held carefully and pushed on to the needle with the point entering the dorsal surface at the suture between the second and third abdominal segments and moving towards the posterior, parallel to the cuticle, into the haemocoel (Fig. 2.3). Care must be taken not to puncture the intestine. For comparative tests of *B. popilliae* isolates, a dose of approximately 1×10^6 spores is applied in 2–3 µl of suspension. Infection with *B. popilliae* can also be obtained with as few as 10–100 vegetative cells.

Incubation Conditions

Conditions of incubation can have a marked effect on the outcome of the bioassay. The thermal requirements for infection may vary. If the bioassay is carried out below the activation temperature of the pathogen, little effect will be recorded. Temperature may be chosen to minimize the effect of other pathogens. We have found that the fungus *Metarhizium anisopliae* is often the cause of death and mycosis among field-collected insects in long assays conducted at >20°C. In bacterial assays conducted at 15°C, occurrence of this fungus is seldom a problem. In soil-free assays, maintaining correct moisture level is critical. Desiccation is harmful to both bacteria and the test insects while free water from condensation can cause death by aiding ingress of contaminatory microbes. To control moisture, trays can be

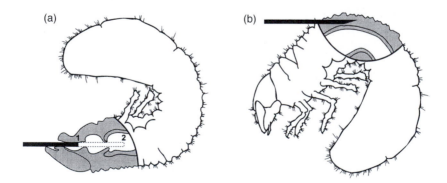

Fig. 2.3. Inoculation of scarab larvae. (a) Oral inoculation of a grass grub larva using a blunted 30-gauge needle inserted into the oral cavity (**1**) or midgut (**2**). (b) Intracoelomic injection of a grass grub larva with a 30-gauge needle inserted dorsally between the second and third abdominal segments.

bound in paper tissue and placed in plastic bags. Larvae produce moisture during respiration which condenses on the sides of the containers. When trays are filled with larvae considerable amounts of moisture can accumulate and we have found it best to transfer larvae to clean dry trays every 3–4 days. Care should be taken to avoid contamination during insect transfer by sterilization of forceps (alcohol or flame) between individuals or treatments.

Moisture and soil type can also affect the survival of insects in soil assay systems. Friable soil, moist but not wet, appears to produce the best results. Small containers, 20–50 ml, should be capped with a loose-fitting lid to prevent excessive evaporation in dry conditions. Soil moisture levels in larger open containers should be maintained by weight.

Screening Bioassays to Find Pathogenic Bacterial Isolates

One of the most common forms of bioassay is screening to detect pathogenic isolates of microorganisms. In any survey for culturable bacteria, a large number of clones can easily be isolated from insect, soil or other materials. If the required numbers of test insects are difficult or costly to obtain, it is important that the desired result, differentiation into pathogenic and non-pathogenic isolates, is obtained with the minimum of effort. In order to establish a screening programme it is important to establish a clear-cut testing protocol. The dose of microbe should be calculated to provide a 'maximum challenge' but with minimal effects caused by non-pathogenic bacteria. The conditions of the assay and expected outcomes should be clearly defined. The design of the assay will then be established to meet these criteria. As an example, the protocol for detection of strains of bacteria that cause amber disease is set out in Box 2.2.

Larvae that do not feed on the fresh untreated carrot offered in the 2–5 day period and show the gut clearance characteristic of amber disease are categorized as diseased, while feeding larvae with a darkened gut are noted as healthy. Dead larvae (usually very few) are omitted from analysis as treatment is not expected to result in death within the time of the experiment.

When large numbers of strains (k) are being screened for pathogenicity the key distinction is between the disease level observed for each strain and that observed in the control treatment. Since the control level is reused many times, multiple replicates of the control treatment should be included in the assay to provide a better estimate of the control level. For optimal efficiency, the multiple is $\sqrt{k}$ (Scheffe, 1959); for example, if $k = 25$ strains to be tested, there should be $\sqrt{25} = 5$ control groups randomized among the 25 treated strain groups within each statistical block.

When there is expectation that there will be few pathogens among the isolates, strains can be grouped, mixed and tested in a single dose. Where there is no evidence of pathogenicity, all isolates in the mix can be designated as non-pathogenic. When a pathogenic response occurs, the strains

Box 2.2. Testing protocol for detection of isolates of bacteria that cause amber disease.

1.   Scarab larvae should be collected from the field (the ideal stage is early 3rd instar, which can be identified by head capsule width, relatively small body size and a darkened midgut).
2.   Larvae are placed in individual compartments (2.5 ml) in trays with small (~20 mg) cubes of carrot for food and maintained in humid conditions in a dark incubator at 15°C for 24–48 h.
3.   Test bacteria are applied to carrot cubes either in suspension by micropipette (5 µl containing at least 1×10^7 cells) or the carrot cubes can be rolled across a growth of bacterial cells on an agar plate to provide a coating of bacteria on the carrot.
4.   Treated and control carrot cubes are placed in trays according to a predetermined randomized block design (see text).
5.   Healthy, feeding larvae (selected after stage 2) are added to cells in the trays in the random order specified by the design.
6.   Assessment of feeding is made after 48 h and larvae transferred to fresh trays with untreated carrot.
7.   After a further 48–72 h, larvae are assessed for feeding and expression of amber disease symptoms.

within the positive group should then be tested individually to determine which are pathogenic.

Experimental Design – Replication and Randomization

Replication and randomization are basic factors to consider when designing a bioassay, but determining the most appropriate layout can be complicated. To take an example (Table 2.2), five strains of bacteria were tested for pathogenicity with 216 larvae available for testing within trays where each larva was contained within an individual cell. Equal numbers of each strain and a control were treated allowing 36 larvae in each test group. These could be arranged in a number of ways varying from no randomization (all larvae in the same treatment group together) to complete randomization (each test insect receiving individual treatment). If the 36 larvae in each treatment are all treated together, there is only one genuine replicate and no valid measure of random variation within experimental treatments against which to assess the variation between treatment means. In practice a randomized block design can be used with experimental 'plots' being groups of insects. The question then arises of how many insects should be contained within each group. Fewer groups are more convenient for the researcher, but a larger number of groups is desirable for assessing the

Table 2.2. Levels of amber disease assessed 2 weeks after treatment with different strains of *S. entomophila*. Larvae were held in individual cells in trays, with the experiment following a randomized block design with an experimental unit of six larvae/treatment/block and six blocks. LSD (5%), least significant difference ($P < 0.05$).

Strain	% Disease
1	83
2	81
3	89
4	83
5	69
Control	17
LSD (5%)	19

variability between groups (i.e. ensuring the degrees of freedom for error is adequate).

In the example under consideration, six groups of six larvae were used for each of the six experimental treatments (five strains plus a control), allocating larvae to the treatments according to a randomized block design. To achieve this, treatment numbers (1–6) were randomly assigned to sets of six cells within each of the six blocks. The cells were then filled by allocation of larvae into cells within each block in turn, with selection from the insect pool sorted on size of larvae.

After 1 month, larvae were assessed and disease levels following bacterial treatment ranged from 69 to 89%, with 17% apparently diseased in the control. Analysis produced a least significant difference ($P < 0.05$) of 19, suggesting that disease levels were significantly ($P < 0.01$) higher than the control for all strains and that disease level following application of strain 3 was significantly ($P < 0.05$) higher than strain 5.

In the analysis, there was a suggestion of variation between blocks ($P < 0.10$), which may indicate an effect of larval size. There was no significant variation between the groups within each statistical block over and above the variation from larva to larva within each group. This suggests that the number of larvae per group was not crucial in this particular study (keeping to a total of 36 larvae per treatment), so larger groups could have been used as long as the degrees of freedom for error remained adequate. In general, we recommend a minimum of four groups per treatment.

A mistake commonly made by researchers is to not recognize the importance of having replicated experimental units that are treated independently of one another. When each insect within an experimental group is considered as an independent replicate, this can give rise to a spuriously low estimate of the experimental variation. For a statistically valid result, it is necessary to apply treatments to experimental units in a random, repli-cated way (e.g. a randomized block design), and statistically analyse the resulting group means.

While a single experiment can provide excellent information on a num-ber of bacterial test preparations, it does not necessarily provide robust infor-mation on the intrinsic variation between strains. Bacterial cultures may be affected by a number of factors including growth medium, culture methods, time to harvest, storage and handling. The effect of any organism will be influenced by conditions during the assay and the susceptibility of the particu-lar batch of test insects. When first examining bacterial pathogens against new targets, subtle interactions will not be known. It is therefore useful to replicate assays to increase the level of confidence that the experimental differences are indeed intrinsic strain effects rather than experimental aberrations.

During bioassays for evaluation of transgenic strains of bacteria (usu-ally 15–20 strains), strains A1MO2 and M4 were regularly used as internal controls. In these experiments, treatment was by presenting larvae with carrot rolled in overnight plate cultures of bacteria. Experimental design was randomized block, with 12 larvae in two groups of six for each treat-ment at each time. Larvae were assessed after 3 days for amount of feeding on the carrot. Table 2.3 presents the feeding results for each individual bioassay (analysed by analysis of variance as two blocks of three treat-ments) and combined over the bioassays by carrying out an analysis of variance of the means in the table as three treatments and eight blocks. The difference in feeding between A1MO2 and control was significant at $P < 0.01$ in two out of the eight assays, at $P < 0.05$ in two additional assays and was significant at $P < 0.01$ in the combined analysis. The difference in feeding between M4 and control was significant at $P < 0.01$ in one assay, at $P < 0.05$ in two additional assays and was significant at $P < 0.01$ in the combined analysis. The difference between the two strains, A1MO2 and M4, was significant at $P < 0.01$ in only one assay out of the eight but was significant at $P < 0.05$ in the combined analysis. In general, the results from the combined analysis can be expected to show more clearly any consis-tent trends in the data; in this example, both A1MO2 and M4 have been shown to have an antifeeding effect ($P < 0.01$), with the effect of A1MO2 significantly greater than the effect of M4 ($P < 0.05$).

When the difference in antifeeding effect between A1MO2 and M4 is compared with the date of assay it is clear that differences occurred only early in the year. Further regression analysis shows that the difference between the two strains is related to size of larvae (Table 2.3; $r = -0.738$; $P < 0.05$). Thus the difference in antifeeding effect is most clearly observed among smaller larvae which show very little feeding after being dosed with A1MO2. The larger larvae are more voracious and less affected by bacteria. In summary, pooling the assays allows greater confidence to be applied to the interpretation of results and elucidation of effects which were not obvious on analysis of the individual assays.

Table 2.3. Replicated tests for comparison of the effects of *S. entomophila* strains A1MO2 (wild-type strain) and M4 (tnPhoA mutant) on level of feeding (% carrot consumed) among 3rd instar grass grub larvae in the first 3 days from treatment.

Date	Bioassay								Mean
	7/3/98	27/3/98	16/4/98	1/7/97	8/8/97	30/7/97	16/10/98	23/10/98	
Larval wt (mg)	122	112	112	164	148	158	145	145	138
Feeding %									
Control	100	100	83	100	76	92	93	100	92.9
A1MO2	8	18	1	17	52	75	38	93	37.6
M4	28	92	43	42	52	68	51	100	59.4
LSD (5%)	28	29	77	51	175	23	79	26	21.7
Significance of differences									
A1MO2 versus Control	**	**	*	*	ns	ns	ns	ns	**
M4 versus Control	**	ns	ns	*	ns	*	ns	ns	**
A1MO2 versus M4	ns	**	ns	ns	ns	ns	ns	ns	*

Difference significant at * $P < 0.05$ and ** $P < 0.01$.

Quantification of Pathogenicity

Calculation of the acute lethal dose is common in toxicology and has often been applied to pathology. This is not without problems, especially when the test material is a replicating bacterium. Bacteria applied to food can grow, equilibrating the doses (see above), and even when applied directly into the gut, growth can occur over time. Hence the concept of infective dose is only really useful in a comparative sense where experimental parameters are clearly defined.

Quantitative assays with chemicals or non-replicating biologicals will lead to a clear-cut dose response. Insects receiving a lethal dose will be killed within a short period of time while those receiving sublethal doses will usually recover. However, many species of bacteria will replicate within the gut and provide a lethal dose as the bacteria increase in numbers. Insects treated with high numbers of bacteria will succumb first, but expression of disease will increase with time at the lower doses and the IC_{50} (a measure of the concentration of cells required to produce 50% infection among the test larvae) will decrease over time during a period of continued daily assessment. This effect is demonstrated in Fig. 2.4, where grass grub larvae were treated with single doses of *S. proteamaculans* over a range of serial dilutions and assessed for expression of amber disease symptoms over a period of 18 days. Disease symptoms appeared after 2–3 days at the higher dose rates and disease levels gradually increased at the lower dose levels over the first 8 days. After 8 days the IC_{50} dose of 6×10^4 bacteria could be calculated. Beyond 14 days, disease and death increased in the control and low-dose treatments causing the calculated IC_{50} to rise. Thus, the optimum time for calculation of the effect of disease and comparison between strains is between 8 and 14 days from treatment, when there is minimum variation in the IC_{50} value.

While the concentration of cells required for acute infection (a single, measured dose over a short time period) can be useful for comparisons between strains and genotypes, it will be considerably higher than the chronic dose acquired in the field while feeding. For example, in acute tests for disease-causing *Serratia* spp. the IC_{50} is approximately 5×10^4 bacteria. In the field, disease epizootics appear to be initiated at lower densities of bacteria as soil is ingested throughout the life of the larvae during consumption of roots and organic matter.

Discussion

The major problems in carrying out and interpreting bioassays for soil-dwelling pests with replicating bacteria can be overcome with attention to some basic principles: quality of test insects, good statistical design and reproducibility of results. If care is taken with selection and handling of

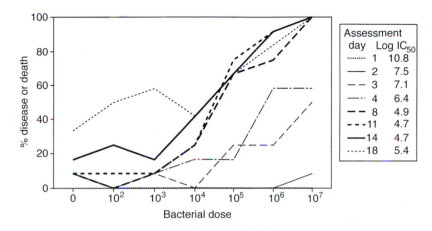

Fig. 2.4. Change in expression of amber disease with varying dose rates of *Serratia proteamaculans* (L2) at different time intervals over a period of 18 days. For each assessment date, a probit model which incorporated a parameter to allow for a non-zero control value (C), was fitted to the binomial data as a generalized linear model using the 'Fitnonlinear' directive of the GENSTAT statistical computing package. The 50% infective concentration (IC_{50}) was defined as the concentration which corresponded to a fitted infection level of (50 + C/2), as in Abbott's formula.

insects for collection and testing, it should be possible to keep non-treatment mortality down to a minimum and thus reduce problems with analysis. The penalty for lack of care in handling or selection of a weak population for testing is obvious – the need for a much greater number of replicates.

For statistically defensible results, each bioassay treatment needs to be replicated more than once, and treatments need to be applied and assessed (preferably blind) in a predetermined random order. This is less convenient for the researcher than a single replicate, but necessary for an assessment of the accuracy of the results. However, in the case of a series of bioassays which are to be pooled, it is sufficient to have just one replication as long as the treatments are applied in a random order.

When analysing bioassay data, care must be taken to avoid the use of the wrong error term. For determining the correct unit for statistical analysis, the rule of thumb is that the experimental unit is the 'item which can be randomized'. For example, if the larvae are in groups of ten which are randomized to treatments, then the experimental unit is the group of ten larvae, and the group means are the entities which should be subjected to statistical analysis. In such a case, the invalid use of the individual larvae as replicates may cause the error term to be spuriously low, with ensuing difficulties of interpretation and reproducibility.

For soil-dwelling pests, standardization of conditions between assays will always be a problem. The test insects are almost inevitably variable and uncertainty is compounded where the pathogens cannot be cultured *in vitro*. Given this uncertainty, greatest knowledge will be generated from simple bioassays that are repeated on a number of occasions rather than large complex experiments on a single population. If inconsistencies occur between assays, it will be necessary to look for interacting factors (stage and condition of pest population, other diseases, etc.). It is far better to find out that there are limitations on the use of a microbial control agent in the laboratory rather than after commercial release in the field.

Most tests with bacteria against soil-dwelling pests have been carried out with relatively few strains of *B. popilliae* or *Serratia* spp. To meet the challenge proffered by emerging pest species, new species and strains of microbe are needed for evaluation. Given the variety and number of diseases associated with the Scarabaeidae and other soil pests, it is certain that there will be new, useful strains waiting to be discovered. Effective bioassays are the first step in the process that can lead to successful microbial controls.

Acknowledgements

To the Microbial Control team at AgResearch, Lincoln, for data (especially Mark Hurst and Maureen O'Callaghan), analysis and discussion of the development of bioassays for soil-dwelling pests.

References

Barkley, W.E. and Richardson, J.H. (1994) Laboratory Safety. In: Gerhardt, P., Murray, R.G.E., Wood, W.A. and Krieg, N.R. (eds) *Methods for General and Molecular Bacteriology.* American Society for Microbiology, Washington, DC, pp. 715–734.

Boucias, D.G., Stokes, C., Storey, G. and Pendland, J.C. (1996) The effects of imidacloprid on the termite *Reticulitermes flavipes* and its interaction with the mycopathogen *Beauveria bassiana. Pflanzenschutz-Nachrichten Bayer* 49, 103–144.

Chilcott, C.N. and Wigley, P.J. (1993) Isolation and toxicity of *Bacillus thuringiensis* from soil and insect habitats in New Zealand. *Journal of Invertebrate Pathology* 61, 244–247.

Dingman, D.W. (1996) Description and use of a peroral injection technique for studying milky disease. *Journal of Invertebrate Pathology* 67, 102–104.

Gerhardt, P., Murray, R.G.E., Wood, W.A. and Krieg, N.R. (eds) (1994) *Methods for General and Molecular Bacteriology.* American Society for Microbiology, Washington, DC, 791 pp.

Glare, T.R. (1994) Stage-dependant synergism using *Metarhizium anisopliae* and *Serratia entomophila* against *Costelytra zealandica. Biocontrol Science and Technology* 4, 321–329.

Glare, T.R., Hurst, M.R.H. and Grkovic, S. (1996) Plasmid transfer among several members of the family Enterobacteriaceae increases the number of species capable of causing experimental amber disease in grass grub. *FEMS Microbiology Letters* 139, 117–120.

Gol'berg, A.M., Itzchaki, J., Nuriel, E. and Keren, S. (1991) The scarabaeid beetle, *Maladera matrida* Argaman, its damage to the fruit trees and its rearing technique. *Alon Hanotea* 45, 281–284 (in Hebrew).

Haseman, J.K. (1978) Exact sample sizes for use with the Fisher–Irwin test for 2 × 2 tables. *Biometrics* 34, 106–109.

Jackson, T.A. (1996) Soil dwelling pests – is specificity the key to successful microbial control? In: Jackson, T.A. and Glare, T.R. (eds) *Proceedings of the 3rd International Workshop on Microbial Control of Soil Dwelling Pests*. AgResearch, Lincoln, New Zealand, pp. 1–6.

Jackson, T.A. and Glare, T.R. (eds) (1992) *Use of Pathogens in Scarab Pest Management*. Intercept Ltd, Andover, 298 pp.

Jackson, T.A., Huger, A.M. and Glare, T.R. (1993) Pathology of amber disease in the New Zealand grass grub *Costelytra zealandica* (Coleoptera: Scarabaeidae). *Journal of Invertebrate Pathology* 61, 123–130.

Kaya, H.K. and Gaugler, R. (1993) Entomopathogenic nematodes. *Annual Review of Entomology* 38, 181–206.

Klein, M.G. (1992) Use of *Bacillus popilliae* in Japanese beetle control. In: Jackson, T.A. and Glare, T.R. (eds) *Use of Pathogens in Scarab Pest Management*. Intercept Ltd, Andover, pp. 179–190.

Klein, M.G. (1997) Bacteria of soil-inhabiting insects. In: Lacey, L.A. (ed.) *Manual of Techniques in Insect Pathology*. Academic Press, San Diego, pp. 101–107.

Koch, A.L. (1994) Growth measurements. In: Gerhardt, P., Murray, R.G.E., Wood, W.A. and Kreig, N.R. (eds) *Methods for General and Molecular Biology*. American Society for Microbiology, Washington, DC, pp. 248–277.

Scheffe, H. (1959) *Analysis of Variance*. John Wiley & Sons, New York, 477 pp.

Smits, P.H., Vlug, H.J. and Wiegers, G.L. (1993) Biological control of leatherjackets with *Bacillus thuringiensis*. *Proceedings of the Netherlands Entomological Society* 4, 187–192.

Suzuki, N., Hori, H., Ogiwara, K., Asano, S., Sato, R., Ohba, M. and Iwahana, H. (1992) Insecticidal spectrum of a novel isolate of *Bacillus thuringiensis* serovar *japonensis*. *Biological Control* 2, 138–142.

Theunis, W. and Aloali'i, I. (1998) Selection of a highly virulent fungal isolate, *Metarhizium anisopliae* Ma TB 101, for control of taro beetle, *Papuana uninodis* (Coleoptera: Scarabaeidae). *Biocontrol Science and Technology* 8, 187–195.

Thurston, G.S., Kaya, H.K., Burlando, T.M. and Harrison, R.E. (1993) Milky disease bacterium as a stressor to increase susceptibility of scarabaeid larvae to an entomopathogenic nematode. *Journal of Invertebrate Pathology* 61, 167–172.

Trought, T.E.T., Jackson, T.A. and French, R.A. (1982) Incidence and transmission of a disease of grass grub (*Costelytra zealandica*) in Canterbury. *New Zealand Journal of Experimental Agriculture* 10, 79–82.

Wigley, P.J. (1992) Development of an artificial diet and laboratory handling methods for the New Zealand grass grub, *Costelytra zealandica*. In: Jackson, T.A. and Glare, T.R. (eds) *Use of Pathogens in Scarab Pest Management*. Intercept, Andover, pp. 11–20.

Bioassays of Entomopathogenic Viruses

3

K.A. Jones

Natural Resources Institute, University of Greenwich, Chatham, Kent, UK

Introduction

Virus diseases of insects have been recognized for many years, although it is only in the last 40 years that significant interest has been shown in controlling insect pests with these viruses (Entwistle and Evans, 1985). At least 11 families and several other groups of viruses have been identified infecting insects. The major families, which include those used for pest control, are shown in Table 3.1. More detailed listings are given by Jones (1994) and, definitively, by Murphy *et al.* (1995). Of the 1200 virus–insect associations listed by Martignoni and Iwai (1981), 71% were found in Lepidoptera, 14% in Diptera, 7% in Hymenoptera and 5% in Coleoptera (Entwistle and Evans, 1985). Baculoviruses have, by far, received the most attention for development of pest control agents. This is primarily due to their inherent safety to humans and other non-target organisms, as well as their high pathogenicity to susceptible insects. Also, baculoviruses are, by far, the largest group of viruses that have been isolated to date, accounting for more than 50% of virus–insect associations described (Entwistle and Evans, 1985). For this reason this chapter will concentrate primarily on baculoviruses, but will mention other virus groups where relevant.

In nature baculoviruses normally infect insects following ingestion by the host. In the midgut, the occlusion body (OB) is dissolved to release infective virions. The replication cycle for nucleopolyhedrovirus (NPV) has been summarized by Winstanley and Rovesti (1993). The envelope of the virion fuses with the membrane of a microvillus and the nucleocapsid of the virus enters a midgut epithelial cell. Here the virus goes through a primary infection cycle. The nucleocapsid is transported to the nucleus of the midgut

Table 3.1. Main groups of viruses pathogenic to insects.

Virus family		Inclusion body	Main arthropod orders attacked	Other phyla attacked
Baculoviridae	Nucleopoly-hedrovirus	+	L, H, D, Col, Cr	None
	Granulovirus	+	L	None
Non-occluded rod-shaped virus	Oryctes-type	−	L, Col	None
Ascoviridae		+/−	L	—
Reoviridae	Cypovirus	+	L, H, D, Col, Cr	Vertebrates, Plants
Poxviridae	Entomopoxviridae	+	L, O, Col, D	Vertebrates
Parvoviridae	Densovirus	−	L, O, D	Vertebrates
Rhabdoviridae	Sigma virus	−	D	Vertebrates, Plants
Picornaviridae		−	L, H, Col, D, Hem	Vertebrates, Plants
Iridoviridae	Iridovirus	−	L, H, Col, D, E, Hem	Vertebrates

L, Lepidoptera; H, Hymenoptera; D, Diptera; Col, Coleoptera; Hem, Hemiptera; O, Orthoptera; E, Ephemoptera; Cr, Crustacea.

cell where the nucleocapsid uncoats to release the nucleic acid. Progeny nucleocapsids are assembled within the cell nucleus and migrate to the distal surface of the midgut cell; here they pick up a viral envelope as they bud from the cell into the haemolymph of the host to give budded virus. This primary infection therefore amplifies the ingested virions. Budded virions are taken up from the haemolymph by susceptible cells by absorptive endocystosis and are transported to the nucleus where they undergo secondary viral replication. During this replication cycle some of the progeny nucleocapsids migrate to the cell membrane and bud from the surface, as previously described. In contrast, the membrane of some of the progeny nucleocapsids is synthesized in the nucleus and these are subsequently occluded. The occluded virus is released on cell lysis and eventual death of the host. From the point of view of bioassay of baculoviruses it is important to note that budded virus is more infectious *via* the haemolymph than are virions obtained from occlusion bodies (the latter sometimes being referred to as polyhedra-derived virus). However, virions from occlusion bodies are more infectious to midgut epithelial cells.

Worldwide use of baculoviruses for pest control has been reviewed by Winstanley and Rovesti (1993) and more recently by Hunter-Fujita *et al.* (1998). Jones (1988a) and Jones *et al.* (1993b) have reviewed their production and use in developing countries. Winstanley and Rovesti (1993) list some 54 insect species that show good potential for control with baculoviruses. The majority (87%) are Lepidoptera, but the list also includes Coleoptera and Hymenoptera (2% and 11%, respectively). The authors also list 31 viral insecticides that are registered or close to registration, for control of 24 pests. The latter list does not include several products that are currently being used in developing countries, such as *Spodoptera exigua* NPV which is used in Thailand, *Spodoptera litura* NPV which is used in India, *Helicoverpa armigera* NPV which is used in India and Thailand, *Errynis ello* granulovirus (GV) in Brazil and *Phthorimaea operculella* GV in central and south America and Africa (Jones *et al.*, 1993b; Hunter-Fujita *et al.*, 1998).

Production of Baculoviruses

To date the large-scale production of insect viruses has only been carried out by infection and harvesting of insects in a factory or by field collection of infected larvae. Whilst *in vitro* cell culture is possible with some baculoviruses this has not yet been expanded to a commercial scale and is more expensive than *in vivo* techniques. Copping (1993) noted that the cost of *in vitro* production needs to fall (or productivity rise) between ten- and 100-fold for it to become commercially attractive. It is worth noting, however, that continued development of *in vitro* production technologies will occur as a result of a move toward greater automation, a closer control of production conditions, as well as the need to have such technology for the production of genetically manipulated viruses and virus expression-vector systems.

On a smaller laboratory scale, production by *in vivo* methods is also the norm. The basic methodology is similar for large- and small-scale production; production of NPV in lepidopterous larvae will be taken as an example.

In vivo production

In vivo production of baculoviruses has been reviewed by Shapiro (1982, 1986). In summary, susceptible insects are reared to an optimum stage and then infected with virus. The insects are then reared for a further period and harvested just prior to, or after, death. Research on optimizing production yields has concentrated on determining the appropriate virus dose administered, age/instar/weight of host insect, and length and temperature of incubation, as well as automating some steps (e.g. dispensing diet, egg placement). Normally, the amount of virus produced per insect is positively correlated to larval weight. Thus conditions are optimized so that larvae reach maximum weight before dying from viral infection.

Detailed descriptions of virus production for different insect species are given by Podgwaite (1981), Smits (1987), Im *et al.* (1989), McKinley *et al.* (1989), Bell (1991), Hughes (1994), Cherry *et al.* (1997) and Grzywacz *et al.* (1998) amongst others. Alternatively, virus can be obtained by field collection of virus-killed insects or production 'in the field' by farmers (Jones *et al.*, 1996).

Harvested larvae contain a mixture of virus, insect debris and contaminant microorganisms (bacteria, fungi, protozoa, etc.). Insect debris and other contaminants can alter the results of a bioassay either by affecting the virus directly, resulting in partial or total inactivation, antagonism or synergism, or by affecting the test insect/cell, resulting in death, or interfering in virus infection/replication or reducing insect feeding and hence virus uptake. These effects are often unpredictable and variable, particularly as the amount and type of contaminants can vary. Many production techniques have been designed to minimize contamination by microorganisms, for example harvesting of infected insects whilst still alive results in reduced numbers of spore-forming bacteria in comparison with harvesting after death (McKinley *et al.*, 1989). Production within a closed automated system also minimizes contamination. It is generally important to minimize potential sources of contamination through proper preparation of insect diets, promotion of a high standard of operator hygiene and selection and maintenance of healthy insect colonies. These are also essential to ensure predictable and even growth of insects, which is essential for optimum production and accurate bioassays.

In vitro production

As mentioned above, the large-scale production of virus *in vitro* is too expensive at present, although rapid advances have been made in recent years. *In vitro* production involves the production and infection of a susceptible insect cell line in a bioreactor. Ignoffo and Hink (1971) summarized the requirements of successful *in vitro* production as: (i) the development of robust, prolific insect cell lines that yield high pathogen titres; (ii) the availability of simple and cheap culture media; and (iii) development of plant-scale equipment and efficient, routine production procedures. Numerous cell lines are now available, along with suitable simple and serum-free media, as well as improved bioreactors and procedures (Monnet *et al.*, 1994; Vaughn, 1994; Weiss *et al.*, 1994). In many systems, however, there are still problems of production reverting to mutants with only budded virus. Also, further bioreactor improvements are required to achieve oxygen levels required in vessels larger than 250 litres. Cell culture of NPV is the best established, with a number of cell lines capable of supporting the replication of *Spodoptera* and *Heliothis* NPVs. Cell culture of GV is less well advanced, with only a few cell lines available that are capable of supporting virus replication (e.g. Winstanley and Crook, 1993). A necessary feature of cell culture systems is a high level of sterility, so the contamination problems encountered with virus produced *in vivo* do not occur, but of course this requires the availability of facilities that allow sterile handling of equipment. For further information of cell culture techniques the reader is referred to King and Possee (1992) and O'Reilly *et al.* (1994).

Purification of Virus Suspensions

Ideally, virus suspensions that are to be bioassayed should be purified so that other material or microorganisms present do not interfere with the infection process. However, in a number of cases the aim of a bioassay is to test the effect of these materials on viral potency. Even in such cases, it is desirable, if not essential, to also include a purified virus sample in the assay for comparison. A number of purification techniques are available which have different efficiencies, some of these methods themselves may also affect virus activity.

Acetone coprecipitation

This was first developed for *Bacillus thuringiensis* (Dulmage and Rhodes, 1971), but was subsequently adapted for use with viruses (e.g. Ignoffo and Shapiro, 1978). Aqueous lactose and acetone solution is slowly added to a virus suspension. This causes the virus to precipitate from the suspension, although bacteria will also precipitate. The suspension is then filtered and washed with sterile water. This technique removes, for instance, insect protein, as well as killing some vegetative bacteria. However, a number of workers have found that it also reduces the potency and shelf-life of baculoviruses (McGaughey, 1975; Hunter *et al.*, 1977; Ignoffo and Couch, 1981).

Density-gradient centrifugation

This has been described by Harrap *et al.* (1977) and Hunter *et al.* (1984) for baculoviruses, and is the most often-used method for producing highly purified suspensions. Infected larvae are macerated in 0.1% (w/v) sodium dodecyl sulphate (SDS), filtered through a double layer of muslin and centrifuged at 100 *g* for 30 s to remove gross debris. The supernatant is then centrifuged at 2500 *g* for 10 min for NPV and 10,000 *g* for 30 min for GV, to remove soluble material, lipids, and other contaminants. The resulting pellet is resuspended in 0.1% SDS, layered on a 45–60% (w/v) sucrose gradient and centrifuged at 50,000 *g* for 1 h for NPV and at 90,000 *g* for 2 h for GVs. The purified virus forms an opaque band at a sucrose density of 54–56%. The band is removed with a syringe or pipette, diluted in 10 times the volume of 0.1% SDS in sterile water, repelleted, as described above, and finally washed in distilled water by suspension and repelleting three times. Extra centrifugation steps may be included to improve purity (see Hunter-Fujita *et al.*, 1998). Although this process results in a highly purified virus, it also results in almost half of the virus being lost (Cherry *et al.*, 1997) and, if used for commercial production, can increase costs fourfold (Jones, 1994); it is therefore more suitable for small-scale laboratory applications. Purification techniques for other virus groups, based on centrifugation, are summarized by Evans and Shapiro (1997).

Semi-purification

Semi-purification of virus, which removes large insect debris and some contaminant microorganisms, can be used if the equipment is not available for density-gradient centrifugation. Semi-purification is also more suitable for large-scale production procedures, as long as the number and type of contaminant microorganisms is monitored (see Grzywacz *et al.*, 1997; Grzywacz, 1998). The following methodology for semi-purification of NPV has been successfully used on a laboratory scale by F.R. Hunter-Fujita and K.A. Jones in Thailand and results in a suspension that can be quantified using a haemocytometer (see 'Using a haemocytometer').

1. Macerate larvae and filter through a double layer of muslin.
2. Dilute 1:4 with 0.1% SDS in distilled water.
3. Centrifuge at 600 rpm in a bench-top centrifuge with a fixed-angle rotor to pellet insect debris. Discard the pellet.
4. Centrifuge the supernatant at 3150 rpm for 10 min and discard the supernatant. The virus will be pelleted in two layers, a darker layer on the bottom and a lighter layer on top.
5. Collect the darker layer and resuspend in 0.1% SDS. Store in a refrigerator.
6. Resuspend the lighter layer in 0.1% SDS and centrifuge at 4050 rpm for 12 min. Discard the supernatant. Again the virus pellet will be in two layers, a darker lower layer and a lighter upper layer.
7. Collect the darker layer and add to the suspension collected earlier in step 5.
8. Resuspend the lighter layer in sterile distilled water and centrifuge at 4050 rpm. Discard supernatant and resuspend the pellet in sterile distilled water and repeat this step. This should be repeated at least twice. The virus should now be stored in the refrigerator or freezer.
9. Take the 'dark' suspension and centrifuge at 4050 rpm for 10 min and discard the supernatant. Remove any virus in a light layer and treat as in step 8. For the dark layer repeat from step 5. This can be repeated as necessary. Finally, wash the remaining dark layer in distilled water as described in step 8 and store separately, or discard.

Other methods

More novel methods have also been studied which generally result in reduced contamination by microorganisms or kill unwanted microorganisms without removing them. These methods include ultra-high-pressure treatment (Butz *et al.*, 1995).

Virus Identification

Although the virus type can be determined visually, accurate identification is not possible unless carried out at the molecular level. This is most graphically illustrated with baculoviruses, which are named after the host from which they are originally isolated, for example *Helicoverpa armigera* NPV. However, *H. armigera* NPV can infect other *Heliothis/Helicoverpa* spp., and vice versa. Moreover, both single-embedded and multiple-embedded *H. armigera* NPV have been isolated (HaSNPV and HaMNPV). Finally, there are different strains of *H. armigera* NPV that have been isolated in different regions, which have different potencies (e.g. Hughes *et al.*, 1983; Ignoffo *et al.*, 1983; Williams and Payne, 1984).

Restriction-endonuclease analysis

Restriction-endonuclease (REN) analysis of viral DNA is by far the most commonly used method. REN analysis relies on the use of specific enzymes (restriction endonucleases), which recognize and cleave specific nucleotide sequences. These cut the viral DNA into fragments of different length, which can be separated through electrophoresis, to visualize and produce a characteristic profile (or 'fingerprint') for each virus (Nathans and Smith, 1975). REN analysis of baculoviruses has been described in detail by several authors, including Smith and Summers (1978) and Harrap and Payne (1979), with a general description of the methodology, including purification of viral DNA, being given by Hunter *et al.* (1984) and Hunter-Fujita *et al.* (1998). REN analysis requires relatively large amounts of virus, which should be available if a subsequent bioassay is planned. However, if only a small amount of virus is available, detection and identification is possible using the polymerase chain reaction (PCR) technique. PCR uses DNA polymerase to synthesize many complementary bands of DNA from a small amount of DNA template (amplification). These bands can be visualized through electrophoresis. Specific DNA sequences can be amplified, which allow identification of virus species or strains, through choice of appropriate primers. The methodology is described in detail by Brown (1998).

Other techniques

Other techniques that have been used include characterization of viral proteins through SDS–PAGE electrophoresis (Laemmli, 1970) and serological techniques, such as immunodiffusion/precipitation (Kalmakoff and Wigley, 1980), which can be used to detect virus in contaminated samples or in individual larvae. More novel methods such as dissolution characteristics of baculovirus occlusion bodies (Griffiths, 1982) have also been suggested. At best, these latter techniques can only distinguish between species and are unable to distinguish between strains. Serological techniques can be highly sensitive when using immunofluorescence, which is also able to detect virus within cells, and both sensitive and specific when monoclonal antibodies are used with radio-immunoassay, enzyme-linked immunosorbent assay (ELISA) or immunoaffinity chromatography (see Volkman, 1985); both sensitivity and specificity differ with the technique used.

Purification of Single-strain Isolates

A single virus-infected insect can often contain a mixture of virus strains. It is often desirable to obtain single-strain isolates for bioassay, for example, in order to determine which is the most infective or productive, or to determine the effect of interaction between strains. Isolation of single strains is possible through *in vitro* plaque assay (see '*In vitro* assay', p. 131); however, this is not possible for viruses for which there is no established cell culture system. Moreover, *in vitro* culture requires expensive facilities. A simple but effective alternative is to use the *in vivo* dilution technique described by Smith and Crook (1988). With this technique a low dose of a virus isolate (approximately LD_{10} or less) is given to early instar insects. Virus is collected individually from the insects that die and the process repeated preferably at least three times. At each stage the REN profile of the virus collected from each insect is obtained (see 'Restriction-endonuclease analysis', p. 101). The concept here is that at a low dose, a certain percentage of young larvae will be infected by a single virus particle, which will lead to the insect dying from a single strain (assuming that the insect is not already infected by an occult or latent virus).

The Bioassay Procedure

Viruses are intracellular, obligate parasites and therefore, unlike fungi and bacteria, which can be grown on artificial media, their potency can only be tested using living cells. This can either be with live insects (*in vivo* testing) or in cell culture systems (*in vitro* testing). Both have their advantages. Cell culture systems provide the opportunity for more precise control of assay conditions, as well as easier automation and scale-up. They also eliminate the possibility of the presence of occult or latent viruses affecting the assay. However, they do not reflect all the processes required for virus infection of insects, such as initial dissolution of OB of baculoviruses in the insect midgut; or may require initial pretreatment which may influence infectivity – for example, with baculoviruses, budded or non-occluded virus must be used for initial infection, which for the latter will require dissolution of occlusion bodies. Moreover, cell culture systems are currently not available for many viruses. Bioassay in live insects will test for all stages involved in the infection process and therefore can be more readily related to infection in nature; however, it is more difficult to control assay conditions, and the natural variability of the insect host necessarily leads to more variation in results. It is generally labour intensive and does not easily lend itself to automation.

A major aim of developing good bioassay methodology is to try and minimize variation whilst ensuring that the assay accurately reflects the conditions and effects one is trying to measure. The choice of bioassay methodology is dependent on facilities available as well as the effects one is trying to measure or observe (e.g. relative potency between different samples or exact measure of the number of virus particles required for infection).

Quantification of virus dose

It is important that the concentration of virus suspensions used in bioassay is accurately known. For occluded viruses, such as NPVs, GVs, cytoplasmic polyhedrosis viruses (CPVs) and entomopox viruses (EPVs), quantification is generally measured in terms of the number of OB. With some viruses – in particular NPVs – there are a variable number of virus particles per OB and therefore quantification of OB is not an accurate measure of the number of infectious units and, ideally, quantification of the number of virus particles should be made. However, in most situations, quantification of OB is sufficient and more practical. Moreover, this more accurately mimics the normal infection process, where an insect ingests an OB, which may be regarded as a concentrated package of infectious units.

There are several methods of quantification of virus suspensions; the method chosen depends on virus type (essentially size of the virus) and purity of the suspension. The OB of NPVs, CPVs and EPVs are large enough to be easily seen under a light microscope and can be counted using a haemocytometer and phase-contrast optics (×400). To ensure that only OB are counted, this method is generally restricted to purified suspensions, although, with practice, it is possible to enumerate OB in semi-purified suspensions. Due to the considerably smaller size of GVs, accurate enumeration using this method is very difficult and dark-field optics should be used. Alternatively, electron microscopy can be used (see 'Proportional count', p. 109).

Mixing of viral suspensions

Accurate quantification, and subsequent bioassay, of any suspension requires that the suspension particles are evenly distributed in the carrier liquid, without any aggregation or clumping. This is normally achieved through thorough mixing of the suspension by agitation for at least 30 s using a 'whirlimix' or similar device. McKinley (1985) reported that after agitation no significant sedimentation of NPV occurs for 10 min. Aggregates can be difficult to disperse and in such cases sonication can be used; preferably using an ultrasonic water bath to minimize contamination. However, it is possible with damaged or delicate viruses that long periods of sonication might disrupt the virus. To minimize the risk of such a possibility, it is suggested that long periods of sonication be avoided. Hunter-Fujita *et al.* (1998) recommend a period of 30 s to 1 min. A 'low-tech' alternative method for breaking up aggregates is to force the suspension through a hypodermic needle by sucking up and expelling the suspension three or four times with a hypodermic syringe. With NPV, McKinley (1985) reported that aqueous suspensions tended to form aggregates after repeated freezing and thawing, but the problem was much less if the suspension was prepared in 0.02 M phosphate buffer (pH 6.9)[1] (see 'Notes', p. 134). Hughes *et al.* (1983) added Darvan No. 2 (10 mg ml^{-1}; R.T. Vanderbilt Co., Los Angeles, California) to doses of NPV isolates to prevent aggregation.

Direct visual enumeration

These techniques are the most often used as they are simple and do not require expensive equipment.

USING A HAEMOCYTOMETER

This is most suitable for purified viruses. With NPVs, CPVs and EPVs a 0.1-mm deep chamber with improved Neubauer ruling is suitable. The use of this method for counting insect viruses has been described in detail by Wigley (1980) and Hunter-Fujita *et al.* (1998). In summary, improved Neubauer ruling consists of two grids of 25 squares, which are further subdivided into 16 smaller squares, each with an area of 0.0025 mm^2. When a coverslip is firmly placed on the slide, such that Newton's rings are visible (i.e. an effective seal exists), a gap of 0.1 mm is left between the top surface of the slide and the undersurface of the coverslip. The small square thus marks the area above which is a volume of 0.1 × 0.0025 mm^3 or 0.00025 µl. The methodology employed is described below.

1. Carefully clean the haemocytometer stage and coverslip with alcohol, using non-fibrous tissue or cloth.
2. Breathe gently on the slide surface and, with even pressure, firmly press

the coverslip on the slide. The coverslip should cover the grids drawn on both sides of the slide. Check that Newton's rings are visible.

3. Dispense the virus suspension at the edge of the coverslip and the surface of the slide; the suspension is drawn beneath the coverslip by surface tension until the grid surface is completely covered.

4. Place the haemocytometer on the microscope stage and leave undisturbed for 10 min to allow the OB to settle and reduction in Brownian motion.

5. The number of OB is then counted under phase-contrast or dark-field optics (×400). With the former, OB appear as bright spherical objects, often with a black dot in the centre. Counts are made of the number of OB contained within each small square. OB touching the top and right hand (central) line are included in the squares, those touching bottom and left are not. It is normally not necessary to count the number in all the squares on the slide.

As a maximum, 20 large squares (320 small squares, 160 from each of the two grids drawn on the haemocytometer) are viewed in a predefined pattern (e.g. Fig. 3.1); however, for accuracy a total of 300 OB should be included in the count. An optimum figure is around six OB per small square; above this counting becomes difficult, below this the statistical variation is high. Thus, concentrated suspensions should be diluted. Dilute suspensions should be concentrated by centrifugation (pellet at high speed in a microfuge and resuspend in a smaller volume of liquid).

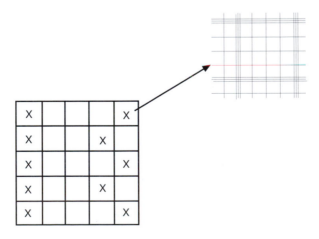

Fig. 3.1. Position of counts on an improved Neubauer haemocytometer.

The concentration of the virus suspension is then calculated as:

$$\text{OB ml}^{-1} = \frac{D \times X}{N \times K}$$

Where D = dilution of suspension dispensed into the haemocytometer, X = number of OB counted, N = number of small squares counted, and K = volume above a small square (in ml) (for a 0.1-mm-deep improved Neubauer haemocytometer this is 2.5×10^{-7} ml).

This can be alternatively written as:

$$M \times D \times V$$

Where D = dilution of suspension dispensed into the haemocytometer, M = mean number of OB per small square, and V = reciprocal of the volume (ml) above a small square (for an improved Neubauer haemocytometer this is 4×10^{6}).

It is important that the OB are randomly distributed and not clumped, as this will lead to an inaccurate count. The OB should follow a Poisson distribution and therefore the standard error of the mean number per small square is:

$$\text{SE} = \sqrt{M/N}$$

Ideally, the number of OB in each of the small squares should be recorded and the mean and standard error compared with that for an ideal Poisson distribution. In practice, the presence of clumping can be visually observed; if not present, a running total of the count can be made. Each suspension should be quantified at least twice; if the results obtained differ by more than the standard error the suspension was poorly mixed and further counts should be made.

Although this method is recommended only for purified virus suspensions, in practice skilled personnel are able to count suspensions accurately that have a low level of particulate contamination, such as semi-purified suspensions following filtration and low-speed centrifugation (see 'Semi-purification', p. 100).

From the calculation shown above it can be realized that the minimum suspension concentration that can be accurately quantified is approximately 1×10^{7} OB ml^{-1} – lower concentrations will have a high standard error. Suspensions with fewer OB should be concentrated by centrifugation, if possible. Granulosis viruses can also be quantified using a haemocytometer. In this case a 0.01- or 0.02-mm depth haemocytometer and dark-field optics are normally used. Moreover, as GV OB are much smaller they are more difficult to count accurately and distinguish from impurities.

DRY FILMS

It is possible to quantify unpurified suspensions of occluded viruses by counting dried stained films on microscope slides. One method has been described by Wigley (1980), in which known volumes of a virus suspension are evenly spread in circles of known size on a glass microscope slide and, after drying, stained with Giemsa (Gurr's improved R66), although Evans and Shapiro (1997) recommend staining with Buffalo Black (= Naphthalene Black B12, see 'Impression-film technique', p. 109). Using a light microscope and an oil-immersion objective (×1000), the number of OB are counted at specific points along the radius of the stained sample. The methodology is as follows:

1. Mix a sample of the test suspension with an equal volume of 0.1% (w/v) gelatin in distilled water.

2. Between 5 and 20 μl of the resulting mixture is spread in a 14-mm diameter circle on a microscope slide. This is achieved by placing an alcohol-rinsed glass slide on top of a template in which 14-mm diameter holes have been drawn or drilled. Four such circles can fit in a line within the area of a standard microscope slide. The suspension is dispensed in the centre of the circle and carefully spread from the centre in a single spiral motion, with a blunt seeker, to cover the whole circle.

3. Repeat with the same suspension for each of the four circles.

4. Leave the slides to dry.

5. Fix and stain the slides as follows:

- Fix by placing for 2 min in Carnoy's fixative[2].
- Wash in absolute alcohol for 30 s and allow to dry.
- Stain in 7% (v/v) Gurr's improved R66 Giemsa stain (BDH Ltd, Poole, Dorset) in 0.02 M phosphate buffer (see note 1) for 45 min.
- Rinse twice in tap water, drain by placing the edge of the slide on tissue paper and allow to dry.

The slide is then viewed under a microscope using oil immersion (×1000). Occluded viruses appear as clear, round objects, whereas bacteria and other contaminants are stained purple. Counts are made along a radius of each stained circle at predetermined positions from the edge of the circle. The number of OB visible at each position, or contained within the area defined by an eyepiece graticule, is noted. Normally ten counts are made along each circle radius, the direction of the counts being different for each circle (see Fig. 3.2). As the thickness of the stained film increases towards the centre of the circle, a weighting system has to be used for each count position. The weighting factor is calculated as:

$$\frac{Or - Ir}{R^2}$$

Where Or = Outer radius of count position (mm from centre of circle), Ir = Inner radius of count position (mm from centre of circle), and R = circle radius (mm).

Fig. 3.2. Direction of counts along radii of stained dried films.

Table 3.2. Count positions and weight factors for the dry film counting technique.

Count no.	Count position (mm from edge of stained circle)	Outer and inner radii (from circle centre)	Weight factor
1	0.25	7.0–6.5	0.138
2	0.75	6.5–6.0	0.128
3	1.25	6.0–5.5	0.117
4	1.75	5.5–5.0	0.107
5	2.25	5.0–4.5	0.097
6	2.75	4.5–4.0	0.087
7	3.25	4.0–3.5	0.076
8	3.75	3.5–3.75	0.096
9	4.75	2.75–1.75	0.092
10	5.75	1.75–0	0.062

The counting position and weight factors for a 14 mm circle are shown in Table 3.2. The mean count per circle is calculated as:

$$\Sigma w \times n$$

Where w = weight factor, and n = number of OB counted at each position. The mean count per sample is obtained as the sum of the counts for each radius divided by four (the number of circles counted). The concentration of the virus suspension is then estimated as:

$$\text{No. inclusion bodies ml}^{-1} = \frac{F}{M} \times C \times \frac{1000}{V} \times D$$

where F = area (mm^2) of each stained circle (e.g. for 14-mm diameter circle = $\pi \times 7^2$), M = area viewed/counted at each count position (area viewed by the eyepiece or eyepiece graticule), V = volume (in µl) dispensed for each circle, D = dilution factor of sample (including the twofold dilution in gelatin), and C = mean number of OB per circle.

The standard error is calculated as:

$$\text{SE} = \sqrt{\Sigma w^2 \times (y/n)}$$

where w = weight factor, y = mean count at each count position, and n = number of circles counted.

From the results the concentration of the virus suspension can be accurately estimated, along with the standard error of the estimate. Evans and Shapiro (1997) state that the most efficient suspension concentration for counting is 5×10^8 OB ml^{-1}; lower concentrations have high standard errors, higher concentrations are difficult to count due to the large number of OB. Suspensions with as little as 1×10^7 OB ml^{-1} can be quantified, but in this case the standard error is very high.

PROPORTIONAL COUNT

A third visual method suitable for purified or unpurified, occluded viruses is the proportional count described by Wigley (1980) where the virus suspension is mixed with a known concentration of polystyrene beads (or in the case cited, puffball spores) and smeared on to a microscope slide. The slide is then stained (as necessary for unpurified virus, e.g. using Giemsa, see 'Dry films', p. 106) and counted. Normally several counts (up to 40) are made per slide. The concentration of the virus suspension is calculated as:

$$R \times S \times \frac{V_1}{V_2}$$

where R = mean ratio of spores or beads to virus, i.e.

$$\frac{\Sigma p \text{ (individual counts of virus)}}{\Sigma b \text{ (individual counts of beads/spores)}}$$

S = no. of spores/beads per ml, V_1 = volume of bead/spore suspension, V_2 = volume of virus suspension.

The standard error of the count is calculated from the standard error of the ratio estimate:

$$\text{SE of } R = \sqrt{y^2/n}$$

where $y = R^2 \times [(1/\Sigma p) + (1/\Sigma b)]$, and n = number of counts.

Again, the limit of sensitivity for this method is approximately 1×10^7 OB ml^{-1}.

A variation of this method, suitable for viruses of any size including non-occluded viruses, is to use an electron microscope. The method has been described by Hunter-Fujita *et al.* (1998). Briefly, a virus suspension, prepared in water or 2% ammonium acetate, is mixed with an equal volume of a suspension of polystyrene beads that are of similar size to the virus and of known concentration. After thorough mixing the resultant suspension is sprayed using a nebulizer on to carbon-coated Formvar grids. After drying and treatment with 2% phosphotungstic acid (pH 7.2, 2–5 µl drop placed on grid, excess removed from edge with filter paper and left to dry for 10 min), the grids are viewed under the electron microscope and the ratio of virus to beads determined. The limit of sensitivity for this method is approximately 2×10^6 virus particles or OB ml^{-1}.

IMPRESSION-FILM TECHNIQUE

The method, described by Elleman *et al.* (1980), can be used to estimate the number and distribution of occluded virus on leaf surfaces. The technique uses clear, double-sided adhesive tape to remove the virus, which is then stained and counted.

1. Cut a 2 cm length of clear, double-sided adhesive tape, remove paper from one side and firmly attach to an alcohol-cleaned glass microscope slide. Care should be taken to ensure no air bubbles are trapped beneath the tape.
2. Repeat for as many samples as required. The slides can be stored, stacked in boxes.
3. When required the paper is removed from the adhesive tape. The plant surface is then evenly pressed on to, and then carefully pulled from, the tape. The pressure used is determined by trial and error. It should be firm enough to strip surface particles, including virus, from the surface, but not so as to exude sap or strip off the epidermis of the plant.
4. The slide is then labelled and placed in a microscope slide box.
5. When all samples have been collected, the slides are stained with Buffalo Black[3] for 5 min at 40–45°C.
6. The slides are then viewed with a light microscope fitted with an eye-piece graticule, under oil immersion ($\times$1000). OB appear as round, black objects, being distinguished from other such particles by their size. The number per unit area can then be determined.

This technique has been shown to be reasonably accurate, although Buffalo Black will stain many different proteinaceous objects. However, the maximum sensitivity for this method, which Entwistle and Evans (1985) put at 3.4×10^4 OB cm^{-2}, may not be sufficient for quantification of virus found on leaves sprayed at normal field rates. For example, when *S. littoralis* NPV was applied to cotton at a rate of 5×10^{12} OB ha^{-1} in 120 l, the mean number of OB cm^{-2} obtained on the lower surface of leaves in the upper plant canopy (where the target young instar larvae are normally located) was 3.85×10^3 (Jones, 1988b).

It should also be noted that the adhesive properties of double-sided tape differ between manufacturers, as well as with the age of the tape. Some tape will always stick too strongly to the leaf, thereby tearing off the epidermis when it is removed. The best tape will need to be determined by trial and error for each leaf species used.

Indirect methods of enumeration

Virus suspensions can also be enumerated using indirect or non-visual methods.

PROTEIN ASSAY

Hunter-Fujita *et al.* (1998) describe the enumeration of GV by determining protein concentration using a commercially available kit (Pierce (UK) Ltd). With this technique a calibration curve of protein concentration is obtained for a standard solution of bovine plasma albumin by measuring the absorbance of solutions of known concentration in a spectrophotometer (at 540 nm). Similarly, the absorbance of the test virus suspensions (note: the virus OB will have been dissolved by reagents in the kit) is determined and the protein concentration determined from the calibration curve. The relationship between protein concentration and virus concentration can be determined for different viruses by measuring the absorbance of virus suspensions of known concentrations. For example, Hunter-Fujita *et al.* (1998) state that for *Plodia interpunctella* GV a protein concentration of 10 mg ml^{-1} is equivalent to 5.03×10^{11} granules ml^{-1}.

ENZYME-LINKED IMMUNOSORBENT ASSAY

Serological techniques provide a method of virus detection and quantification which is both sensitive and specific and enzyme-linked immunosorbent assay (ELISA) has been used to detect and, in some cases, quantify insect viruses (Kelly *et al.*, 1978; Crook and Payne, 1980; Kaupp, 1981; Volkman and Falcon, 1982; Ma, 1985). ELISA uses an antibody to detect the virus and a subsequent enzyme-mediated reaction to quantify it. The end result is a colour reaction, the intensity of which is proportional to the amount of virus present. Jones and McKinley (1983) and Jones (1988b) used a double-antibody sandwich methodology (Voller *et al.*, 1976; methodology described in detail by Voller *et al.*, 1979) to quantify *Spodoptera littoralis* NPV. This technique used antibodies raised against the purified polyhedral protein of the NPV. Methodologies for purification of the polyhedrin (individual virus polypeptide) and antibody preparation are described by Jones (1988b) and Hunter-Fujita *et al.* (1998). The ELISA methodology is summarized below in which the various reagents are introduced into a polyethylene tube or well (Fig. 3.3). Commercially available 96-well microtitre plates are most frequently used, allowing several samples to be tested.

With most viruses a simple calibration curve of colour intensity versus virus concentration can be obtained by using suspensions of known concentration. Test suspensions are then quantified by comparison with the calibration curve. However, with occluded viruses the OB need to be solubilized to facilitate the antibody–antigen reaction. This can normally be achieved by addition of a low concentration (0.05 M) of Na_2CO_3. However, with unpurified and semi-purified suspensions of *S. littoralis* NPV, solubilization of OB was found to require harsher conditions (0.5 M Na_2CO_3 + 0.5 M NaOH + 0.1% (v/v) Tween 20). This has further implications resulting from the need to return the suspension to a neutral pH, and the subsequent high ionic concentration of the suspensions. Highly ionic solutions interfere with the antibody reactions. Moreover, interference of the reaction occurs at high concentrations of antigen; this results in a bell-shaped curve, rather than a straightforward calibration curve. Finally, the presence of impurities such as insect protein, formulation additives or plant material can affect the results, either through non-specific reactions (minimized by the addition of bovine albumin and polyvinylpyrrolidone) or enhancing sensitivity. The net

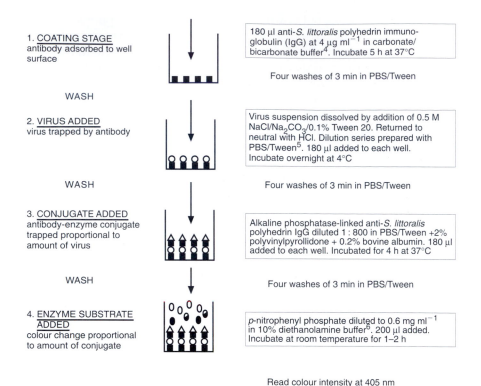

1. COATING STAGE
antibody adsorbed to well surface

180 µl anti-*S. littoralis* polyhedrin immuno-globulin (IgG) at 4 µg ml^{-1} in carbonate/bicarbonate buffer[4]. Incubate 5 h at 37°C

WASH

Four washes of 3 min in PBS/Tween

2. VIRUS ADDED
virus trapped by antibody

Virus suspension dissolved by addition of 0.5 M NaCl/Na$_2$CO$_3$/0.1% Tween 20. Returned to neutral with HCl. Dilution series prepared with PBS/Tween[5]. 180 µl added to each well. Incubate overnight at 4°C

WASH

Four washes of 3 min in PBS/Tween

3. CONJUGATE ADDED
antibody-enzyme conjugate trapped proportional to amount of virus

Alkaline phosphatase-linked anti-*S. littoralis* polyhedrin IgG diluted 1 : 800 in PBS/Tween +2% polyvinylpyrollidone + 0.2% bovine albumin. 180 µl added to each well. Incubated for 4 h at 37°C

WASH

Four washes of 3 min in PBS/Tween

4. ENZYME SUBSTRATE ADDED
colour change proportional to amount of conjugate

p-nitrophenyl phosphate diluted to 0.6 mg ml^{-1} in 10% diethanolamine buffer[6]. 200 µl added. Incubate at room temperature for 1–2 h

Read colour intensity at 405 nm

Fig. 3.3. Summary of the double-antibody sandwich method of ELISA used for quantification of *S. littoralis* NPV.

result is that one standard calibration curve is not sufficient, and a standard dilution series of known concentration and containing the same additives as the test suspension should be included in each test; moreover, a dilution series of the test suspensions should also be made so that comparison can be made on the basis of plotted curves, rather than a single point. Using this technique suspensions containing as little as 1000 OB ml^{-1} can be quantified (Jones, 1988b).

ELISA can be used to detect virus particles, rather than occlusion body proteins. Hence it is suitable for non-occluded viruses. Arguably, with occluded viruses, quantification of virus particles rather than OB should be made, particularly as with some viruses, such as NPVs, the number of virus particles contained within OB varies; also the size of the individual OB varies. However, as mentioned above, quantification of occluded virus is predominately on the basis of OB; moreover several hundred OB are contained within each sample tested, which will tend to average out variation.

Coulter counter

Highly purified suspensions of occluded virus can be enumerated automatically using a Coulter counter. Burges and Thompson (1971) found this to be the most accurate method of a number tested.

Larval equivalents

A number of reports in the literature (mainly referring to field trials), particularly early papers, quantify virus in terms of 'larval equivalents', which is taken as the amount of virus produced by a virus-killed insect (normally a final-instar larva). Generally this has been used for baculoviruses. Its advantage is that it gives an idea of the production requirements for field application. For NPV, one larval equivalent is taken to be approximately $1-2 \times 10^9$ OB; however, the amount of virus produced by a single larva varies both within and between species. This leads to inaccuracies and in some cases confusion, thus the use of larval equivalents should be discouraged, being used only as a last resort to give a very approximate indication of virus concentration.

Insect supply

Insects used for bioassay need to be as uniform as possible, and should be defined in terms of age, instar and, ideally, weight. All of these factors can influence susceptibility. A primary requirement is that the insects should be taken from a colony that is healthy and free of disease. This is achieved through rearing under controlled conditions (e.g. temperature, humidity, day length, food quality) and strict hygiene. Availability of a semi-synthetic diet gives more controlled and uniform food quality than a natural diet. Separate staff should be used for insect rearing and handling of virus, and insect rearing and virus handling/bioassay rooms (including post-treatment incubation facilities) should be well separated.

Principles and problems of mass rearing of insects have been described by Singh (1980) and Singh and Moore (1985), to which the reader is referred. Protocols for individual species have been published for a number of major insect pests, such as *Spodoptera littoralis* (McKinley *et al.*, 1984), *Helicoverpa armigera* (Armes *et al.*, 1992), locusts (Harvey, 1990) and so on. Insect age should be defined from hatch, which can be synchronized by collection of newly laid eggs daily; closer synchronization can be obtained by storage of eggs in the refrigerator and removal prior to being needed. For example, *S. littoralis* eggs normally hatch on the fifth day after laying at 26°C and hatching can occur over several hours. Placement of these eggs in a refrigerator at 4°C on the third day after laying for a period of 16 h resulted in hatching of all the eggs within a 2 h period 4 days later (McKinley, 1985).

Bioassay techniques

Bioassays are used to measure dose–response relationships or time–response relationships. The methodologies employed for these measures are essentially the same, although there are some important differences. The following sections describe the techniques used to measure dose–response relationships; the measurement of time–response relationships is described in 'Lethal-time and survival-time assay' (p. 130).

All *in vivo* dose–response assays should aim to contain a dose range that results in insect mortalities between 10 and 90%. In general, five doses should be used, with an absolute minimum of 20 insects treated at each dose. There is normally no advantage of treating more than 50 larvae per dose. A negative control, consisting of the carrier used for preparation of the virus doses, should be included in all assays. Assays should be repeated at least three times. The requirements for *in vitro* assays are given in the section headed '*In vitro* assay' (p. 131).

Mass dosing bioassays

Mass dosing assays are used to estimate the effect of virus concentration on a test insect, most commonly as LC_{50} – the concentration required to kill 50% of the sample population. The advantage of these assays is that a large number of insects can be dosed, allowing for a direct comparison to be made between several samples, often expressed in terms of relative potency (see 'Analysis of results', p. 127). In general, mass dosing assays are most often used for first-instar or neonate larvae; however, the method can be used with any age of larvae. There is a possibility that larvae may interact/disturb each other, leading to variation in the amount of virus dose ingested and variable growth of larvae. In practice, this is not a problem with many species, but preliminary tests should be carried out to determine this. In contrast, mass dosing assays are essential for insects that feed gregariously. In most mass dosing assays the virus dose is either spread on the surface of the insect diet or incorporated into the insect diet. A third alternative that is being increasingly used is presenting the dose to the insect as drops from which the insect drinks. Examples of each type are presented below.

SURFACE DOSING
This method is most suitable for larvae that feed on the surface of the diet/food; many noctuid larvae, such as *Spodoptera littoralis*, eat only the

leaf epidermis as young (I–III instar) larvae. This method of dosing mimics the natural situation, where virus is picked up from the leaf or food surface. The virus inoculum is evenly spread over the surface of the insect diet on which the insects are placed and feed (leaves may be used, but these need to be held flat). An important assumption here is that the inoculum is evenly spread over the entire surface of the diet; in practice this means that the diet surface must be horizontal and free of holes, where the inoculum can collect, or lumps, which would not be covered by the inoculum. McKinley *et al.* (1984) achieved this by using a diet that contained fewer and more finely ground solids in the diet used for bioassay than in the diet used for normal insect rearing[7].

The method described is one developed for *Spodoptera littoralis* by McKinley (1985).

1. Carefully pour the artificial diet into the bottom of a 30-g (1-oz) plastic pot to form a smooth layer approximately 4-mm thick. It is important not to introduce air bubbles into the diet or splash diet on to the sides of the pot. A plastic water bottle used for household pets is suitable for pouring the diet in this way.
2. Leave the pots undisturbed on a horizontal surface until the diet has cooled and solidified. The pots can then be stacked and stored in the refrigerator until required.
3. A dilution series of the test suspension is prepared in water or phosphate buffer[1]. Normally a fivefold dilution series, with five dilutions is suitable. As mentioned above, the series should result in mortalities ranging between 10 and 90%; an initial assay with tenfold dilutions can be used to estimate at what suspension concentrations this will occur. An untreated control of distilled water or buffer only is also prepared.
4. The virus dose is dispensed in volumes of 100–150 μl on to the surface of the diet. The pot is then gently tilted and rotated to spread the dose over the diet surface. The use of a brush or cloth to spread the virus over the diet surface is not recommended as these will collect or absorb a variable amount of the dose. It should also be noted that different diets (including different batches of artificial diet) might absorb varying amounts of the dose, which can lead to some further variation. It is recommended that pots are treated with the lowest dose first (starting with an untreated control), working up to the highest dose, in order to minimize the risk of contamination of doses.
5. Dosed pots are then placed open on a flat surface for the dose to dry. (Note: open pots should not be left too long as the diet will begin to dry out and shrink from the walls of the pot, leaving an undosed surface.)
6. Ten neonate larvae are then placed in each dosed pot using a small paintbrush. The pots are then closed with a lid labelled with treatment and replicate number. Larvae are placed in the lowest dose pots first. To avoid build up of condensation water in the pots, which can result in

larvae drowning, small holes are made in the lid and a piece of filter paper or tissue placed between the lid and pot.

7. The paintbrush should be sterilized between each dose by dipping in 2% sodium hypochlorite solution, followed by rinsing in distilled water.

8. The pots are then kept in a rearing room or incubator at 26 ± 1°C and a 12-h photoperiod.

9. After 24 h the number of live larvae in each pot is counted in order to account for any deaths due to handling. Deaths are assessed by gently tapping the pots on a bench surface; larvae that do not move or those that disintegrate are counted as dead.

10. The number of live larvae is again counted on the fifth day after dosing. With NPV, the average time taken for the virus to kill larvae is approximately 4 days, so termination of the assay on the fifth day avoids secondary infection of larvae. Moreover, larvae up to 7 days old (third instar) are not cannibalistic. It is possible to assess the assay up to the seventh day after dosing. McKinley (1985) noted that, although total mortality of each test larval population increased with time post-dosing, the relative difference between each group remained the same. If the assay needs to be continued beyond 7 days, larvae must be placed individually in pots. (Note: with genetically engineered viruses where the speed of kill has been enhanced, the time taken to kill the insects is less and therefore the assessment times will need to be adjusted accordingly.)

11. A minimum of 30 larvae are treated at each dose. If ten larvae are placed in each pot, this number is increased to 50 larvae (five pots at each dose). Five doses are included in each assay, along with an untreated control in which distilled water (or buffer or other carrier, if this has been used to prepare the doses) is dispensed into the pots in place of the virus suspension.

DIET INCORPORATION

Essentially the methodology is similar to the surface dosing assay except that the virus dose is evenly mixed throughout the diet. This is normally achieved using a food mixer. With agar-based artificial diet the dose must be incorporated before the agar cools sufficiently for it to gel. Approximately 5 ml of the virus dose is added to 95 ml of diet. However, the virus cannot be added when the liquid diet is very hot, as this could inactivate the virus. Thus the inoculum needs to be added when the diet is at temperatures below 40°C and the diet poured immediately into the appropriate containers and allowed to set. McKinley (1985) noted that during setting of diet with low solid content, the solids tended to sink to the bottom; this may result in uneven distribution of the virus in the diet. Thorough mixing of the virus inoculum and diet does result in an 'even spread' of the dose throughout the media; in contrast the surface dosing technique is more prone to a more uneven spread of the dose. Also, as the inoculum is mixed throughout all the insect diet, the dose is diluted to a greater degree than with surface dosing assays, although this is partly compensated by the fact that all of the food that the larvae consume will contain virus. The diet incorporation methodology is most suitable for species that do not just feed on the surface of the diet but tend to burrow, such as codling moth, *Cydia pomonella*.

Either of the above methods can be used for larger larvae, but normally only one larva would be placed into each pot, as many lepidopteran larvae are cannibalistic. Moreover, those that are not normally cannibalistic will often eat virus-infected larvae. With larger larvae, the larvae should be incubated until death or moth emergence, rather than the assay being terminated after a maximum of 7 days (see 'Diet-plug assay', p. 120).

DROPLET ASSAY

This technique is based on the synchronous peroral method described by Hughes and Wood (1981) and Hughes *et al.* (1986), in which neonate larvae are allowed to drink from drops of virus suspension. The methodology, summarized by Jones *et al.* (1993a), is described below.

1. Five dilutions (normally on a three- to fivefold dilution scale) of each test suspension are prepared in sterile distilled water containing 1% (w/v) Brilliant Blue R (Sigma Chemical Co., Poole, UK). A control treatment, consisting of sterile distilled water containing 1% Brilliant Blue R, is also prepared.
2. For each dilution, six 5 µl drops are dispensed by pipette on a sheet of Parafilm (American Can Co., Greenwich, Connecticut), or similar hydrophobic surface, in a circle approximately 2 cm in diameter. All five dilutions from a test suspension can be dispensed in this way on to a 20 cm × 10 cm sheet.
3. Between 100 and 150 neonate larvae are placed in the centre of each circle of drops by tipping them through a 7-cm diameter plastic funnel.
4. Clean 30-g plastic pots are then inverted over each circle of drops, thereby enclosing the larvae. The pots are labelled with dose or treatment number and left undisturbed for 30 min during which time the larvae approach and drink from the drops. During this period further treatments are prepared in the same way.
5. After 30 min, 90% of the larvae will have drunk from the drops and are distinguishable by the presence of blue dye in their guts. The covering pot is removed, one dose at a time, and the dosed larvae transferred to 30-g plastic pots containing artificial diet[7]; although it is not necessary to have a flat layer of diet poured into the pots, in practice this makes subsequent counting easier. For each dose, ten larvae are transferred to each of five pots. Larvae are always transferred from the lowest dose first, the untreated control being handled before any of the test samples. The pots are closed and labelled as described above for the surface dosing assay.
6. Between samples, the brush used for transferring the larvae is washed in

a 2% sodium hypochlorite solution and rinsed in distilled water in order to minimize any chance of contamination.

7. Mortality is assessed after 24 h and then 5–7 days after dosing as described in 'Surface dosing' above. As explained with the surface dosing assay, if the assay is continued beyond this period, larvae should be placed individually in pots after dosing.

With the methodology described by Hughes *et al.* (1986), Petri dishes containing diet were inverted over the drops and the drops were dispensed by dipping the heads of a circle of pins or nails in the test suspension. One drawback of this is a variable number of larvae migrate to the diet. It is more laborious to remove dosed larvae from the diet (in order to achieve a standard number per dose) than it is to transfer dosed larvae to diet. Not all species will readily drink, although this might be influenced by how the eggs/larvae have been treated prior to assay. Hughes and Wood (1981) added sucrose (5–20% w/v has been used) to encourage drinking with some species and this also delayed sedimentation of the virus; however, the amount of suspension ingested by some species has been shown to be less when sugar is added (Hughes and Wood, 1986). Assays with and without sugar should not be directly compared. Jones (1988b) noted that *S. littoralis* larvae that hatched from eggs which had been surface-sterilized did not drink; the reason for this is unclear, but could be related to a higher water content of the surface-sterilized egg masses or some effect of residual chemicals (formalin or sodium hypochlorite) used for sterilization.

The droplet assay allows for a large number of larvae (and hence samples) to be tested in a single assay – a great advantage when wishing to compare treatments. Jones (1988b) was able to dose over 5000 larvae (equivalent to 20 treatments of five doses, plus an untreated control) in a single assay. Moreover, as dosing is synchronous, it is highly suited to time–effect assays.

It has also been shown that the volume imbibed by neonate larvae is remarkably constant (Table 3.3), which facilitates an estimation of lethal dose from the LC_{50} value obtained from the assay. For example, the LC_{50} of NPV in neonate *S. littoralis* larvae using this method is 3.94×10^5 PIB ml^{-1} or 3.94×10^2 PIB μl^{-1}. The volume ingested by the larvae is 0.015 μl. The 'estimated LD_{50}' is therefore $0.015 \times 394 = 5.91$ PIB per larva; it should be emphasized that this is an estimation based on lethal concentration.

The amount ingested by a larva can be determined as follows.

1. Dose 20 or more larvae with an aqueous solution of 2% (w/v) Tinopal CBX (Ciba-Geigy Ltd) and 0.1% (v/v) Triton X-100 (BDH Chemicals Ltd), using the droplet technique described above, but without the inclusion of Brilliant Blue.

2. Observe the insects so you know when they have ingested the solution. Once they have finished drinking, transfer the larvae to individual 1.5-ml microcentrifuge tubes containing 1 ml distilled water.

Table 3.3. Volumes (μl) ingested by different lepidopterous larvae.

Estigmene acrea	0.049 ± 0.006
Helicoverpa zea	0.014 ± 0.004
Heliothis virescens	0.01114[a]
Leucoma salicis	0.11 ± 0.02
Lymantria monacha	0.18 ± 0.06
Spodoptera exigua	0.023 ± 0.015
Spodoptera frugiperda	0.006 ± 0.002
Spodoptera littoralis	0.015 ± 0.0019
Trichoplusia ni	0.013 ± 0.003

[a] Estimated by increase in weight of 100 fed larvae (see text).
Data taken from Hughes and Wood, 1986; Lameris *et al.*, 1985; Smits, 1987; Jones, 1988b; Barnett, 1992.

3. The larvae are then macerated with a blunt seeker, left for 15 min and then the insect debris is sedimented by centrifugation at 12,000 rpm in a bench-top microcentrifuge.
4. The supernatant is transferred to a 20-ml universal bottle and the volume made up to 5 ml with distilled water.
5. Five control larvae, dosed with distilled water, are treated in the same way.
6. The fluorescence of each sample is measured in a fluorimeter, which has previously been calibrated with two solutions containing 10 and 100 mg ml^{-1} Tinopal CBX and 0.1% Triton X-100.
7. From the calibration it is possible to determine the amount of Tinopal CBX, and hence volume of liquid, ingested.

This same technique can be used with other fluorescent material; however, the material chosen must not repel the larvae or act as an antifeedant. Similarly, water containing ^{32}P has also been used. Alternatively, 100 dosed and undosed larvae are weighed and the weight difference calculated and hence the volume of liquid ingested.

EGG DIPPING
Upon hatching, most neonate larvae of many lepidopteran species will eat the chorion of their egg. This habit has been exploited in bioassay by dipping eggs in a virus suspension. A wetting agent (e.g. 0.1% (v/v) Tween 80 or Teepol) is added to the dose to aid an even coverage of the surface – this is particularly important as many species lay their eggs in groups and cover them with hydrophobic scales. Normally, eggs that have been laid on filter paper or a similar surface are dipped in the suspension, left to dry and the larvae allowed to hatch and eat the chorion. The larvae are then transferred to pots of artificial or natural diet. With this technique, as with the leaf dipping technique (see 'Other methods for individual dosing of large larvae',

p. 122), the amount of virus on the egg surface is variable, particularly when egg masses are dipped. Egg masses may be separated by immersion in a trypsin solution for a few minutes followed by rinsing and filtration with distilled water. A further source of variation is that not all the larvae will completely consume the chorion.

Individual dosing bioassays

With these assays a known amount of virus is ingested by the insect and therefore an accurate assessment of virus dosage, normally medium lethal dose (LD_{50}) – the dose required to kill 50% of the test insects – can be made. There are a number of different techniques, most of which are based on allowing a larva to consume the whole of a substrate on which the virus dose has been placed.

DIET-PLUG ASSAY

With this method the dose is presented on a small plug of artificial diet. The method described is based on that of McKinley (1985) and is suitable for lepidopteran larvae of third instar (6 days old) and above.

1. Five doses (normally on a fivefold dilution scale) are prepared of the test suspension in distilled water or buffer[1]. An untreated control of distilled water or buffer alone should also be prepared.

2. Larvae of known age, preferably from a single night's egg laying, are individually weighed and placed in 30 g pots, which are labelled with the weight of the insect. Enough larvae should be weighed so that at least 20, and preferably 30, larvae are available for treatment at each of the five doses and the untreated control, leaving an extra ten larvae.

3. The labelled pots are then arranged in order of weight and the five heaviest and five lightest larvae removed – these ten insects are used to determine the instar range of the test insects through measurement of their head capsule width (e.g. McKinley et al., 1984).

4. The remaining pots are then labelled with treatment numbers. Starting with the lowest weight, the pots are labelled in order: 1, 2, 3, 4, 5, 6, 6, 5, 4, 3, 2, 1, 1, 2, 3, 4, 5, 6, 6, 5 ... and so on (treatments 1–5 and 6 for the untreated control). The net result is that larvae of different weights are distributed across the treatments – each treatment will have very similar mean larval weight and standard deviation of the mean.

5. Prepare plugs of artificial diet by using a 5-mm diameter cork borer to cut out cores of diet, which are then sliced into 3-mm thick plugs. The plugs should be of a size that can be consumed by a single larva in 24 h, depending on insect species and age/instar. However, they should not be less than 2–3 mm thick as they dry out rapidly and become hard and inedible. The plugs are then arranged on aluminium foil or a plastic or Parafilm sheet, which has been labelled with the treatment number. One plug is prepared for each treatment replicate.

6. Pipette 1 µl of the appropriate virus doses on to each diet plug. The doses should have been prepared at concentrations that have the appropriate concentration of virus in 1 µl. This volume is used as it is difficult to dispense smaller volumes accurately, and larger volumes do not soak quickly into the diet plug and can flow on to the plastic sheet, which will result in inaccurate dosing (note: if larger pieces of diet are used, larger volumes can be dispensed). The plugs are left to dry at room temperature for 15–30 min.

7. For each dilution, the plugs are then carefully placed into small vials or tubes using forceps – microcentrifuge tubes are ideal, but lengths of glass or plastic tubing plugged with cotton wool are also suitable. One diet plug is placed in each tube; as with the other assay techniques the lowest dose is handled first.

8. Between each dose, sterilize the forceps by heating in a flame, or by dipping in 2% sodium hypochlorite solution, followed by rinsing in distilled water.

9. A single larva from the appropriate treatment number is placed in each vial and the vial sealed (lid of microcentrifuge tube closed – the lid should have a small ventilation hole, or tubing plugged with cotton wool). The vial is then labelled with the treatment number, and if necessary insect weight, or placed in a plastic bag labelled with the treatment number.

10. The vials should be stored for 24 h at constant temperature and a reasonably high humidity (>70%), the latter to prevent the diet plug from drying out and becoming inedible.

11. After 24 h those larvae that have completely consumed the diet plug are transferred to labelled 30-g plastic pots containing a cube of artificial diet. The same pots that have been used for weighing the larvae can be used. One larva is placed in each pot. Larvae that have not consumed the diet plug completely are discarded. Extra larvae may be included at each dose in order to compensate for losses in replicate numbers as a result of the discard; however, this should be avoided if possible, because it might result in a non-random selection of insects.

12. The insects are then observed periodically for death, pupation or other effects. Pots containing the insects are opened and the insect touched with a blunt seeker; insects that disintegrate or do not move are recorded as dead. For LD_{50} estimation the insects are normally observed daily and mortality recorded. For LT_{50}, observations are made more frequently (see 'Lethal-time and survival-time assay', p. 130). With larger larvae it is normally not necessary to start observations until the fifth day after dosing, unless the assay is measuring time taken to death or a genetically engineered virus with enhanced speed of kill is being tested. The assay can be terminated at any point (depending on the activity being measured), but normally would proceed until all insects have either died or pupated. With some viruses and insects it is recommended that the assay proceeds until moth emergence, for example McKinley (1985) noted that deaths from NPV occurred with pupae of *S. littoralis.*

OTHER METHODS FOR INDIVIDUAL DOSING OF LARGE LARVAE

Many assay techniques described are essentially a variation on the above. Hunter-Fujita *et al.* (1998) describe an assay for *Plodia interpunctella* in which diet plugs are pushed into capillary tubing. The dose is applied to the plug. A larva is then trapped in the tube between the diet plug and a plug of cotton wool. The capillary tube is then placed in a pot containing artificial diet. The larva eats the diet plug (that represents a dose) to escape from the tube and into the pot.

A number of assays use a dosing substrate other than artificial diet. For example, leaf discs can be used or a single leaflet (e.g. of clover or lucerne). The leaves in this case are normally larger than the diet plugs – approximately 10 mm in diameter, and are placed directly into the pots in which the insects have been weighed. The effect of some leaf surface chemicals can be reduced in this case by immersing the leaves in a solution of 0.1% Teepol in distilled water, followed by blotting with tissue or filter paper. This also improved the wetting properties of many leaves, allowing easier placement of the virus dose. With termites, virus doses have been successfully administered by placing the dose on filter paper, which the termites subsequently eat (K.A. Jones, unpublished). Klein (1978) describes a method where later-instar lepidopterous larvae are allowed to drink from drops, which they entirely consume. However, McKinley (1985) found that many larvae did not completely consume the drop, that some insect strains would not drink at all and, if force fed with a blunt-needle syringe, regurgitated the dose. Dosing via a syringe is also slow, and the dose is not administered synchronously. A number of authors also described methodologies that are similar to the above but rather than known volumes of virus being placed with a pipette on a leaf, the leaf is dipped into a virus suspension and allowed to dry ('leaf dipping technique'). Whilst this is quick and easy, the amount of virus deposited on the leaf is variable and, therefore, even if individual larvae consume whole leaves, this method can only be used for estimation of LC_{50}. Dipping larger leaves and placement of single or several insects per leaf can be regarded as a variation of the surface mass dosing assay described in 'Surface dosing' (see p. 114).

INDIVIDUAL DOSING OF SMALL LARVAE

With smaller insects (e.g. lepidopterous larvae of first and second instar), it is often difficult to provide a small enough dosing substrate so that it will be consumed within 24 h – small pieces of artificial diet and leaves dry out and become inedible in only a short period. Assay techniques have been adapted to overcome this problem. A suitable methodology is adapted from that described by Evans (1981). The apparatus used consists of two metal plates, one with several holes drilled of between 5 mm and 1 cm diameter, and a plate of glass that is the same size as the metal plates. Leaves are placed between the plates such that parts of them are beneath the holes. The virus doses are then applied to the surface of the leaf that is showing and allowed to dry. A single larva is then placed on each dosed leaf area, and enclosed by securing the plate of glass with a rubber band on top of the drilled plate. The insects are left to feed for 24 h and then transferred to pots containing natural or artificial diet. Damp filter paper can be placed between the first plate and the leaves to prevent them drying out rapidly. Evans (1981) used leaves sandwiched between two Perspex plates with opposing holes. Larvae were enclosed by sealing with a plastic film.

Small pieces of leaves have also been placed on a layer of agar in the bottom of the wells of a microtitre plate. After dosing the leaves with virus suspension and allowing the dose to dry, larvae are enclosed by placing a lid over the plate. A final adaptation, which can be used with artificial diet, is to cut the bottom off a microtitre plate. This is then partially pushed into a tray of artificial diet. The virus dose is applied evenly to the surface of the diet in each cell, allowed to dry and single larvae enclosed in each well by fitting a lid on the plate. Thus a small area of diet is dosed, but as it is continuous with a large volume of diet, it will not dry out quickly and become unpalatable. Larvae are transferred to individual pots with diet after 24 h; only those that have completely eaten the surface of the diet should be transferred. The risk here, however, is that some of the dose will have soaked into the diet and remain unconsumed. For very small larvae, microtitre plates can be substituted by glass microtubes; enclosed larvae are allowed to consume all the diet.

HAEMOCOELIC INJECTION

Non-occluded viruses, or virus particles liberated from OB can be injected into the haemocoel of susceptible insects with a microsyringe. Bergold (1951) injected 1-µl doses of virus into the tip of the proleg of *Choristoneura fumiferana*; this part of the insect being chosen as the muscle would close the wound, preventing bleeding and presumably reducing risk of bacterial infection. This technique is time-consuming and does not mimic the normal method of infection in nature for most viruses (infection *via* parasitoids being an exception, although these do not parasitize larvae *via* the proleg). It is also very disruptive to larvae. Hughes and Wood (1986) suggest that this method is useful for studies on host range and barriers to infection.

Field and glasshouse assays

Field assays are where the virus is applied to a crop or other target site and its effect measured by collection of leaves, or similar target, which are fed to insects in the laboratory, or from which the virus is recovered and assayed. Alternatively, insects can be collected after a predefined period and reared in the laboratory. Arguably, a classical field trial, where treatments are assessed by measuring crop/commodity damage and insect populations is also a field assay, but this is considered out of the scope of this chapter and the reader is referred to several texts (Fisher and Yates, 1953; Fisher, 1966; Gomez and Gomez, 1976). However, it is worth mentioning here that dose-dependent responses can be obtained from virus field trials based on simple damage assessment (e.g. Jones *et al.*, 1994), and such results based on damage or yield loss of a crop (or equivalent) are the ultimate test of effectiveness of a viral insecticide. Field assays are subject to considerably more variation than laboratory assays. However, the methodology adopted should ensure that where possible variation is minimized. For example, the variation in coverage of leaves by spray application could be estimated by sampling the spray coverage with water/oil-sensitive cards (Ciba-Geigy Ltd). The amount and distribution of virus can also be determined by the impression-film technique described above (note: leaves sampled in this way should be removed so that they are not subsequently selected for bioassay). The variability could be completely removed by recovery of the virus from leaves by washing and quantifying the resulting suspension. Once samples are within the laboratory, whether insect, leaves or other substrate, the assay technique should be subject to the same control criteria as for the laboratory assays. However, the number of replicates should be increased to account for the increased variation. A protocol for conducting a field assay is given below.

1. Mark out spray plots in the field. Five replicates need to be included for each treatment to be tested and an untreated control. The plots should be assigned to each treatment randomly. The size of the plots will depend on available space and conditions, as well as type of sprayer and possibilities of drift between plots. It should also take into account the migratory tendencies of the target insect. With lepidopterous larvae, aim for plots of about 5 m × 5 m; if possible, leave spaces between plots and stagger adjacent plots to avoid contamination by drift. In some cases individual plants can be regarded as replicates; under such circumstances only one plot per treatment is required.
2. Attach water- or oil-sensitive paper in the crop (randomly assigned, approximately 20 cards to the spray plots; placement of some in the untreated control plots can be used to test for drift).
3. Mark a predetermined number of leaves in each plot with coloured sticky tape placed around the leaf stem – this will be used to identify which leaves

have been sprayed if samples are collected over a period of time (some plants can grow by a considerable amount in a short time, resulting in new growth which has not been treated). The number of marked leaves depends on the total number to be collected.

4. Apply treatments, starting with lowest dose first.

5. Collect spray cards and assess evenness of spray; these can be kept for further analysis.

6. Collect leaves from plots; marked leaves do not need to be collected on the day of application. Leaves should not be collected from plants within 1 m of the edge of the plot. The number collected should reflect the degree of variation in the coverage observed on the spray cards (as should the region of the plant canopy from which the leaves are collected). Normally about five leaves are collected from each plot, giving a total of 25 leaves per treatment.

7. Leaves are taken to a 'laboratory' site where they are placed in pots and larvae added. For neonate larvae and second instar, several larvae can be placed on a single leaf. An excess number of larvae can be placed, allowing for easy transfer later. For larger larvae, or larvae that will not feed gregariously, one larva is placed per leaf.

8. After 24 h transfer larvae to pots containing undosed artificial diet/leaves. With neonate larvae, ten larvae can be transferred from each leaf and placed together in a pot.

9. Assess mortality as described above for surface or individual dosing.

10. If required, leaves can be collected at subsequent intervals after spraying – normally a log scale such as 0, 1, 3 and 7 days post-spray is used. After the 'day 0' collection, marked leaves are collected. The leaves are treated in the same way as above.

11. It is recommended that the following measurements of crop/environmental conditions be taken in order to aid interpretation of the results.

- Wind speed and direction, temperature and relative humidity during virus application. This will aid interpretation of droplet coverage and distribution data.
- Plant/row spacing, height of plants, number of leaves and leaf area (average of at least 20 measurements, the leaf area measurements should be taken from all parts of the plant canopy); this should be done prior to virus application and, with the exception of the plant spacing, at each subsequent collection time. This will aid interpretation on initial coverage and droplet distribution data, plus subsequent 'growth dilution' effects.
- Ultraviolet light intensity, preferably at different parts of the plant canopy. Readings should be taken at regular intervals, at specific times of the day, or using a methodology that gives a cumulative reading (e.g. Jones and McKinley, 1987; Jones, 1988b; Jones *et al.*, 1993a). This is important if the experimenter is looking at persistence of the virus by collection of samples after specific time intervals; UV

inactivation is in many field situations the most important factor. Temperature, relative humidity and leaf/substrate surface chemicals can also be measured. Methodologies that have been used in the field are described by Jones and McKinley (1987) and Jones (1988b).

The methodology of feeding the larvae field-collected leaves or virus dose can be varied. Leaves, stems or pine needles may be fitted into cork, which is then fitted into glass vials. The leaf stem can then be placed in water to prevent the leaf drying out. As an alternative, virus can be recovered from the collected leaves by taking leaf discs or placing whole leaves in bottles and washing. Several leaves/discs can be placed in a single bottle. The suspensions are then quantified before assay. For example, Jones (1988b) placed ten 1-cm diameter cotton leaf discs (obtained using a cork borer) into an 8-ml glass vial containing 2 ml distilled water. The bottles were placed on a blood suspension mixer at 30 rpm for 4 h. The control was virus suspension only. The amount of virus recovered is shown in Table 3.4. Efficiency of recovery may vary between plant species/surfaces and therefore needs to be determined for each surface type. Formulation additives, such as stickers, will also influence the efficiency of recovery. Addition of a surfactant to the wash bottles may be necessary, but should be avoided where possible as it may affect assay results in some cases. If recovery is low, the experimenter should be aware of the possibility that a 'non-random' recovery may be made, for instance partially dissolved/inactivated virus may be preferentially recovered. This would render this technique unsuitable. A variation of recovery from the leaf is to attach polythene slides to leaves by stapling. These can then be collected and the virus recovered. This has the advantage of allowing the coverage on the upper and lower leaf surfaces to be assessed separately and also allows leaf surface effects to be separated from other environmental effects.

The same general method can be used for insect collections. Plots are treated, after which insects are collected directly from the field into pots of undosed artificial diet or leaves. As they are collected, a note of instar/size group is made. In this case a pre-spray collection is often made, and to reduce possibilities of larvae wandering between plots, plot sizes should be bigger and collection made only from the centre of the plots.

Greenhouse assays are in many ways similar to field assays, except that conditions are more controlled. Individual plants can be sprayed, using a

Table 3.4. Recovery of *S. littoralis* NPV from cotton leaves.

Surface	Concentration of virus (PIB ml^{-1})
Control	$1.25 \pm 0.04 \times 10^7$
Cotton leaf, upper surface	$1.24 \pm 0.08 \times 10^7$
Cotton leaf, lower surface	$1.17 \pm 0.06 \times 10^7$

track sprayer for example, and more even coverage obtained. Leaves can be collected, or a known number of insects placed and enclosed in cages on the plants.

Confirmation of virus

It is recommended with all assays that samples of virus are taken from randomly selected larvae that have died and the identity of the virus confirmed (see 'Virus identification', p. 101) to be the same as the virus used for inoculation. This rules out the possibility of the presence of latent or occult virus being triggered by the inoculum, as well as 'contamination' resulting from poor experimental technique, or with mortality induced by naturally occurring virus. With well-established laboratory assay systems using an established insect culture the frequency of virus identification can be reduced over time.

Analysis of results

The statistics of bioassay analysis is discussed in Chapter 7. Here a few comments relating to bioassay of viruses will be made. As mentioned earlier, unless the actual dose consumed by (or injected into) an insect is known, lethal concentration rather than lethal dose can be estimated. In some cases, the presence of virus such as OB in tissues is determined (by making insect squashes on microscope slides and staining with an appropriate stain, e.g. Giemsa) rather than mortality; this is expressed as the effective dose (e.g. ED_{50}), or similar designation. When assessing viral insecticides, it is always preferable to assess in terms of mortality, unless another effect is being assessed, or there is high confidence that infection will inevitably lead to death. Generally, the LC_{50} or LD_{50} of a suspension is quoted. Most often this is calculated using probit analysis (Finney, 1971), although other transformations, such as logits (Ashton, 1972) have been used. The use of probit analysis is not strictly correct as the statistical theory assumes that the effects of increasing dosage are cumulative, and therefore there is a threshold dose above which the insect is certain to die. However, as discussed by Ridout *et al.* (1993), with viruses each virus particle is capable of independently causing infection and, therefore, increasing dose does not result in an additive effect within the insect but an increased probability of virus particles causing infection. This situation is more properly interpreted through an exponential model (Hughes and Wood, 1986). Moreover, the log-probit model assumes that the active ingredient is evenly distributed between subsamples that are presented to the test insects. However, with particulate active ingredients, such as viruses, the distribution (at low dose concentrations) follows a Poisson distribution and therefore some subsamples presented (or consumed by) a test insect will not contain any particles. In this case logit transformation may be a more appropriate methodology. Hughes and Wood (1986) discuss the relative merits of different methods of analysis for dose–mortality assays. In practice, results obtained with probit analysis and other methodologies are almost identical and most laboratories have access to computer programs that use probits. Normally, treatment mortalities are corrected for control mortalities using the method of Abbott (1925); distinction can be made between control mortalities resulting from virus infection and those resulting from other causes.

When comparing virus suspensions on the basis of LC_{50} or LD_{50} it is important to consider the slope of the probit mortality–log dose line. These should be the same, within acceptable limits, for each suspension compared. If they are not, the relative difference between suspensions will vary at different mortality levels. If lines are not parallel, LD_{50} and slope (plus possibly LD_{90}) should be quoted. Particularly with mass dosing assays, comparison of samples is made on the basis of relative potency, normally of a test suspension to a standard (see 'Assay variation', p. 129). This, again, can only be done on the basis of comparison of parallel probit mortality–log dose lines.

All assays should be repeated three to five times, if possible. Average results then need to be found by combining the results of the assays. Inclusion of a standard suspension in the assay aids this process and reduces variation between assays. There are a number of methodologies for combining bioassay results, discussed by Finney (1971); most are based on weighting the individual estimates of LC/LD_{50} or relative potency on the basis of the inverse of the variance of the estimate prior to combination.

Finally, with assays that have been used to determine virus infectivity over time (e.g. assessing persistence of the virus in the field), a number of authors have used a single dose to obtain a percentage mortality and determined the percentage original activity remaining (%OAR) or percentage of original activity (Ignoffo and Batzer, 1971; Jaques, 1972, 1985; Ignoffo *et al.*, 1977; Lewis and Yendol, 1981). This is defined as:

$$\% \text{ OAR} = \frac{\% \text{ mortality after exposure} \times 100}{\% \text{ mortality before exposure}}$$

Although simple and convenient, this method is subject to flaws. First, basing the calculation on a single dose rather than several doses results in a high degree of error (although this can be reduced through replication). Second, the methodology does not take into account the probit mortality–log dose line, which may vary with different samples (see above). Finally, as pointed out by Richards and Payne (1982), although this method is used to calculate half-lives (time taken for 50% of the virus to be inactivated), this is not a true estimate of virus half-life, as the relationship between percentage mortality and virus dose (when not transformed to probits and logs, respectively) is not a linear one and hence a halving of mortality does not represent a halving in the amount of infective virus. Therefore, if possible, it is recommended that persistence bioassays are based on multidose assays where the virus has been quantified. With some field assays this may not be possible and single dose assays are used; in this case half-lives of virus should not be quoted directly on the basis of %OAR. Moreover, some studies where %OAR has been calculated have had an initial mortality of 100% – in this case there is no way of knowing whether the amount of active virus is 1, 10 or 100 times greater than that required to give a 100% mortality, thus making interpretation of results very difficult.

Assay variation

Although the methodologies outlined above are designed to minimize variation, the inherent heterogeneity of live assay systems inevitably means that variations will occur. Moreover, different assay techniques will have different amounts of variation. This has been discussed by Entwistle and Evans (1985). A reflection of heterogeneity is the slope of the transformed percentage (probit) mortality and the log virus dose line. In general, slope values of between 1 and 2 have been obtained with virus assays, although higher values of 3–4 have been reported under specific conditions (e.g. Wigley, 1976; Evans, 1981). However, the independent action theory (see 'Analysis of results', p. 127) states that dosage–mortality slopes cannot be greater than 2; Hunter-Fujita *et al.* (1998) speculate that the higher slopes that have been reported in some assays are a result of intrinsic variability. For comparison, slopes of 5 or more can be expected with topically applied chemical pesticides.

Susceptibility of insects to virus varies with age and instar – referred to by some authors as 'maturation resistance' (e.g. Whitlock, 1977). For example, the LD_{50} of one isolate of *Spodoptera exigua* NPV ranged from 4 PIB per larva for first-instar larvae to 1.3×10^6 PIB per larva for fifth-instar larvae (Smits, 1987). However, 90% of the observed variation can be removed by plotting as $logLD_{50}$ mg^{-1} body weight (Entwistle and Evans, 1985). Some variation may still occur within instars – McKinley (1985) noted larvae to be more susceptible immediately after a moult and, although able to relate LD_{50} of an NPV to age and weight of *S. littoralis* larvae, the regression obtained:

$$Log_{10} LD_{50} = 1.416 \, log_{10} \, weight - 0.007 \, age^2 + 1.03$$

accounted for only 73% of the variation.

Variation may also occur between different batches of insects, even though rearing conditions have been standardized. Jones (1988b) reported that the LC_{50} obtained with the droplet bioassay of a standard suspension of *S. littoralis* NPV against neonate larvae varied from 0.96×10^5 to 7.67×10^5 PIB ml^{-1} over a 16 month period. This effect can be largely removed by including a standard set of doses in each assay and expressing the results of test suspensions in terms of relative potency to the standard; the standard should be homologous with the test suspension (i.e. at least the same organism) (Dulmage, 1973). Inclusion of a standard also allows results from different assays to be combined (see Finney, 1971; Jones et al., 1993a).

There can also be a great variation between the results obtained from assays using different substrates to present the dose. This may be a result of leaf surface or other substrate chemicals affecting the virus, for example the salts found on cotton leaves (Young et al., 1977; Elleman and Entwistle, 1985) or other leaf chemicals that may affect the virus. A number of authors have recorded different LD_{50} for a virus when presented on different leaves or when compared with artificial diet (e.g. Richter et al., 1987; Jones, 1988b; Santiago-Alverez and Ortez-Garcia, 1992). Richter et al. (1987) speculated that differences were related to stress of larvae on unsuitable food sources. However, Jones (1988b) found that rearing food made little difference and the important factor was the dosing substrate. These effects can be avoided by using an inert substrate or droplet method, although this does not then reflect the field situation. Inclusion of a standard set of doses and expression of results as potency ratios will also take these effects into account.

The presence of formulation materials may interfere with an assay. For example, oil formulations cannot be assayed by the droplet technique directly because neonate larvae become trapped and drown in the oil. Similarly, the oil does not dry and so larvae also drown with the surface dosing technique. Thus, an assay system must be chosen that avoids this problem, such as individual dosing of larger larvae, or if there is a need to test large numbers of samples, the virus must be recovered and resuspended in water. For oil, this can be achieved through filtering and rinsing in acetone. However, as mentioned previously (see p. 99), this can result in inactivation of some viruses. The method used depends on the formulation material. With aqueous formulations the virus can be recovered through centrifugation and resuspension (similar to the purification processes described above). Again the inclusion of a standard set of doses allows for formulation effects to be taken into account.

Variation also occurs between assays carried out in different laboratories and by different researchers (Burges and Thompson, 1971); inclusion of a standard suspension in all assays will allow accurate comparison between results. Hunter-Fujita et al. (1998) also report that between- and within-assay variation is less when researchers carry out assays more frequently.

Lethal-time and survival-time assay

Increasing interest is being shown in determining the time that a virus takes to kill or affect the host, particularly with regard to genetic engineering of baculoviruses to increase speed of effect. The bioassay methodology used to determine this is essentially the same as for dose–mortality assays. However, there are some important additional requirements for accurate results. It is desirable that all test insects have, as far as possible, hatched synchronously (see 'Insect supply', p. 113) to minimize variation. It is also important that the dose be administered synchronously. With mass dosing assays this is a primary advantage of the droplet dosing method (see 'Droplet assay', p. 117). With individual dosing assays, the diet-plug assay (see p. 120) should be used in preference to haemocoelic injection or drop dosing, although the dosing substrate should be as small as possible so that the dose is consumed over a short period. Overall the droplet dosing method is considered the most suitable for time–effect assays as dosing is synchronous.

Normally, a single dose is administered for each sample being tested. The assay should then be checked at regular intervals. With unmodified viruses, the assay is checked daily until the first death occurs and then at 6–8 h intervals. With viruses that are genetically modified to enhance speed of kill, more frequent observations should be made starting from initial dosing. Medium lethal or medium survival times (LT_{50} and ST_{50}) are then calculated.

Hughes *et al.* (1983) described the use of the droplet assay to assess time–mortality response of *Helicoverpa zea* to several NPV isolates. Each isolate was diluted to 1×10^6 OB ml^{-1}. The average amount imbibed by a neonate larvae was determined to be 0.11 µl, and therefore each larvae will on average receive a dose of 11 OB; this is sufficiently high to result in few larvae surviving. Thirty larvae, between 4 and 8 h old, were treated for each isolate, using the droplet assay described by Hughes and Wood (1981; modified version described in 'Droplet assay', p. 117). Mortality was recorded at 6-h intervals, from just before onset of the first mortality until all infected larvae had died (the assay should not be terminated at insect pupation). As the dose was high enough to kill most larvae, survivors were considered to have escaped infection and were not included in the data analysis. The assay was replicated two to five times.

LT_{50} and ST_{50} should be calculated from life-tables (Hunter-Fujita *et al.*, 1998), however, in practice probit analysis is used. With the assay described above, Hughes *et al.* (1983) calculated maximum likelihood ST_{50} values based on a logit version (logit = slope $\times$ log[ST/ST_{50}]) of the probit model described by Bliss (1937). A computer program used to estimate this is given by Hughes and Wood (1986).

In vitro *assay*

With *in vitro* assays the researcher has more control over assay conditions. However, there is a need to have access to a sterile working environment, plus the availability of suitable cell lines (see '*In vitro* production', p. 98).

The methodology used is based on being able to count the number of infection foci or plaques that develop at each dilution in susceptible cells attached in a layer to a solid surface. Plaque assays have been described by a number of authors for NPVs in lepidopteran cell lines (e.g. Hink and Strauss, 1977; Knudson, 1979; Hunter-Fujita *et al.*, 1998). For each assay system developed it is important to determine the most suitable state of the cells used, the appropriate composition of diluent, the density of cells and cell culture medium (e.g. TC100 with 10% fetal bovine serum). All of these affect the efficiency of cell attachment, viral replication and development of foci (Hughes and Wood, 1986). The general methodology is as follows:

1. Susceptible cells must be available in the logarithmic phase of growth, when they are most susceptible to virus infection.

2. Cells are seeded into appropriate containers – normally multicellular or Petri plates. The cell densities used depend on the cell growth rate (doubling time) and virus growth rate (both dependent on assay conditions – see above), the faster the cell doubling rate the lower the initial level required; however, higher cell densities are required for viruses that replicate faster.

3. Leave the cells to incubate for 2–24 h, during which time they attach to the plate surface. Attachment efficiency can be increased by centrifugation, but this is not suitable for all cell lines.

4. Carefully remove tissue medium with a pipette or aspirator.

5. Add virus inoculum – this must be a suspension of non-occluded virus, obtained by dissolution of the occlusion body, or preferably budded virus obtained by extraction from the haemolymph of an infected insect. The inoculum is added in a volume of medium sufficient to cover the attached cells and which will not dry out during the assay period.

6. Place the plates on a rocker platform for 1–2 h. If a platform is not available, the plates should be gently tilted every 15 min. With some cell lines the plates can be centrifuged to concentrate virus and cells in the bottom of the plate, which can increase sensitivity tenfold (Wood, 1977).

7. Carefully remove as much of the inoculum medium as possible using an aspirator. An aspirator with a small orifice placed at the edge of the plate where the meniscus of the medium is higher is recommended by Hughes and Wood (1986). Care should be taken not to allow the cells to dry out.

8. An overlay is now applied to the cells. This amplifies the infection foci making them readily visible and delineates the infection site allowing plaque purification of virus isolates. The overlay is prepared by dissolving agarose in the basic salt solution of Grace's medium by autoclaving. The tubes of agarose–salt solution are gently shaken to dissolve any remaining solids and left to cool.

9. The agarose solution ($2\times$ concentration) is then melted in a boiling water bath and mixed with complete medium and placed in a water bath at 37°C. The final concentration of agarose thus obtained depends on the agarose used, as well as the plaque assay cell-line, but is normally between 0.3 and 2.0% (w/v).

10. The overlay is applied to cover all the cells, after which the assay plates are sealed to avoid desiccation and left to incubate for 5 or more days.

11. After the incubation period the number of plaques are counted using a binocular microscope (although they can be viewed with the unaided eye). Plaques are recognized by the presence of OB in the cells, or by alterations in cell morphology (including in some cases cell lysis) and inhibition of cellular division.

12. Stains can be used to aid visualization (see Hughes and Wood, 1986), although this is not possible if the plates are to be observed on successive days.

From the results of a number of plaques obtained at different virus dilutions the ED_{50} can be estimated. This methodology is outlined by Hughes and Wood (1986). A suitable method, recommended by Finney (1971), is the Spearmen–Karber analysis. The method requires that the data span the full range of response from 0 to 100%. The ED_{50} is calculated as:

$$Log_{10}ED_{50} = X_{p=1} + (\tfrac{1}{2}d) - d\Sigma p$$

where $X_{p=1}$ = the highest log_{10} dose giving 100% positive response, d = log_{10} dilution factor (i.e. for a tenfold dilution this = 1), p = proportion of positive responses at a given dose, and Σp = sum of p for $X_{p=1}$ and all higher dilutions.

The standard error is calculated as:

$$SE = \sqrt{d^2 \times \Sigma[p(1-p)/n-1]}$$

where n = the number of inoculated samples.

Concluding Remarks

This chapter has outlined the many different bioassay techniques that are used with insect viruses. The choice of technique used depends on what the experimenter wishes to find out, on available time and resources. Thus, the need to find out an accurate estimate of viral infectivity requires a more labour-intensive approach than the need to compare the activity of different viral formulations. However, all the techniques can give statistically valid results if applied correctly. There are many variations on the described methodologies. Further variations will be developed for new insect–virus systems. Future techniques will also involve adaptations of bioassay techniques used for other stomach poisons or microbes – for example tests for mosquitoes and blackfly described in Chapter 1. The basic need of standardization, however, will remain the same.

The importance of bioassay in the study and development of insect viruses and viral insecticides cannot be overstated – this represents the only accurate measure of the efficacy of a virus.

Notes

1   Solution A: 28.39 g NaH_2PO_4 made up to 1 litre in distilled water. Solution B: 31.21 g $Na_2HPO_4.2H_2O$ made up to 1 litre in distilled water. Add 55 ml solution A to 45 ml solution B and make up to 1 litre with distilled water.
2   100 ml glacial acetic acid, 300 ml chloroform, 600 ml ethanol.
3   1.5 g Naphthalene black B12, 40 ml glacial acetic acid, 60 ml distilled water.
4   1.5 g Na_2CO_3, 2.93 g $NaHCO_3$, 0.2 g NaN_3 made up to 1 litre in distilled H_2O.
5   8 g NaCl, 2.9 g $Na_2HPO_4.12H_2O$, 0.2 g KH_2PO_4, 0.2 g KCl, 0.2 g NaN_3 made up to 1 litre in distilled H_2O, plus 0.5 ml Tween 20 (pH 7.4).
6   97 ml diethanolamine, 800 ml distilled H_2O, 0.2 g NaN_3 adjusted to pH 9.6 with HCl and made up to 1 litre in distilled H_2O.
7   557 ml distilled H_2O, 1.8 g methyl-4-hydroxybenzoate, 37 g finely ground dry solids (18 g wheatgerm, 16 g dried baker's yeast, 2.9 g casein, 2.9 g sucrose), 6.0 g agar.

References

Abbott, W.S. (1925) A method for computing the effectiveness of an insecticide. *Journal of Economic Entomology* 18, 265–267.

Armes, N.J., Bond, G.S. and Cooter, R.J. (1992) The laboratory culture and development of *Helicoverpa armigera*. *Report OB57*, Natural Resources Institute, Chatham, UK.

Ashton, W.D. (1972) *The Logit Transformation with Special Reference to its Uses in Bioassay*. Griffin, London.

Barnett, A.L. (1992) An investigation into the encounter of a baculovirus deposit on a cotton leaf surface by *Heliothis virescens* (Lepidoptera: Noctuidae) larvae. MSc thesis, University of London.

Bell, M.R. (1991) *In vivo* production of a nuclear polyhedrosis virus utilizing tobacco budworm and a multicellular larval rearing container. *Journal of Entomological Science* 26, 69–75.

Bergold, G.H. (1951) The polyhedral disease of the spruce budworm *Choristoneura fumiferana* (Clem) (Lepidoptera: Totricidae). *Canadian Journal of Zoology* 29, 17–23.

Bliss, C.I. (1937) The calculation of the time–mortality curve. *Annals of Applied Biology* 24, 815–852.

Brown, M. (1998) Detection and identification of virus DNA using the polymerase chain reaction (PCR). In: Hunter-Fujita, F.R., Entwistle, P.F., Evans, H.F. and Crook, N.E. (eds) *Insect Viruses and Pest Management: Theory and Practice*. John Wiley & Sons, Chichester, UK, pp. 425–429.

Burges, H.D. and Thompson, E.M. (1971) Standardization and assay of microbial insecticides. In: Burges, H.D. and Hussey, N.W. (eds) *Microbial Control of Insects and Mites*. Academic Press, London, pp. 591–622.

Butz, P., Fritsch, E., Huber, J., Keller, B., Ludwig, H. and Tauscher, B. (1995) Ultra-high pressure decontamination of insect biocontrol preparations. *Biocontrol Science and Technology* 5, 243–246.

Cherry, A.J., Parnell, M.A., Grzywacz, D. and Jones, K.A. (1997) The optimization of *in vivo* nuclear polyhedrosis virus production in *Spodoptera exempta* (Walker) and *Spodoptera exigua* (Hübner). *Journal of Invertebrate Pathology* 70, 50–58.

Copping, L.G. (1993) Baculoviruses in crop protection. *Agrow Report DS 85*. PJB Publications, Richmond, Surrey, UK.

Crook, N.E. and Payne, C.C. (1980) Comparison of three methods of ELISA for Baculoviruses. *Journal of General Virology* 46, 29–37.

Dulmage, H.T. (1973) Assay and standardization of microbial insecticides. In: Bulla, L.A. Jr (ed.) *Regulation of Insect Populations by Microorganisms. Annals of the New York Academy of Sciences* 217, 187–199.

Dulmage, H.T. and Rhodes, R.A. (1971) Production of pathogens in artificial media. In: Burges, H.D. and Hussey, N.W. (eds) *Microbial Control of Insects and Mites*. Academic Press, London, pp. 507–540.

Elleman, C.J. and Entwistle, P.F. (1985) The effect of magnesium ions on the solubility of polyhedral inclusion bodies and its possible role in the inactivation of a nuclear polyhedrosis virus of *Spodoptera littoralis* by cotton leaf exudate. *Annals of Applied Biology* 106, 93–100.

Elleman, C.J., Entwistle, P.F. and Hoyle, S.R. (1980) Application of the impression film technique to counting inclusion bodies of nuclear polyhedrosis viruses on plant surfaces. *Journal of Invertebrate Pathology* 3, 129–132.

Entwistle, P.F. and Evans, H.F. (1985) Viral control. In: Kerkut, G.A. and Gilbert, L.I. (eds) *Comprehensive Insect Physiology, Biochemistry and Pharmacology*, Vol. 12, *Insect Control*. Pergamon Press, Oxford, pp. 347–412.

Evans, H.F. (1981) Quantitative assessment of the relationships between dosage and response of the nuclear polyhedrosis virus of *Mamestra brassicae*. *Journal of Invertebrate Pathology* 37, 101–109.

Evans, H. and Shapiro, M. (1997) Viruses. In: Lacey, L.A. (ed.) *Biological Techniques: Manual of Techniques in Insect Pathology*. Academic Press, London, pp. 17–53.

Finney, D.J. (1971) *Probit Analysis*, 3rd edn. Cambridge University Press.

Fisher, R.A. (1966) *The Design of Experiments*, 8th edn. Hafner Publishing Company, New York.

Fisher, R.A. and Yates, F. (1953) *Statistical Tables for Biological, Agricultural and Medical Research*, 4th edn. Oliver Boyd, Edinburgh.

Gomez, K.A. and Gomez, A.A. (1976) *Statistical Procedures for Agricultural Research with Emphasis on Rice*. The International Rice Research Institute, Los Baños, Manila.

Griffiths, I.P. (1982) A new approach to the problem of identifying baculoviruses. In: Kurstak, E. (ed.) *Microbial and Viral Pesticides*. Marcel Dekker, New York, pp. 507–583.

Grzywacz, D. (1998) Microbiological examination of virus produced *in vivo*. In: Hunter-Fujita, F.R., Entwistle, P.F., Evans, H.F. and Crook, N.E. (eds) *Insect Viruses and Pest Management: Theory and Practice*. John Wiley & Sons, Chichester, UK, pp. 523–541.

Grzywacz, D., McKinley, D.J., Jones, K.A. and Moawad, G. (1997) Microbial contamination in *Spodoptera littoralis* nuclear polyhedrosis virus produced in insects in Egypt. *Journal of Invertebrate Pathology* 69, 151–156.

Grzywacz, D., Jones, K.A., Moawad, G. and Cherry, A. (1998) The *in vivo* production of *Spodoptera littoralis* nuclear polyhedrosis virus. *Journal of Virological Methods* 71, 115–122.

Harrap, K.A. and Payne, C.C. (1979) The structural properties and identification of insect viruses. *Advances in Virus Research* 25, 273–355.

Harrap, K., Payne, C.C. and Robertson, J. (1977) The properties of three baculoviruses from closely related hosts. *Virology* 79, 14–31.

Harvey, A. (1990) Rearing and breeding of locusts in the laboratory. *Report MSC17*, Natural Resources Institute, Chatham, UK.

Hink, W.F. and Strauss, E.M. (1977) An improved technique for plaque assay of *Autographa californica* nuclear polyhedrosis on TN-368 cells. *Journal of Invertebrate Pathology* 29, 390–391.

Hughes, P.R. (1994) High density rearing system for larvae. US patent number 5,351,643.

Hughes, P.R. and Wood, H.A. (1981) A synchronous peroral technique for bioassay of insect viruses. *Journal of Invertebrate Pathology* 37, 154–159.

Hughes, P.R. and Wood, H.A. (1986) *In vivo* and *in vitro* bioassay methods for baculoviruses. In: Granados, R.R. and Federici, B.A. (eds) *The Biology of Baculoviruses*, Vol. 2. *Practical Application for Insect Control*. CRC Press, Boca Raton, Florida, pp. 1–30.

Hughes, P.R., Getting, R.R. and McCarthy, W.J. (1983) Comparison of the time-mortality response of *Heliothis zea* to 14 isolates of *Heliothis* nuclear polyhedrosis virus. *Journal of Invertebrate Pathology* 41, 256–261.

Hughes, P.R., van Beek, N.A.M. and Wood, H.A. (1986) A modified droplet-feeding method for rapid assay of *Bacillus thuringiensis* and baculoviruses in noctuid larvae. *Journal of Invertebrate Pathology* 48, 187–192.

Hunter, D.K., Collier, S.S. and Hoffman, D.F. (1977) Granulosis virus of the Indian meal moth as a protectant for stored in-shell almonds. *Journal of Economic Entomology* 70, 493–494.

Hunter, F.R., Crook, N.E. and Entwistle, P.F. (1984) Viruses as pathogens for the control of insects. In: Grainger, J.M. and Lynch, J.M. (eds) *Microbiological Methods and Environmental Biotechnology*. Society for Applied Bacteriology, London, pp. 323–347.

Hunter-Fujita, F.R., Entwistle, P.F., Evans, H.F. and Crook, N.E. (1998) *Insect Viruses and Pest Management*. John Wiley & Sons, Chichester, UK.

Ignoffo, C.M. and Batzer, O.F. (1971) Microencapsulation and ultraviolet protectants to increase sunlight stability of an insect virus. *Journal of Economic Entomology* 6, 850–853.

Ignoffo, C.M. and Couch, T.L. (1981) The nucleopolyhedrosis virus of *Heliothis* species as a microbial insecticide. In: Burges, H.D. (ed.) *Microbial Control of Pests and Plant Diseases 1970–1981*. Academic Press, London, pp. 330–362.

Ignoffo, C.M. and Hink, W.F. (1971) Propagation of arthropod pathogens in living systems. In: Burges, H.D. and Hussey, N.W. (eds) *Microbial Control of Insects and Mites*. Academic Press, London, pp. 541–580.

Ignoffo, C.M. and Shapiro, M. (1978) Characteristics of baculovirus preparations processed from living and dead larvae. *Journal of Economic Entomology* 7, 186–188.

Ignoffo, C.M., Hostetter, D.L., Sikorowski, P.O., Sutter, G. and Brooks, W.M. (1977) Inactivation of representative species of entomopathogenic viruses, a bacterium, fungus and protozoan by an ultraviolet light source. *Environmental Entomology* 6, 411–415.

Ignoffo, C.M., McIntosh, A.H. and Garcia, C. (1983) Susceptibility of larvae of *Heliothis zea, H. virescens* and *H. armigera* (Lep: Noctuidae) to three baculoviruses. *Entomophaga* 2, 1–8.

Im, D.J., Choi, K.M., Lee, M.H., Jin, B.R. and Kang, S.K. (1989) *In vivo* mass production of *Spodoptera litura* nuclear polyhedrosis virus. *Korean Journal of Applied Entomology* 28, 82–87.

Jaques, R.P. (1972) The inactivation of foliar deposits of viruses of *Trichoplusia ni* and *Pieris rapae* and tests on protectant activities. *Canadian Entomologist* 10, 1985–1994.

Jaques, R.P. (1985) Stability of insect viruses in the environment. In: Maramorosch, K. and Sherman, K.E. (eds) *Viral Insecticides for Biological Control.* Academic Press, London, pp. 285–360.

Jones, K.A. (1988a) The use of insect viruses for pest control in developing countries. *Aspects of Applied Biology* 17, 425–433.

Jones, K.A. (1988b) Studies on the persistence of *Spodoptera littoralis* nuclear polyhedrosis virus on cotton in Egypt. PhD thesis, University of Reading, UK.

Jones, K.A. (1994) Use of baculoviruses for cotton pest control. In: Matthews, G.A. and Tunstall, J.P. (eds) *Insect Pests of Cotton.* CAB International, Wallingford, UK, pp. 477–504.

Jones, K.A. and McKinley, D.J. (1983) The use of enzyme-linked immunosorbent assay (ELISA) for quantification of virus spray deposits. In: *Proceedings of the 10th International Congress of Plant Protection*, Brighton, 1983, Vol. 2. British Crop Protection Council, Croydon, UK.

Jones, K.A. and McKinley, D.J. (1987) The persistence of *Spodoptera littoralis* nuclear polyhedrosis virus on cotton in Egypt. *Aspects of Applied Biology* 14, 323–334.

Jones, K.A., Moawad, G., McKinley, D.J. and Grzywacz, D. (1993a) The effect of natural sunlight on *Spodoptera littoralis* nuclear polyhedrosis virus. *Biocontrol Science and Technology* 3, 189–197.

Jones, K.A., Westby, A., Reilly, P.J.A. and Jeger, M.J. (1993b) The exploitation of microorganisms in the developing countries of the tropics. In: Jones, D.G. (ed.) *Exploitation of Microorganisms.* Chapman & Hall, London, pp. 343–370.

Jones, K.A., Irving, N.S., Grzywacz, D., Moawad, G.M., Hussein, A.H. and Fargahly, A. (1994) Application rate trials with a nuclear polyhedrosis virus to control *Spodoptera littoralis* (Boisd.) on cotton in Egypt. *Crop Protection* 13, 337–340.

Jones, K.A., Grzywacz, D. and Cherry, A. (1996) Appropriate methods of insect virus production in resource-poor regions. Abstracts: Technology Transfer in Biological Control: From Research to Practice, Montpellier, France, September 9–11, 1996. *IOBC wprs Bulletin* 19 (8), 137.

Kalmakoff, J. and Wigley, P.J. (1980) Practical: purification of NPV from *Heliothis armigera* larvae and serological detection of infection. In: Kalmakoff, J. and Longworth, J.F. (eds) *Microbial Control of Insect Pests. DSIR Bulletin 228*, Department of Scientific and Industrial Research, Wellington, New Zealand, pp. 26–29.

Kaupp, W.J. (1981) Studies on the ecology of a nuclear polyhedrosis virus of the European pine sawfly, *Neodiprion sertifer* Geoff. D.Phil thesis, University of Oxford.

Kelly, D.C., Edwards, M.L., Evans, H.F. and Robertson, J.S. (1978) The use of enzyme-linked immunosorbent assay to detect a nuclear polyhedrosis virus in *Heliothis armigera* larvae. *Journal of General Virology* 40, 465–469.

King, L.A. and Possee, R.D. (1992) *The Baculovirus Expression System.* Chapman and Hall, London.

Klein, M. (1978) An improved peroral administration technique for bioassay of nucleopolyhedrosis viruses against Egyptian cotton worm, *Spodoptera littoralis. Journal of Invertebrate Pathology* 31, 134–136.

Knudson, D.L. (1979) Plaque assay of baculoviruses employing an agarose-nutrient overlay. *Intervirology* 11, 40–46.

Laemmli, U.K. (1970) Cleavage of structural proteins during assembly of the head of bacteriophage T4. *Nature* 227, 680–685.

Lameris, A.M.C., Ziemnicka, J., Peters, D., Grijpma, P. and Vlak, J.M. (1985) Potential of baculoviruses for control of satin moth, *Leucoma salicas* L. (Lepidoptera: Lymantriidae). *Mededelingen van de Rijksfaculteit Landbouwwetschappen te Gent* 50, 431–438.

Lewis, F.B. and Yendol, W.G. (1981) Gypsy moth nucleopolyhedrosis virus: efficacy. In: Doanne, C.C. and McManus, M.L. (eds*) The Gypsy Moth: Research Toward Integrated Pest Management.* USDA Technical Bulletin 1584, pp. 503–512.

Ma, M. (1985) Enzyme immunoassays for the detection of gypsy moth nuclear polyhedrosis virus. In: *Proceedings, Symposium: Microbial Control of Spruce Budworms and Gypsy Moths, April 10–12 1984.* Forest Service USDA, Northeastern Forest Experiment Station, Broomall, PA 19008, pp. 125–131.

Martignoni, M.E. and Iwai, P.J. (1981) A catalogue of viral diseases of insects, mites and ticks. In: Burges, H.D. (ed.) *Microbial Control of Pests and Plant Diseases 1970–1980.* Academic Press, London, pp. 897–911.

McGaughey, W.H. (1975) A granulosis virus for Indian meal moth control in stored wheat and corn. *Journal of Economic Entomology* 68, 346–349.

McKinley, D.J. (1985) Nuclear polyhedrosis virus of *Spodoptera littoralis* Boisd. (Lepidoptera, Noctuidae) as an infective agent in its host and related insects. PhD thesis, University of London.

McKinley, D.J., Smith S. and Jones, K.A. (1984) The laboratory culture and biology of *Spodoptera littoralis* Boisduval. *Report L67,* Natural Resources Institute, Chatham, UK.

McKinley, D.J., Moawad, G., Jones, K.A., Grzywacz, D. and Turner, C. (1989) The development of nuclear polyhedrosis virus for control of *Spodoptera littoralis* (Boisd.) in cotton. In: Green, M.B. and de Lyon, D.J. (eds) *Pest Management in Cotton.* Ellis Horwood, Chichester, UK, pp. 93–100.

Monnet, S., Gazon, M., Deramoudt, F.-X., Quiot, J.-M. and Devauchelle, G. (1994) Production of polyhedral inclusion body in a high density insect cell bioreactor. In: *Proceedings VIth International Colloquium on Invertebrate Pathology and Microbial Control,* Montpellier, France, 28 August–2 September, 1994. Society for Invertebrate Pathology, pp. 447–451.

Murphy, F.A., Fauquet, C.M., Bishop, D.H.L., Ghabrial, S.A., Jarvis, A.W., Martelli, G.P., Mayo, M.A. and Summers, M.D. (eds) (1995) Virus taxonomy: Classification and nomenclature of viruses, the sixth report of the international committee on taxonomy of viruses. *Archives of Virology Supplement* 10, Springer-Verlag, Vienna, Austria.

Nathans, D. and Smith, H.O. (1975) Restriction endonucleases in the analysis and restructuring of DNA molecules. *Annual Review of Biochemistry* 44, 73–293.

O'Reilly, D.R., Miller, L.K. and Luckow, V.A. (1994) *Baculovirus Expression Vectors, a Laboratory Manual.* Oxford University Press, Oxford.

Podgwaite, J.D. (1981) NPV production and quality control. In: Doanne, C.C. and McManus, M.L. (eds) *The Gypsy Moth: Research Toward Integrated Pest Management.* USDA Technical Bulletin 1584, pp. 461–464.

Richards, M.G. and Payne, C.C. (1982) Persistence of baculoviruses on leaf surfaces. In: *Invertebrate Pathology and Microbial Control, Proceedings of IIIrd International Colloquium on Invertebrate Pathology,* 6–10 September 1982, University of Sussex, Brighton, UK. Society for Invertebrate Pathology, pp. 296–301.

Richter, A.R., Fuxa, J.R. and Abdel-Fattah, M. (1987) Effect of host plant on the susceptibility of *Spodoptera frugiperda* (Lepidoptera: Noctuidae) to nuclear polyhedrosis virus. *Environmental Entomology* 16, 1004–1006.

Ridout, M.S., Fenlon, J.S. and Hughes, P.R. (1993) A generalized one-hit model for bioassays of insect viruses. *Biometrics* 49, 1136–1141.

Santiago-Alverez, C. and Ortez-Garcia, R. (1992) The influence of host plant on the susceptibility of *Spodoptera littoralis* (Boisd.)(Lep., Noctuidae) larvae to *Spodoptera littoralis* NPV (Baculoviridae, Baculovirus). *Journal of Applied Entomology* 114, 124–30.

Shapiro, M. (1982) *In vivo* mass production of insect viruses. In: Kurstak, E. (ed.) *Microbial and Viral Pesticides.* Marcel Dekker, New York, pp. 463–492.

Shapiro, M. (1986) *In vivo* production of baculoviruses. In: Granados, R.R. and Federici, B.A. (eds) *The Biology of Baculoviruses,* Vol. II. CRC Press, Boca Raton, Florida, pp. 31–62.

Singh, P. (1980) Problems of large-scale insect rearing for pathogen production. In: Kalmakoff, J. and Longworth, J.F. (eds) *Microbial Control of Insect Pests. DSIR Bulletin 228,* Department of Scientific and Industrial Research, Wellington, New Zealand, pp. 13–17.

Singh, P. and Moore, R.F. (1985) *Handbook of Insect Rearing,* Vol. II. Elsevier, Amsterdam.

Smith, G.E. and Summers, M.D. (1978) Analysis of baculovirus genomes with restriction endonucleases. *Virology* 89, 519–527.

Smith, I.R.L. and Crook, N.E. (1988) *In vivo* isolation of baculovirus genotypes. *Virology* 166, 240–244.

Smits, P.H. (1987) Nuclear polyhedrosis virus as a biological control agent of *Spodoptera exigua.* PhD thesis, Wageningen Agricultural University.

Vaughn, J.L. (1994) Large volume insect cell culture, selection of cells and medium. In: *Proceedings VIth International Colloquium on Invertebrate Pathology and Microbial Control,* Montpellier, France, Society of Invertebrate Pathology, pp. 432–437.

Volkman, L.E. (1985) Classification, identification and detection of insect viruses by serological techniques. In: Maramorosch, K. and Sherman, K.E. (eds) *Viral Insecticides for Biological Control.* Academic Press, London, pp. 27–53.

Volkman, L.E. and Falcon, L.A. (1982) Quantitative use of monoclonal antibody in enzyme-linked immunosorbent assay to detect the presence of *Trichoplusia ni* (Lepidoptera: Noctuidae) nuclear polyhedrosis virus polyhedrin in *T. ni* larvae. *Journal of Economic Entomology* 75, 868–871.

Voller, A., Bidwell, D.E. and Bartlett, A. (1976) Enzyme immunoassays in diagnostic medicine: theory and practice. *Bulletin of the World Health Organization* 53, 55–65.

Voller, A., Bidwell, D.E. and Bartlett, A. (1979) *The Enzyme-linked Immunosorbent Assay (ELISA). A Guide with Abstracts of Microplate Applications.* Dynatech Ltd, Billingshurst, UK.

Weiss, S.A., Dunlop, B.F., Georgis, R., Thomas, D.W., Vail, P.V., Hoffman, D.F. and Manning, J.S. (1994) Production of baculoviruses on industrial scale. In: *Proceedings VIth International Colloquium on Invertebrate Pathology and Microbial Control,* Montpellier, France, 28 August–2 September, 1994, Society for Invertebrate Pathology, pp. 440–446.

Whitlock, V.H. (1977) Effect of larval maturation on mortality induced by nuclear polyhedrosis and granulosis infections of *Heliothis armigera. Journal of Invertebrate Pathology* 30, 80–86.

Wigley, P.J. (1976) The epizootiology of a nuclear polyhedrosis virus disease of the winter moth, *Opherophtera brumata* L., at Wistman's Wood, Dartmoor. D.Phil thesis, University of Oxford.

Wigley, P.J. (1980) Practical: Counting micro-organisms. In: Kalmakoff, J. and Longworth, J.F. (eds) *Microbial Control of Insect Pests. DSIR Bulletin 228,* Department of Scientific and Industrial Research, Wellington, New Zealand, pp. 29–35.

Williams, C. and Payne, C.C. (1984) The susceptibility of *Heliothis armigera* larvae to three nuclear polyhedrosis viruses. *Annals of Applied Biology* 104, 405–412.

Winstanley, D. and Crook, N. (1993) Replication of *Cydia pomonella* granulosis virus in cell cultures. *Journal of General Virology,* 74, 1599–1609.

Winstanley, D. and Rovesti, L. (1993) Insect viruses as biocontrol agents. In: Jones, D.G. (ed.) *Exploitation of Microorganisms.* Chapman & Hall, London, pp. 105–136.

Wood, H.A. (1977) An agar overlay plaque assay method for *Autographa californica* nuclear-polyhedrosis virus. *Journal of Invertebrate Pathology* 29, 304–307.

Young, S.Y., Yearian, W.C. and Kim, K.S. (1977) Effect of dew from cotton and soybean foliage on activity of *Heliothis* NPV. *Journal of Invertebrate Pathology* 29, 105–111.

Bioassays of Entomogenous Fungi

4

T.M. Butt[1] and M.S. Goettel[2]

[1]School of Biological Sciences, University of Wales, Swansea, UK; [2]Lethbridge Research Centre, Agriculture and Agri-Food Canada, Alberta, Canada

Introduction

Many entomogenous fungi are relatively common and often induce epizootics and are therefore an important factor regulating insect populations. Most species attacking terrestrial insects belong to the Hyphomycetes and Entomophthorales while those attacking aquatic insects are generally from the Chytridiomycetes and Oomycetes. The host is usually invaded through the external cuticle, although infection through the digestive tract occurs with some species. Spores attach to the cuticle, germinate, and penetrate the integument by means of a combination of physical pressure and enzymatic degradation of the cuticle. The mycelium then ramifies throughout the host haemocoel. Host death is usually due to a combination of nutrient depletion, invasion of organs and the action of fungal toxins. Hyphae usually emerge from the cadaver and, under appropriate conditions, produce spores on the exterior of the host.

The importance of entomogenous fungi as biological control agents has been reviewed by Latge and Moletta (1988), McCoy *et al.* (1988), McCoy (1990), Ferron *et al.* (1991), Roberts and Hajek (1992), Tanada and Kaya (1993), and Hajek and St Leger (1994). Examples of common entomogenous fungi, including those of commercial importance, are given in Table 4.1, but a more detailed list is provided by Roberts (1989).

The search for commercially viable entomogenous fungi for use in integrated pest management programmes entails several steps. Fungal species and isolates must first be obtained from diseased insects or the environment, and identified. They must then be evaluated under laboratory conditions to identify the most promising candidates. Concomitantly, several problems

Table 4.1. Some common entomogenous fungi and their hosts.

Entomogenous fungus	Invertebrate host
Division Zygomycotina	
Conidiobolus obscurus	Aphids
Entomophaga aulicae	Lepidopteran insects
Entomophaga grylli	Orthopteran insects
Entomophthora muscae	Dipteran insects
Entomophthora thripidum	Thrips
Erynia neoaphidis	Aphids
Massospora cicadina	Cicada
Neozygites fresenii	Aphids
Zoophthora radicans	Certain Hemiptera and Lepidoptera
Division Deuteromycotina	
Aschersonia aleyrodis	Whiteflies, scales
Beauveria bassiana	Wide host range
Beauveria brongniartii	Cockchafers and sugarcane borer
Culicinomyces spp.	Mosquitoes
Hirsutella thompsonii	Spider mites, citrus mites
Metarhizium album	Homopteran insects
Metarhizium anisopliae	Wide host range
Metarhizium flavoviride	Orthopteran insects
Nomuraea rileyi	Lepidoptera
Paecilomyces farinosus	Coleoptera, Lepidoptera
Paecilomyces fumosoroseus	Wide host range
Tolypocladium cylindrosporum	Mosquitoes
Verticillium lecanii	Wide host range

have to be addressed. The selected isolate must be economically mass produced, have adequate storage properties, and it must be efficacious under field conditions. Formulation is an important factor that can affect many of these properties. For instance, formulations can improve storage time and field efficacy by protecting against desiccation and harmful UV radiation. Some formulations can enhance fungal virulence by improving spore attachment to the host surface, diluting the fungistatic compounds in the epicuticular waxes and stimulating germination. Rapid germination and infection are a hallmark of virulent isolates. Finally, the inoculum must be targeted effectively because mortality is dose related.

Well designed bioassays are central to the successful development of entomogenous fungi. There exists a wide range of attributes among fungal isolates and species (Table 4.2). Bioassays are the tools for identifying the following key parameters: (i) host range, (ii) virulence, (iii) ecological competency (i.e. performance under field conditions), (iv) conditions impeding/enhancing epizootics, and (v) barriers to infection.

The development of bioassays requires a thorough understanding of both host and pathogen requirements. Failure to understand these can lead

Table 4.2. Comparison of the attributes of specialist and generalist insect pathogenic fungi.

Specialist	Generalist
Narrow host range	Wide host range
Mostly biotrophs	Mostly hemibiotrophs
Usually produce relatively few, large conidia	Produce copious small conidia
May produce more than one type of spore	Produce single type of spore
Conidia may be forcibly discharged from conidiophores	Conidia usually passively dispersed
Conidia coated in mucilage	Hydrophobic conidia
Few conidia required to cause rapid infection	Many conidia required to cause rapid infection
Subtilisins not detected	One or more subtilisins secreted
Little evidence of toxicosis	Toxins may play important role in host death
Colonize haemocoel as protoplasts or thin-walled hyphal bodies	Colonize haemocoel as blastospores or filamentous hyphae
Induce spectacular epizootics	Natural epizootics usually less obvious
Diseased insects usually located on aerial parts of plant	Diseased insects mostly found in the soil
Difficult to culture	Easy to culture

to inconsistent results, high control mortality, and poor assessment of fungal virulence. The production, formulation and application methods employed can also influence fungal viability, virulence and efficacy. Methods for isolation, cultivation and storage of entomogenous fungi and important aspects of host–pathogen relationships are briefly reviewed, focusing on specific factors which could influence the results of laboratory and field-based bioassays. The methods used for bioassay of these fungi against insects are discussed and examples are used to illustrate the different methods used, augmenting the recent reviews on bioassay techniques by Goettel and Inglis (1997), Kerwin and Petersen (1997) and Papierok and Hajek (1997).

Isolation of Entomogenous Fungi

Details on initial handling and diagnosis of diseased insects have been recently reviewed by Lacey and Brooks (1997). Pathogens can be retrieved directly from the surface of cadavers if the fungus has already sporulated. Most Hyphomycetes can be scraped directly off the cadaver (Goettel and Inglis, 1997), while insects infected with entomophthoralean fungi may be positioned to shower their conidia directly on to a nutrient surface

(Papierok and Hajek, 1997). If sporulation or external hyphal growth has not yet taken place, diseased insects can be incubated in a humid chamber such as a Petri dish lined with moist filter paper to encourage sporulation. Sporulating cadavers can be placed whole or dabbed on a selective medium for isolation of the pathogen.

Most selective media contain either a fungicide and/or antibiotics which encourage growth of entomogenous fungi and discourage growth of saprophytic fungi and bacteria. Some entomopathogenic fungi can be isolated indirectly from the soil by live baiting with insects such as larvae of *Galleria* spp. (Zimmermann, 1986), or directly by extraction using an aqueous solution, often in conjunction with a selective medium (e.g. Beilharz *et al.*, 1982; Appendix 4.1) or discontinuous density gradients (Hajek and Wheeler, 1994). Aquatic fungi can be baited using a variety of substrates such as hemp seed (Kerwin and Petersen, 1997).

Once isolated, many fungi, especially from the Hyphomycetes, can be maintained *in vitro* on several media. Conidia and mycelium should be stored in cryovials under nitrogen, or freeze-dried and stored in sterile glass ampoules (Humber, 1997). Freshly harvested conidia can also be air dried and stored in a desiccator at 4°C or room temperature. Several hyphomycete fungi (e.g. *Verticillium lecanii, Metarhizium anisopliae*) can be stored as conidia bound to silica gel at −40°C. More details on isolation and storage of entomopathogenic fungi are given by Goettel and Inglis (1997), Humber (1997), Kerwin and Petersen (1997) and Papierok and Hajek (1997).

Production and Formulation

Once isolates have been identified, the next step is the production of stable, non-attenuated inoculum for use in evaluation bioassays. Minor changes in production, storage or formulation can greatly influence bioassay results. The amount of inoculum required for most bioassays is minimal and can sometimes even be obtained from cadavers. Although this is sometimes the only source of inoculum in fastidious fungi which do not readily grow on artificial media, it is preferable to obtain inoculum cultured on an artificial medium. The method of culture will largely depend on fungal species and the type of propagule required. More details on laboratory-scale production of entomopathogenic fungi are given by Goettel and Inglis (1997), Kerwin and Petersen (1997) and Papierok and Hajek (1997). More recent general reviews on mass production and formulation are those of Bartlett and Jaronski (1988), Baker and Henis (1990), Auld (1992), Bradley *et al.* (1992), Goettel and Roberts (1992), Feng *et al.* (1994), Jenkins and Goettel (1997) and Moore and Caudwell (1997).

Attenuation of virulence

Successive subculturing on artificial media often results in attenuation of virulence. Therefore, when possible, large quantities of inoculum should be produced using the initial isolate and stored (e.g. as dry conidia) for use in successive bioassays and studies. The rate of attenuation clearly depends on the isolate and species of pathogen.

Some fungal pathogens retain their virulence even after prolonged culture *in vitro* (e.g. *Culicinomyces clavisporus*, Sweeney, 1981; *Beauveria bassiana*, Samsinakova and Kalalova, 1983; *V. lecanii*, Hall, 1980). In contrast, some isolates rapidly loose virulence after only a few subcultures on artificial media. For instance, Nagaich (1973) noted that an isolate of *V. lecanii* pathogenic to aphids lost its virulence after the second or third subculturing. *Lagenidium giganteum* progressively lost the ability to form oospores and zoospores and to infect *Aedes aegypti* larvae after prolonged culture on a sterol-free agar medium (Lord and Roberts, 1986).

Virulence of attenuated isolates can often be regained with passage through an appropriate host. This has been demonstrated with several pathogens including *M. anisopliae* (Fargues and Robert, 1983), *Nomuraea rileyi* (Morrow *et al.*, 1989), *Paecilomyces farinosus* (Prenerová, 1994), *B. bassiana* (Hall *et al.*, 1972; Wasti and Hartmann, 1975), *Lagenidium giganteum* (Lord and Roberts, 1986) and *Conidiobolus coronatus* (Hartmann and Wasti, 1974).

The effects of culture history on virulence poses a special problem if bioassays are used to compare virulence among isolates obtained from various sources and culture collections, as the precise culture history of such isolates is seldom known. In an attempt to address this problem, each isolate can first be passed through an insect host prior to culture on artificial media and use in bioassays. Vidal *et al.* (1997) first passed 30 isolates of *Paecilomyces fumosoroseus* through nymphs of *Bemisia argentifolii*, then carried out bioassays to compare their virulence against this host. Fargues *et al.* (1997b) passed isolates through a non-host prior to bioassays against a host; isolates of *Metarhizium flavoviride* were first passed through the waxmoth, *Galleria mellonella*, by injecting conidia into seventh-instar larvae prior to use in bioassays against the desert locust, *Schistocerca gregaria*. Although the waxmoth is not known to be a natural host of this pathogen, it was felt that growth and sporulation of the fungus on a natural substrate would help restore virulence. Although original, the utility of this technique in comparative bioassays of fungal isolates has not yet been determined. Further studies in this regard are warranted.

Production of infection propagules

There are four general methods for the production of fungal propagules on artificial media: (i) surface culture on solid media, (ii) fermentation on

Table 4.3. Production and storage information on selected entomogenous fungi.

Pathogen	Production method	Media	Form of inoculum
Aschersonia aleyrodis	1, 3	PDA or chopped millet	DM, B, C
Beauveria bassiana	1, 3	Most nutrient agar and liquid media	DM, B, C
Beauveria brongniartii	1, 3	Most nutrient agar and liquid media	DM, B, C
Coelomomyces spp.	2	Host rearing medium	S, infected copepods
Culicinomyces clavisporus	1, 3	Most nutrient agar and liquid media	DM, B, C
Entomophaga spp.	1, 2	Sabouraud dextrose, egg yolk, milk agar (SEMA)	C
Erynia neoaphidis	1, 2, 3	SEMA	DM, C
Hirsutella spp.	1, 3	Most nutrient agar and liquid media	Submerged C, B, C
Lagenidium giganteum	1	Different solid or liquid media that include a sterol	Z
Metarhizium anisopliae	1, 3	Most nutrient agar and liquid media	DM, B, C
Metarhizium flavoviride	1, 3	Most nutrient agar and liquid media	DM, B, C
Nomuraea rileyi	1, 3	Most nutrient agar and liquid media	DM, B, C
Paecilomyces farinosus	1, 3	Most nutrient agar and liquid media	DM, B, C
Tolypocladium spp.	1, 3	Most nutrient agar and liquid media	DM, B, C
Verticillium lecanii	1, 3	Most nutrient agar and liquid media	DM, B, C
Zoophthora radicans	1, 2, 3	SEMA	DM, C

Production method is: 1, surface or submerged culture; 2, live host; 3, semi-solid or diphasic culture. Form of inoculum: C, conidia; DM, dry mycelium; B, blastospores; Z, zoospores; S, sporangia.

semi-solid media, (iii) submerged fermentation and (iv) diphasic fermentation. Although production on solid media is considered as the most expensive, it is also the simplest and usually suffices for the production of the relatively small amounts of inoculum required for laboratory bioassays. The production methods of some important fungi are summarized in Table 4.3. Because few generalizations can be made regarding culture and production of propagules of the more fastidious fungi, we focus our attention here on those fungi that are more amenable.

Surface culture on solid media

Most facultative entomogenous fungi will grow on one or more defined or semi-defined agar-based medium (e.g. Czapek-Dox, Sabouraud) or on natural substrates (e.g. wheat, bran, rice, egg yolk, potato pulp). Specialist fungi are usually fastidious on artificial media and are usually best maintained on their respective hosts. A few can be grown *in vitro* but require a complex medium.

For example, *Lagenidium giganteum* can be cultivated on simple medium but requires sterols to induce oosporogenesis (Kerwin *et al.*, 1991). Entomophthoralean fungi grow well on Sabouraud dextrose or maltose agar fortified with coagulated egg yolk and milk (Papierok, 1978; Wilding, 1981).

Petri dishes and autoclavable plastic bags are recommended for small- and larger-scale production, respectively. However, other containers such as pans, glass bottles and inflated plastic tubing have been used (Samsinakova *et al.*, 1981; Goettel, 1984; Jenkins and Thomas, 1996). Agar-based media are usually used for routine culture. Alternatively, cheaper substrates such as rice or shelled barley can be used in autoclavable bags or other containers, especially when larger amounts of inoculum are required (Aregger, 1992; Jenkins and Thomas, 1996). Once the fungus has sporulated, conidia are harvested either by washing off using water or a buffer, direct scraping from the substrate surface (e.g. agar), or by sieving (e.g. rice). For some entomophthoralean fungi, the forcibly discharged conidia are allowed to shower directly on the host (Papierok and Hajek, 1997).

To obtain conidia virtually free of nutritive substrate contamination, non-cellulolytic fungi can be grown on a semi-permeable membrane such as cellophane (Goettel, 1984). Pans containing a nutritive substance such as bran are lined with the cellophane, placed in sterile bags, autoclaved, inoculated and incubated. After sporulation has taken place, the membrane with the adhering sporulating fungus is lifted from the nutritive substrate. Conidia can then be scraped from the cellophane surface.

Fermentation in semi-solid media

Production of fungi on semi-solid media involves impregnation of small particles with nutrients. Typically wheat bran is mixed with an inorganic substance such as vermiculite, although other substances can be used to provide a large surface area for growth. The mixture is then steam sterilized and the moisture content adjusted to 50–70%. The fermentation process takes place either in a bin or a rotating drum through which sterile, moist air is passed. Primary inoculum is usually grown in liquid medium. Toward the end of the fermentation cycle, the moist air is replaced by dry air to reduce the moisture content of the bran and to encourage sporulation. The temperature is controlled by regulating the circulating air temperature.

More recently, nutrient-impregnated membranes have been shown to reduce production costs of *M. anisopliae* conidia (Bailey and Rath, 1994). A range of membranes impregnated with skimmed milk were screened including blotting paper, fly screen, hessian, and gauze-type fabrics. Sporulation was profuse on Superwipe (an absorbent fibrous material) soaked in skimmed milk (20 g l^{-1}) supplemented with sucrose (2 g l^{-1}) or dextrose plus potassium nitrate. Spores could be washed off in a similar way to removal of conidia from grain.

Submerged and diphasic fermentation

Submerged fermentation can be used for production of blastospores and submerged conidia of selected isolates of entomogenous fungi. Dimorphic filamentous fungi like *M. anisopliae, B. bassiana, Beauveria brongniartii, V. lecanii, Paecilomyces farinosus* and *Nomuraea rileyi* produce relatively thin-walled blastospores in submerged culture that are infectious but difficult to preserve (Adamek, 1965; Samsinakova, 1966; Blanchere *et al.*, 1973; Ignoffo, 1981). Blastospores are produced in relatively large quantities during the log phase of growth. Most often they are spherical, oval or rod-shaped single cells which usually germinate within 2–6 h. Although several species of entomogenous fungi produce blastospores, there is considerable intraspecific variation. Some isolates produce blastospores more readily than others. The culture medium has a profound influence on blastospore production. There are several recipes for blastospore production (Appendix 4.2).

Blastospores sometimes are indistinguishable from submerged conidia. For example, some isolates of *M. flavoviride, M. anisopliae,* and *Hirsutella thompsonii* will produce conidia-shaped cells in submerged culture occasionally from phialide-like structures (van Winkelhoff and McCoy, 1984; Jenkins and Prior, 1993; T.M. Butt, unpublished observations). Van Winkelhof and McCoy (1984) noted that of 14 isolates of *H. thompsonii* only one produced true conidia. The others produced conidia-like cells.

Diphasic fermentation entails growth of fungi in liquid culture to the end of log phase followed by surface conidiation on a nutrient or inert carrier. This method has been developed for mass production of *B. bassiana* (Bradley *et al.*, 1992) and *M. flavoviride* (Jenkins and Thomas, 1996). A similar approach was used in the production of dry marcescent entomphthoralean mycelium (McCabe and Soper, 1985).

Dry marcescent mycelium

The development of the dry marcescent process (McCabe and Soper, 1985) provides a convenient method for production of fungi, especially fastidious species like *Zoophthora radicans*. This process entails the production of the mycelium by submerged fermentation, harvesting by filtration, coating the harvested mycelium with a protective layer of sugar solution and then drying under controlled conditions. When hydrated, the mycelium quickly sporulates to produce infectious conidia. The dry marcescent process has been used successfully as a source of inoculum for *M. anisopliae* (Pereira and Roberts, 1990; Krueger *et al.*, 1992), *C. clavisporus* (Roberts *et al.*, 1987), *B. bassiana* (Rombach *et al.*, 1988), *Z. radicans* and *Erynia neoaphidis* (Wraight *et al.*, 1990; Li *et al.*, 1993).

Effects of culture conditions on virulence and ecological fitness

Culture conditions can greatly influence the virulence, longevity and ecological fitness of the resultant propagules. For example, St Leger *et al.* (1991) found that levels of enzymes on conidia from infected *Manduca sexta* larvae were higher than those cultured on an agar medium. Papierok (1982) found that conidia of four isolates of *Conidiobolus obscurus* produced *in vitro* were less virulent against aphids than those produced *in vivo*. Hallsworth and Magan (1994a, b) found that *B. bassiana*, *M. anisopliae* and *P. farinosus* accumulate polyols when grown on media with increasing ionic solute concentration and with different carbohydrate types at different concentrations. Inoculum with high reserves of polyols was shown to germinate and grow more rapidly at much lower water activities (A_w 0.90 = 90% RH) than those with small reserves of these polyols (Hallsworth and Magan, 1994c, 1995). Furthermore, in bioassays with *G. mellonella* larvae at different RHs, conidia with large amounts of glycerol and erythritol were more virulent than conidia grown on rich nutrient substrates (Hallsworth and Magan, 1994c). Culture conditions can also influence thermal tolerance. Increasing the sucrose content of the growth medium from 2 to 8% resulted in a reduction of thermal tolerance by conidia of *M. flavoviride* (McClatchie *et al.*, 1994).

Postharvest storage

The postharvest storage conditions greatly affect fungal viability and efficacy. Conidial moisture content is an important factor with respect to temperature tolerance and viability. Zimmermann (1982) showed that the tolerance of *M. anisopliae* for high temperatures increases with increasing desiccation, whilst Daoust and Roberts (1983) showed that at 37°C, two isolates of *M. anisopliae* retained most viability after long-term storage at either 0 or 96% RH. Drying conidia in the presence of desiccating agents like silica gel and $CaCl_2$ appears to improve their viability but direct contact with the desiccant can be detrimental (Daoust and Roberts, 1983).

Moore *et al.* (1995) found that dried conidia stored in oil formulations remained viable longer than those stored as a dried powder, especially if stored at relatively low temperatures (10–14°C compared with 28–32°C). Addition of silica gel to oil-formulated conidia appears to prolong their shelf life. Undried conidia of *M. flavoviride* lose viability rapidly, with germination dropping below 40% after 9 and 32 weeks at 17°C and 8°C, respectively. After 127 weeks in storage, germination remained at over 60 and 80% for the dried formulations at 17°C and 8°C, respectively (Moore *et al.*, 1996). These conidia were found to have retained virulence similar to that of freshly prepared formulations. Furthermore, conidia dried to 4–5% moisture content showed greater temperature tolerance than conidia with a higher moisture content (McClatchie *et al.*, 1994; Hedgecock *et al.*, 1995).

Formulation

Formulations can greatly improve the efficacy of entomopathogens both in protected and field crops. The type of formulation and selection of additives for a given formulation are critical to their stability. The basic components of most formulations include, in addition to the active ingredient (i.e. fungal spore), one or more of the following: a carrier, diluent, binder, dispersant, UV protectants and virulence-enhancing factors (Moore and Caudwell, 1997).

The most widely used carriers are oil and water. Because of their hydrophobic nature, conidia of some hyphomycete fungi readily suspend in oils, but oil itself can be toxic, especially when applied against small insects. Oils are reasonably effective in sticking spores to insect and plant surfaces (Inglis *et al.*, 1996a). In contrast, surfactants (e.g. Tween) need to be added to water to ensure conidial suspension, but these are toxic to conidia if used at high concentrations (e.g. >0.1% v/v). Incorporation of humectants (e.g. Silwet) can improve infection by providing moisture for germination and infection.

Recent studies show that more than 60% of the fungal inoculum can be removed from leaf surfaces by rain (Inglis *et al.*, 1995c; T.M. Butt, unpublished observations). Compounds increasing adhesion of spores to insect and plant surfaces need to be evaluated. Equally important, the formulation must not interfere with the infection process, and at best it should enhance disease transmission.

Photoinactivation has emerged as one of the major environmental factors affecting persistence and thus efficacy of entomogenous fungi. Ultraviolet radiation can sterilize surfaces of plants and insect cuticle (Carruthers *et al.*, 1992; Inglis *et al.*, 1993). Incorporation of UV blockers (e.g. Tinopal) in formulations can offer some protection against harmful UV radiation (Inglis *et al.*, 1995b).

Other Factors Affecting Virulence

How the culture, storage and formulation of fungi can influence their viability, virulence and field efficacy has been summarized in the previous section. In this section, other factors which could influence the results of laboratory and field bioassays are considered.

Most entomopathogenic fungi gain entry to the haemocoel by penetrating the host cuticle using a combination of hydrolytic enzymes and mechanical force (Goettel *et al.*, 1989; St Leger *et al.*, 1989a, b; Butt *et al.*, 1990, 1995; Schreiter *et al.*, 1994). The speed of kill, and to some extent the host range, are influenced by the number of infection propagules in contact with the cuticle. Mortality is dose related. There are vulnerable sites on the cuticle, such as the intersegmental membranes and sites under the elytra of certain beetles. Basking in the sun, preening and ecdysis reduce the amount of viable inoculum on the insect surface. Handling of insects, rearing conditions and insect vigour influence their susceptibility to fungal infection. Fungal pathogens are greatly affected by abiotic factors such as temperature, light and humidity.

Dose-related mortality

Susceptibility of most insects is dependent on spore dosage. It is presumed that a threshold exists whereby a certain number of propagules are necessary to overcome the host, however, the exact nature of this relationship has not been determined. A positive correlation between the number of infective spores and mortality by mycosis has certainly been established for most insect/pathogen combinations, but there are exceptions. For instance, Goettel *et al.* (1993) reported a negative correlation between dose and mortality at concentrations greater that 10^4 ascospores of *Ascosphaera aggregata* per leaf-cutting bee, *Megachile rotundata,* larva. Therefore, care must be taken when interpreting results of very high application rates in some systems as the possibility of self-inhibition exists.

The dose–mortality relationship is the principal component in many bioassay designs (Chapter 7). Insects are treated at several increasing doses and the LD_{50} and its fiducial limits are then used to compare virulence or 'potency' against other isolates. The slope of the dose–mortality curve is also very useful when comparing virulence amongst different isolates.

Vulnerable sites on the cuticle

Not all areas of the insect cuticle are equally vulnerable to penetration by propagules of entomopathogenic fungi. The intersegmental membranes (Wraight *et al.*, 1990), areas under the elytra (Butt *et al.*, 1995) and the buccal cavity (Schabel, 1976) can be preferential sites of infection. Therefore, the location where the inoculum lands on the cuticle can also influence the probability of infection and the speed of kill. Consequently, targeting of the inoculum is an important consideration in the development of bioassay protocols.

Insect behaviour may affect ultimate sites of penetration. For instance, results of laboratory studies demonstrated that the most sensitive sites for penetration of *Beauveria brongniartii* on larvae of the cockchafer, *Melolontha melolontha,* were the mouth and anus (Delmas, 1973). However, Ferron (1978) found that in larvae of the same species collected in nature, the most frequent sites of infection occurred on the membranes between the head capsule and thorax or between the segments on appendages. This apparent contradiction is possibly due to the larval behaviour of burrowing in soil; particles continuously scrape infectious inoculum off the exposed cuticle whereas the intersegmental membrane is protected from this mechanical action. This is a good example of how results obtained from laboratory bioassays must be treated with caution if used to predict the situation in the field.

Ecdysis and developmental stage

Not all stages in an insect's life cycle are equally susceptible to infection by entomogenous fungi. Pupal stages are often the most resistant stage, while adults can be the most susceptible. For instance, larvae of the thrips, *Frankliniella occidentalis* were found less susceptible to *V. lecanii* and *M. anisopliae* than adults, while later instars were less susceptible than earlier instars (Vestergaard *et al.*, 1995). Larvae of *Ostrinia nubilalis* were found to be most susceptible to infection by *B. bassiana* when exposed as first-instar larvae, while fourth instars were most tolerant (Feng *et al.*, 1985). Fransen *et al.* (1987) found that older instars of *Trialeurodes vaporariorum* were less susceptible to *Aschersonia aleyrodis*, while adults were seldom infected. Within adult stages, there could also be differences in susceptibility between different sexes and forms such as aphid alates and apterae (Oger and Latteur, 1985).

The time of inoculation prior to ecdysis, and the length of the inter-moult period are important factors that may significantly affect bioassay results. Moulting may remove the penetrating fungus prior to the colonization of the insect, if it occurs shortly after inoculation (Vey and Fargues, 1977; Fargues and Rodriguez-Rueda, 1979). In contrast, Goettel (1988) found that larvae of the mosquito, *Aedes aegypti*, were more susceptible to *Tolypocladium cylindrosporum* during their moulting period.

Effect of diet on susceptibility

Successful infections are also dependent upon the host diet. For example, some insects maintained on artificial diet can be more susceptible to infection than insects fed a natural diet (Boucias *et al.*, 1984; Goettel *et al.*, 1993). Likewise, laboratory-reared insects can be more susceptible than field-collected ones (Bell and Hamalle, 1971). Insects which have been starved can also differ in their susceptibility compared with well-fed ones (Milner and Soper, 1981; Butt, unpublished observations).

Tritrophic interactions between host plants, insect pests and ento-mopathogens have been reported for fungi. The pathogenicity of the ento-mogenous fungus *B. bassiana* mediated by host plant species has been reported for both the Colorado potato beetle (Hare and Andreadis, 1983) and the chinch bug (Ramoska and Todd, 1985). Presumably, larvae growing on more favourable plant species are better able to mount a successful defensive reaction to pathogens (Hare and Andreadis, 1983) or have a shorter intermoult period (see previous section). For chinch bugs, adults feeding on wheat or artificial diet and inoculated with *B. bassiana* demonstrated higher mortality and greater fungal development than adults feeding on maize or sorghum. These data were interpreted as showing that insects are benefiting from the fungistatic secondary chemicals in maize and sorghum (Ramoska and Todd, 1985).

An association has also been demonstrated between volatiles and fungal development. Crucifers contain glucosinolates, nitrogen- and sulphur-containing secondary metabolites. These are hydrolyzed by an enzyme to release biologically active compounds which, in addition to playing a major role in defending the plants from herbivores and fungal pathogens (Chew, 1988), also appear to interfere with the infection processes of insect-pathogenic fungi (Inyang *et al.*, 1999).

Sublethal Effects and Other Attributes

Measurement of the effectiveness of a pathogen against a host insect must be based on many factors, in addition to virulence. Not all insects treated with a fungus succumb to infection. Sublethal effects of entomopathogenic fungi have been insufficiently studied. It is usually presumed that those insects that do not succumb to infection do so at no expense. However, this is not necessarily so. For instance, Fargues *et al.* (1991) demonstrated that the fecundity of the Colorado potato beetle, *Leptinotarsa decemlineata*, surviving treatment was much lower than in beetles that were not treated. This study demonstrates that survival does not necessarily come without its price.

Many attributes of a pathogen are important in determining its ecological fitness. There exists a wide range of tolerance among fungal isolates to environmental factors such as sunlight (Fargues *et al.*, 1996) and temperature (Fargues *et al.*, 1997a, b) biotic attributes such as speed of germination (Papierok and Wilding, 1981) and ability to sporulate on the host cadaver (Hall, 1984). In addition to virulence, these are some of the important aspects that need to be considered when determining the effectiveness of an isolate for development as a microbial control agent. Determination of sublethal effects and other attributes is an important, yet much neglected area, which warrants further study.

Bioassay Procedures

Use of bioassay to assess the effects of entomopathogenic fungi in insects is essentially limitless. This, combined with the fact that there is a vast array of entomopathogenic fungi with a great variety of hosts, means that there are no standardized bioassay methods as far as entomopathogenic fungi are concerned. Consequently, bioassays must be tailored according to host, pathogen and bioassay objective.

Bioassays can be used to determine and quantify host–pathogen relationships and the effect of biotic and abiotic parameters on these. Bioassays of entomopathogenic fungi have been used extensively in five important applications: (i) determination of virulence, (ii) comparison of virulence among isolates, (iii) determination of host range, (iv) determination of epizootic potential, and (v) studies on effects of biotic and abiotic factors such as host age, host plant, temperature, humidity and formulation.

The objective of a bioassay must be well defined before a bioassay protocol is adopted. Although bioassay procedures must be as efficacious as possible, they must also be designed to address the objectives and provide as meaningful results as possible to meet these. Choice, rearing and developmental stage of the host, infective propagule, formulation and inoculation method, conditions of post-inoculation incubation, method of mortality assessment (including mortality in controls), bioassay design and statistical analyses must all be carefully considered.

Special care must be taken when bioassays with non-target organisms (NTO) are used for risk assessment. It is common for entomopathogenic fungi to infect hosts in the laboratory which are never infected in the field. For instance, Hajek *et al.* (1996) demonstrated that data on host range of *Entomophaga maimaiga* from laboratory bioassays gave poor estimates for predicting non-target impact; the host range under field conditions was much narrower than that predicted from laboratory results. Laboratory assays demonstrated an LD_{50} of 2.2×10^5 conidia of *B. bassiana* per honey bee worker, whereas subsequent whole-hive exposures resulted in less than 1% mortality (M.R. Loeser, S.T. Jaronski, and J.M. Bromenshenk, cited in Goettel and Jaronski, 1997). The US Environmental Protection Agency (EPA) now accepts that infectivity tests with caged honey bees can be misleading and recommends that 30-day whole-hive tests be used instead (Goettel and Jaronski, 1997). Development of laboratory bioassays which better simulate the environment to which bees are exposed (e.g. internal hive temperatures are commonly held between 32 and 36°C) may provide a better and cheaper alternative to whole-hive assays.

Low mortalities in the field can be obtained even after application of highly virulent propagules. Inglis *et al.* (1997a) obtained low efficacy after applying conidia of *B. bassiana* on to native rangeland against grasshoppers, despite excellent targeting. Results of laboratory assays demonstrated high virulence, and high levels of infection were observed in caged field-collected grasshoppers maintained under glasshouse conditions with similar temperature and humidity to those experienced in the field. Subsequent studies revealed the importance of thermoregulation; grasshoppers bask in the sun, elevating their temperature to levels that prevent disease progress

(Inglis *et al.*, 1996b, 1997b). The development of a bioassay design whereby grasshoppers were allowed to thermoregulate allowed for more meaningful prediction of virulence under field conditions (Inglis *et al.*, 1996b, 1997b, c).

In bioassays designed to predict performance of the pathogen under field conditions, as many pertinent environmental (e.g. temperature, photoperiod) and other (e.g. inoculation method) parameters as possible must be taken into account and incorporated into the bioassay design. Unfortunately most bioassays are performed under static conditions (e.g. constant temperature, RH, photoperiod). Although such assays may be useful in comparing activity of different isolates, they often provide misleading information as far as performance under field conditions is concerned. However, bioassays performed under constant conditions studying single parameters can be very useful in identifying the pertinent parameters that need to be considered.

All bioassays should include a non-treatment control in order to monitor survival of insects under the post-inoculation incubation conditions. Such control insects should be treated with a carrier used for application of the inoculum. In dose–mortality assays, control mortality is then corrected for in the statistical analyses (Chapter 7). If bioassays are used for host range or safety to non-target organisms, it is imperative that a known susceptible host is also treated in parallel with the non-target organisms (i.e. positive control). Otherwise negative results are difficult to interpret unless evidence is provided regarding the virulence of the inoculum against a susceptible host under the same bioassay conditions.

Choice of sample size and range of doses is usually difficult when dealing with fungal pathogens due to the great variability in responses between different isolates and hosts. For dose–mortality assays, preliminary bioassays should be first conducted using a wide range of doses and relatively small numbers of hosts. A range of doses that would result in mortalities between 25 and 75% should then be chosen. The choice of sample size may be more problematic and will depend very much on the pathogen–host system. For instance, Oger and Latteur (1985) determined that, in bioassays of *Erynia neoaphidis* against the pea aphid, *Acyrthosiphon pisum*, the factor that most affected precision was the number of replicated assays. They found that a sample size of ten aphids for each of 10–20 doses replicated three or four times gave an adequate precision for comparative assays. However, more commonly, five doses should suffice, especially if at least three or four of the doses fall in the 25–75% mortality range. As in any scientific study, the whole bioassay must be repeated at a later date, preferably using another batch of insects and inoculum preparation in order to ensure reproducibility of results and thereby substantiate the conclusions. More discussion on choice of doses, sample size, bioassay design and repetition of experiments is presented in Chapter 7.

Inoculation

Method of inoculation is influenced primarily by the form of the inoculum and the size and fragility of the insect. Inoculum is most commonly administered to the surface of the cuticle either through direct methods such as dusting, dipping or spraying or through indirect methods such as the use of baits. Whatever the inoculation method, it is imperative that the viability of the inoculum be determined as close to the treatment time as possible. Otherwise, it is not possible to determine if lack of efficacy is due to low viability of the inoculum. If at all possible, viability assessments of the formulated product should be made. It may be necessary to compare viabilities between the active ingredient and formulated product in order to determine if the formulation adjuvants have a detrimental effect. Methods for viability assessments are summarized by Goettel and Inglis (1997).

Entomophthoralean fungi differ from hyphomycete fungi in several characteristics relevant to the development of bioassays. The former usually produce comparatively few, large, forcibly discharged, sticky conidia. The latter generally produce numerous, small, dry conidia. The methods for inoculating insects with entomophthoralean fungi are usually limited to either showering conidia on to anaesthetized insects or the host's food source such as leaf surfaces. The inoculum may be showered from mycosed cadavers, sporulating cultures or marcescent mycelium (Papierok and Hajek, 1997). In contrast, inoculum of hyphomycete fungi may be applied by the methods noted above.

The most important factor in choosing an inoculation method is to ensure presentation of a precise dose which will reduce variability and help ensure repeatable results. Consequently, crude methods of inoculation such as allowing an insect to walk on the surface of a sporulating culture should be avoided, although such methods may be useful in certain studies whose aims are, for instance, to establish new host records *per se*. It is preferable, however, if the amount of inoculum administered to each insect can be controlled as precisely as possible. This is usually accomplished through enumeration of the inoculum and administering a precise dose to each insect. In situations where it is difficult or impractical to determine the precise dose being administered, it is common practice to obtain estimates of the dose by recovering the propagules after application of the inoculum, either through washing or homogenizing the insects, and then estimating propagule concentrations or through direct enumeration or spread plating (Goettel and Inglis, 1997). Details on methods for enumeration of propagules are presented by Goettel and Inglis (1997) for Hyphomycetes, Kerwin and Petersen (1997) for water moulds, and Papierok and Hajek (1997) for Entomophthorales.

Introduction of inoculum through injection can be used in situations where the importance of the cuticular barrier is not an issue (i.e. immunological assays). A tuberculin syringe attached to a motorized microinjector is usually used to treat many insects rapidly and effectively. The inoculum can be introduced *per os* or directly into the haemocoel by piercing the inter-segmental membrane. Aquatic insects are usually treated by introducing known numbers of propagules into their rearing medium. Specific methods of inoculation are described in the examples presented below.

Bioassay chambers

A wide range of bioassay chambers has been used by various workers for disparate insect species. Bioassay chambers are usually chosen according to availability, price, convenience, ease of cleaning and requirements of the host. With entomopathogens, it is important that the chambers be ade-quately decontaminated prior to reuse. An alternative is to use disposable containers. Some commonly used bioassay chambers include inexpensive plastic or polystyrene coffee cups, ice cream cartons, cigar boxes, glass jars, plastic bins, buckets or bowls, portable cages, and nylon/cotton fine-mesh sleeves.

For assays with small insects, it is often possible to use single leaf peti-oles or excised leaves in small bioassay chambers such as Petri dishes. In such systems, it is important to delay leaf senescence as long as possible, by providing a nutrient or water source for the plant tissue. For instance, stems of single leaf petioles can be immersed in water (Fig. 4.1), kept wet with moistened cotton wool placed on parafilm (Mesquita *et al.*, 1996; Fig. 4.2) or placed directly on to a nutritive substrate. Vidal *et al.* (1997) cut out 3.5-cm diameter discs from ornamental sweet-potato leaves, disinfected them in a series of alcohol and sodium hypochlorite solutions, and placed them in small Petri plates containing a KNOP medium (in g l^{-1} water: 0.25 KCl, 0.25 KH_2PO_4, 0.25 $MgSO_4$, 0.02 $FeSO_4$, 10 agar) (Fig. 4.3).

Choice of bioassay chamber is critical in field-cage bioassays. Although these best 'mimic' field situations, the type of bioassay chamber can greatly influence the climate within. Even screened cages provide shading and pro-tection from wind. In a field-cage experiment using screened cages, Inglis *et al.* (1997b) found minimal differences in temperature and relative humid-ity within and outside the cages, but there was approximately 55% shading within the cage due to the mesh screening.

Post-inoculation incubation conditions

Conditions of humidity, temperature and light can greatly influence bioas-say results. Consequently, after inoculation, the insects should be incubated under controlled environmental conditions or transferred to field bioassay cages. Controlled environmental conditions are usually maintained using

Fig. 4.1. Bioassay chamber containing single leaf with its water supply. Photo by courtesy of Lerry Lacey.

Fig. 4.2. Ventilated bioassay chamber containing single blades of barley leaves. Moisture is provided by soaked cotton battens placed on pieces of Parafilm. Photo by courtesy of Antonio Mesquita and Lerry Lacey.

environment chambers or incubators. Insects can often be pooled in assay chambers according to treatment group; however, it is preferable to incubate cannibalistic insects such as grasshoppers singly. The conditions chosen will vary according to the objectives of the bioassay, but they generally should be favourable for survival of non-inoculated insects.

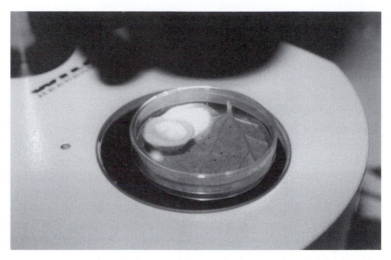

Fig. 4.3. Ventilated bioassay chamber containing an excised sweet-potato leaf disc placed on a nutritive agar medium. Photo by courtesy of Claire Vidal and Lerry Lacey.

Bioassays must often be run for 1–2 weeks or even longer with slower-acting pathogens such as fungi, where LT_{50}s of 4–6 days are common. With such long time spans, control mortalities can often be problematic. Meaningless results are obtained if control mortalities are too high. Control mortalities are usually accounted for in the statistical analyses (Chapter 7), but as a general rule, results must be suspect if control mortalities are higher than 15–20%. If control mortalities greater than 20% are unavoidable, results must be interpreted with caution.

For some fungi, humidities approaching saturation must be maintained in order to obtain infection (Papierok and Hajek, 1997). Saturated and near-saturated conditions are usually provided using water agar or saturated filter paper within the bioassay chambers. Precise conditions of humidity can be maintained in the bioassay chambers by continuously circulating air through a humidifying medium of saturated salt solution (Fargues *et al.*, 1997b). Sealed chambers without air circulation should be avoided as the aerial humidity occurs at equilibrium only at the solution/air interface.

Temperature is one of the most easily controlled factors. Most bioassays comparing virulence among isolates use only one constant temperature, however, the results obtained could provide misleading information. For instance, Fargues *et al.* (1997b) compared the virulence of four isolates of *M. flavoviride* at four constant temperatures. No significant differences in virulence occurred among these isolates at temperatures of 25, 30 and 40°C. In contrast, there were significant differences in virulence among isolates at 35°C. This demonstrates the importance of using several temperatures when screening for the most virulent isolates. Cyclical conditions of temperature approximating as much as possible the natural conditions should be considered for such screenings.

Mortality assessments

Mortality assessments should be made daily. In addition to computing median lethal doses, this allows computation of median lethal times, which can be very useful in making comparisons between different treatments (Chapter 7). If insects are not being incubated singly, dead insects must be removed as soon as possible and certainly prior to sporulation to prevent horizontal transmission and loss due to cannibalism. Incubation time varies according to the insect and fungus being evaluated. Generally, incubation should continue as long as insects continue to succumb to the pathogen.

Mycosis is usually verified by incubating dead insects at high RH (e.g. in Petri dishes containing moistened filter paper or water agar) to allow for fungal colonization and sporulation on the cadaver. It is important to also incubate cadavers from control insects to determine residual infection levels or if accidental contamination occurred.

Some Examples

Here we provide specific examples to illustrate some of the many different bioassay methods used with an array of fungal pathogens and insect hosts. It is not our intent to provide a detailed evaluation or critique on the methods used. Each bioassay needs to be adapted to the specific needs of the host/pathogen combination and according to the objectives of the bioassays themselves. We have divided the sections according to inoculation method.

Spray

Spray bioassays are used extensively, especially against small and fragile insects that are otherwise difficult to treat. Although sophisticated stationary and track sprayers are available for this purpose, simpler and less expensive systems can be developed using an artist's air-brush. Drop size and distribution must be carefully monitored.

Honey bees with M. flavoviride

As mentioned previously, laboratory assays for testing safety against honey bees can provide misleading results (Goettel and Jaronski, 1997). The assay presented here was used to determine the safety of an oil formulation of *Metarhizium flavoviride* to adult honey bee workers (Ball *et al.*, 1994). Similar protocols were used by Vandenberg (1990) and Butt *et al.* (1994) to test the safety of *B. bassiana* and *M. anisopliae* to honey bees.

1. Combs containing mature honey bee pupae were removed from colonies and maintained overnight in an incubator at 35°C. Groups of newly emerged bees (less than 18 h old) were transferred to small cages, supplied with concentrated sucrose, water and pollen, and maintained at 30°C for 1 week before use.

2. For the application of conidial suspensions, bees were briefly anaesthetized with carbon dioxide and transferred to spray cages made from 12.5 cm square perspex 6 mm thick with a hole 10 cm in diameter in the centre. The bees were sandwiched between two layers of 0.71 mm galvanized wire mesh with an aperture of 2.46 mm, 25 bees to each cage.

3. The bees were allowed to recover at room temperature and oil formulations of conidia were applied from a rotary atomizer attached to a track sprayer in a room maintained at 30°C. Sprays were calibrated to simulate field dose levels equivalent to twice and 20-fold expected field application rates. A solution of fluorescent tracer Uvitex OB (Ciba-Geigy) in Ondina oil was used to determine the volume of formulation deposited on the bees. The controls consisted of six cages of bees which were not sprayed and eight cages of bees sprayed with the oil carrier alone. Locust positive controls were also treated in parallel.

4. Immediately after treatment, bees were returned to their original cages without anaesthetization and maintained at 30°C. Bees that died within 24 h of transfer were omitted from the assay.

5. Cages were checked daily for 14 days and dead bees removed and incubated at room temperature on moist filter paper in plastic Petri dishes. Conidia of the fungus appeared within a few days over the surface of infected individuals.

6. A field dose killed 11% of the bees, twice field dose killed 30%, while a 20-fold dose killed 87%.

Whiteflies with Aschersonia aleyrodis

Spore suspensions are often sprayed directly on to leaves containing the host. The assay presented here was used by Fransen *et al.* (1987) to study differential mortality of different life stages of the glasshouse whitefly, *Trialeurodes vaporariorum* treated with conidia of *Aschersonia aleyrodis*. Difficulties with this protocol can be encountered when used with highly mobile insects, as the insects may differentially pick up inoculum post-application depending on their mobility.

1. Spores of *A. aleyrodis* were obtained from a 3-week culture grown on coarse cornflour at 25°C. Spores were suspended in sterile distilled water and enumerated in a counting chamber.

2. Two millilitres containing 4×10^6 spores ml^{-1} were applied with a Potter spray tower at 34.5 kPa to the underside of a cucumber leaf bearing the whiteflies. Ten leaves per age class were treated with the spore suspension and two leaves were treated with distilled water as a control.

3. Spore viability was determined by spraying a spore suspension on to a water agar plate and counting the number of germlings 24 h after spraying at 25°C.

4. After the water had evaporated from the leaves, plants were covered with plastic bags to ensure saturated moisture conditions for the first 24 h at 20°C and 16 h photoperiod. Thereafter, the bags were removed and the plants were kept at 70 ± 10% RH.

5. Adult whiteflies had to be anaesthetized before inoculation, but difficulties were encountered. If adults did not revive quickly, they drowned in the spore suspension. To alleviate this problem, adults were exposed directly to spores on the surface of a sporulating culture for 24 h prior to transfer to leaves in clip cages.

6. Disease progress was recorded at 3- to 4-day intervals until 90% of the survivors had developed into adults.

7. It was found that older instars were less susceptible and that adults were seldom infected.

European corn borer with B. bassiana

Larger insects are often sprayed directly and then transferred to rearing containers. Using the bioassay method described here, Feng *et al.* (1985) determined age-specific dose–mortality effects of *B. bassiana* on the European corn borer (ECB), *Ostrinia nubilalis*.

1. Larvae were collected from overwintering sites (field corn stubble), transferred to a meridic diet and maintained for at least three generations to eliminate weak and diseased individuals.

2. Three isolates of *B. bassiana* obtained from a culture collection were first passaged through ECB larvae. Conidial suspensions were tower-sprayed on to larvae and incubated at 26°C. Conidia were then harvested from cadavers and inoculated on to Sabouraud dextrose agar (SDA). After incubation at 26°C for 20 days, conidia were harvested, dried and stored in vials at 4°C until use.

3. Spore suspensions were prepared and applied to groups of 20 newly moulted larvae of each instar in individual Petri dishes using a spray tower. Dose was assessed by spraying SDA Petri plates and counting colony-forming units (CFU) cm^{-2} that developed. Five doses of between 9.3×10^2 and 2.9×10^6 CFU cm^{-2} were applied for each isolate.

4. The Petri dishes in which the larvae were sprayed were covered. Larvae were fed an artificial diet with all fungicides removed, and incubated at 26°C for 24 h, after which time they were transferred to uncontaminated diet without fungicides and incubated for a further 24 h. After 48 h, standard artificial diet was provided. The larvae were examined daily for mortality.

5. The experiment was replicated four or five times using different batches of larvae and new conidial suspensions.

6. First-instar larvae were found most susceptible, there was little difference between 2nd, 3rd and 5th instars while 4th instars were most tolerant.

Immersion

Insects are often dosed by immersion into suspensions of spores for a specified time. Although this is usually a quick and convenient dosing method, precise measurement of dose is difficult. In order to ensure that each insect from a treatment group receives similar doses, groups of insects should be simultaneously dipped. Care must be taken that each insect remains precisely the same amount of time in the dipping suspension. If insects are dosed singly, then a separate suspension should be prepared for each insect. Otherwise, each subsequent insect dipped could receive less inoculum, especially since conidia of many entomogenous fungi are hydrophobic and therefore adhere preferentially to the insect cuticle.

Aphids with Verticillium lecanii

The difficulty of immersing small insects for a specified time can be overcome by draining off the inoculum rather than dipping the insects into the inoculum *per se*. This method was used by Hall (1976) to bioassay *Verticillium lecanii* against the aphid, *Macrosiphoniella sanborni* (see below). It can be adapted for use against almost any insect which can withstand submergence for a short period. However, this method is difficult to use when attempting to study effects of different carriers and formulations, as the inoculation method does not represent that which would be expected under operational conditions.

This method has been adopted for use with many insects and fungi. Some examples include bioassay of *B. bassiana* against larvae of the curculionid weevils in the genera *Sitona*, *Hylobius*, *Diaprepes*, *Chalcodermus* and *Pachnaeus* (McCoy *et al.*, 1985) and bioassays of *M. anisopliae* against beetle and aphid crucifer pests (Butt *et al.*, 1992, 1994).

1. Conidia were obtained from a single spore isolate of *V. lecanii* grown on SDA at 23°C for 7 days and then stored at −17°C. Conidia for bioassay were produced by spread plating a spore suspension obtained from storage on to SDA and incubating at 23°C. After 7 days, spores were harvested using a bent glass rod and phosphate buffer containing 0.02% Triton X100 as a wetting agent. The spore suspension was purified by filtering through cheesecloth, centrifuging and washing four times in phosphate buffer. The concentration of spores was determined using a haemocytometer. A viability assessment was performed by incubating three drops of a suspension containing approximately 10^6 spores ml^{-1} on a thin layer of sterile SDA on a glass slide and incubating for 12 h at 23°C before examining using phase-contrast microscopy.

2. Adult alate aphids, obtained from a stock culture maintained on potted chrysanthemum plants, were transferred to chrysanthemum leaf discs in breeding cells. After 7 days, apterous progeny from the alatae were transferred to fresh leaf discs and were used for bioassay 8 days later.

3. Batches of mature aphids were placed in glass Petri dishes. Each batch was transferred on to filter paper in a 7.5-cm diameter Büchner funnel. Thirty millilitres of spore suspension were then gently poured over the aphids. After 2 s, the suspension was quickly drained by suction.

4. After inoculation, treated insects were placed singly on leaf discs in high humidity bioassay chambers, incubated at 20°C for 6 days, and examined daily for mortality. Dead insects were examined microscopically for signs of mycosis.

5. The LC$_{50}$ was found to be 2.3×10^5 spores ml^{-1}. It was noted that aphids tolerated transfer to the assay cells much better if preconditioned in breeding cells than direct transfer from plants.

Dusting

Dusting is sometimes used to inoculate insects. Whereas some workers literally dust the insects others may simply allow healthy insects to walk over a sporulating culture (Bidochka *et al.*, 1993). Dusting allows inoculation of large numbers of insects at once. Care must be taken to ensure that the insects can withstand this procedure. If at all possible, attempts should be made to quantify the amount of inoculum received by each insect. Because it may be possible that death could be caused by suffocation due to obstruction of tracheal passages, controls should consist of killed spores plus the carrier. The great variance in dosage acquired should be noted from the example below. For this reason, generally, this method should be avoided for dose–mortality assays unless precautions are taken to ensure that the variation in the amount of dose administered within a dosage group is minimized.

Chinch bug with B. bassiana

Ramoska and Todd (1985) used a dusting method to study the effects of host plant on virulence of *B. bassiana* towards the chinch bug, *Blissus leucopterus leucopterus*. Although dose–mortality assays were not used, dosage levels received by each insect were estimated.

1. A culture of *B. bassiana* was grown on Sabouraud maltose agar for 3 weeks at 27°C. Conidia were harvested, dried, sieved and stored at 4°C until used.

2. Chinch bugs were inoculated by placing 25 adults at a time in a Petri dish containing dry conidia. Petri plates were then shaken to ensure full coverage of the insects. Insects were then removed to incubation chambers.

3. In order to quantify dose levels, two insects from each batch were removed and immersed in 10 ml of 5% v/v aqueous Tween 20 and vigorously agitated. Conidial density in the suspension was calculated using a haemocytometer. Dosage levels ranged from 0.9×10^5 to 1.8×10^8 conidia per insect.

4. Inoculated bugs were transferred on to host plant seedlings and a 43-cm, ventilated, clear plastic collar was placed over the plant which served to cage the insects. Twenty-five insects were placed into each of four replicate chambers.

5. After 2 weeks, the test chambers were dismantled and mortality was assessed. Dead bugs were transferred to Petri dishes containing moist paper towelling to assess fungal growth on the cadavers.

6. Results showed that feeding on sorghum and maize resulted in greater tolerance to the fungus compared with insects feeding on other food sources.

Japanese beetle with B. bassiana *and* M. anisopliae

Although dusting is generally not recommended for dose–mortality assays due to the high variability of inoculum received by insects in a dosage group, as seen in the example above, this inoculation method has been used successfully in dose–mortality assays with some insects. In the example presented here, Lacey *et al.* (1994) determined LT_{50} and LC_{50} estimates for isolates of *B. bassiana* and *M. anisopliae* in adults of the Japanese beetle, *Popillia japonica*.

1. One isolate each of *B. bassiana* and *M. anisopliae* was cultured on SDA, harvested with a rubber spatula, dried in an incubator overnight at 30°C, and then passed through a 250 µm sieve. Spore viability was determined by plating 100 µl of the conidial suspensions on SDA and counting the number of colonies formed after 48 h. Spore counts were estimated as 2×10^7 conidia mg^{-1} for *B. bassiana* and 3.6×10^7 conidia mg^{-1} for *M. anisopliae*.

2. Adult Japanese beetles were field-collected using baited traps and held briefly in the laboratory prior to treatment. One hundred individuals were counted out into 140-ml polystyrene cups, which were closed with perforated screw caps.

3. Five dosages ranging from 0.5 to 10 mg conidia were weighed out and added to the cups containing the adults. In a later modification of the procedure, lower dosages were used in which talcum powder at a ratio of 990 mg to 10 mg conidia was added as a carrier. Five replicate cups were prepared for each dosage. The cups were periodically rotated end-over-end for a 1-h period to help distribute the conidia.

4. Thirty beetles from each treatment were then removed and divided among three holding cages, which consisted of 950-ml plastic containers with perforated lids. Water and humidity was provided by dental wicks which protruded through the bottom of the cage into 100 ml reservoirs below. Adults were incubated at 22–24°C, checked for mortality, and provided with fresh blackberry leaves daily for up to 8 days. Four replicate tests were conducted on each of four separate dates.

5. Dead beetles were transferred to 950-ml plastic tubs containing 500 g moistened sterilized soil, incubated at 22–24°C for 1 week and then were examined for fungal outgrowth.

6. A dose–mortality response was obtained with LC_{50} estimates of 0.7 mg conidia per 100 adults for *M. anisopliae* and 0.026 mg conidia per 100 adults for *B. bassiana*.

Direct deposition on to individual insects

A precise droplet of inoculum can be placed directly on to the surface of the insect. This method can be used with larger insects that can tolerate handling. Often, some form of immobilization of the insect is required. Due to the hydrophobic nature of the cuticle, it is sometimes difficult to administer a drop of aqueous inoculum precisely. In such cases, it may be helpful to choose an inoculation site that would absorb the droplet by capillarity.

Cocoa weevil with B. bassiana

Prior *et al.* (1988) used the direct deposition method to compare the virulence of water and oil formulations of *B. bassiana* against the cocoa weevil pest, *Pantorhytes plutus*. Other examples of direct deposition bioassay include studies with *M. anisopliae* against flea beetles (Butt *et al.*, 1995).

1. *B. bassiana* was cultured for 2 weeks at 28°C on 2% malt agar or on autoclaved brown rice or oat grain in 250-ml conical flasks. Formulations were prepared by adding 100 ml of either filtered coconut oil or distilled water with 0.01 ml of Tween 80 into the cereal cultures, stirring and filtering through a metal strainer. Serial dilutions were then made as necessary and conidial concentrations determined using a haemocytometer. Conidial viability was verified by streaking on to malt agar plates.

2. Adult weevils were field-collected, individually secured by pressing the dorsal abdomen lightly on to Blue tak® adhesive, and inoculated by applying 1 µl to the mouthparts. A Hamilton syringe was used to apply the oil formulation and an Agla microsyringe for the water formulation. For the water formulation, it was necessary to retain the insects secured until the drop of inoculum had dried. Otherwise the drop would run off.

3. Inoculated insects were transferred to 1-l plastic containers and fed every 3–4 days with pieces of cocoa stem. Mortality was checked daily and dead insects were transferred to plastic cups containing damp tissue for verification of fungal outgrowth. Only insects that showed visible outgrowth of the fungus were included in the analyses to determine the LD_{50}.

4. The oil formulation was found to be much more effective than the water formulation. LD_{50} estimates were 1.2×10^3 conidia per insect for the oil formulation and 4.3×10^4 for the water formulation.

Subterranean termite with B. bassiana *and* M. anisopliae

Lai *et al.* (1982) used the direct drop deposition method in bioassays to determine the virulence of six isolates of entomogenous fungi to the subterranean termite, *Coptotermes formosanus.*

1. Cultures of the fungi were kept on SDAY (SDA with yeast). Virulence was maintained by passage through termites every 3 months. For bioassay, 0.1 g of spores were scraped off 20-day-old culture plates. Conidia were then suspended in 10 ml of 0.1% Tween 80 to a final dilution of 1:100 using a magnetic stirrer. The suspension was filtered through two layers of Kimwipes®. A 0.5 µl aliquot of the spore suspension was placed on a microscope slide and the number of conidia in this drop counted under a phase-contrast microscope (Ko *et al.*, 1973). Dilutions were then performed as required.

2. Foraging termite workers were obtained from a field colony. One hundred workers were weighed in groups of ten in glass vials and the mean body weight was used to determine the dosage as expressed by the number of conidia per mg body weight.

3. Termites were anaesthetized with CO_2 for 10 s then transferred to a 100 × 20 mm Petri dish lined with filter paper.

4. A Hamilton microsyringe was used to apply 0.5 µl inoculum to the surface of the prothoracic area. This volume was enough to cover the insect without runoff. Mortality was reduced by avoiding direct contact with the syringe on the termite.

5. Insects were kept in inoculation chambers at 25°C and 56% RH with the filter paper and applicator sticks changed every 4 days to prevent secondary infection by saprophytes such as *Rhizopus* spp.

6. Thirty foraging workers were treated at each dosage level and caged in groups of ten. Three groups constituted a single replication and the experiment was repeated three times. Control groups were inoculated with the carrier alone.

7. Inoculated termites were incubated at room temperature for 15 days. Mortality data at 12 days was used for the probit analysis as control mortalities drastically increased thereafter. Samples of dead termites were homogenized on a microscope slide and examined for the presence of hyphal bodies.

8. Estimates of LD_{50} and LT_{50} revealed differences in virulence among isolates. Overall, isolates of *M. anisopliae* appeared more virulent than those of *B. bassiana*.

Inoculation of soil

Insects which either inhabit or are associated with soil during a part of their life cycle can be exposed to inoculum contained on the surface or within the soil. Soil texture, humidity and microbial flora can affect conidial viability and virulence and need to be considered. A discussion of procedures and precautions is presented by Goettel and Inglis (1997).

Pecan weevil with B. bassiana

Champlin *et al.* (1981) applied an aqueous suspension of *B. bassiana* conidia to the surface of soil to compare virulence of mutants against the pecan weevil, *Curculio caryae*.

1. Five *B. bassiana* mutants were obtained by ultraviolet irradiation of a wild-type culture of *B. bassiana*. Conidia from 14- to 21-day-old cultures grown on SDA + 3% yeast extract were obtained by washing with sterile 0.03% Triton X-100. Conidia were washed twice in sterile distilled water and the concentration estimated spectrophotometrically at 540 nm. Concentrations were determined by plating appropriate dilutions on SDA and counting CFUs.

2. Large plastic cups (14.5 cm deep, 11.5 cm diameter) containing autoclaved soil–sand mixture (10 : 1) were inoculated with 10 ml of spore suspension distributed over the entire 95 cm^2 surface in a dropwise manner using a pipette. After allowing the solution to be absorbed into the soil (to an estimated depth of c. 1.3 cm) 25 4th-instar larvae of field-collected pecan weevils were allowed to burrow down into the soil in each cup. Four dilutions of each mutant were prepared.

3. Cups were covered with Parafilm in which ten holes were punched, and incubated at 25°C. After 7 days, 5 ml of sterile water was added to each cup to maintain moisture. Two to three replicate treatments were performed for each mutant strain. Mortality was assessed 21 days post-inoculation. The percentage of insects that were mummified was used in the LC_{50} assessments.

4. The mutants exhibited different degrees of virulence with LC_{50}s ranging from 9.7×10^6 to 1.0×10^9 conidia ml^{-1}.

Ovipositing grasshoppers with B. bassiana

Conidia can be incorporated directly into the soil to bioassay virulence against insects which burrow or oviposit into soil (Fig. 4.4). Inglis *et al.* (1995a, 1998) used this method to determine the effects of conidial concentration, soil texture and soil sterilization on virulence of *B. bassiana* to ovipositing adults and emerging nymphs.

1. Conidia of *B. bassiana* were obtained commercially. Numbers of conidia g^{-1} were determined using serial dilutions in water and a haemocytometer. Conidia were mixed uniformly into autoclaved sand at a concentration of 10^8 conidia g^{-1} dry-weight sand. Water was added to obtain a water content of 9.2% (w/w) and 880 g of sand was added into each of three plastic containers (10 × 10 × 7 cm) per treatment. A 5 mm layer of dry sterile sand

Fig. 4.4. Grasshopper ovipositing into *B. bassiana*-augmented sand. Photo by courtesy of Doug Inglis.

was then placed on the surface. The moisture level within each cup was monitored daily by weighing and readjusted to 9.2% as necessary.

2. Two cores of sand were removed from each container using a 3.5-mm diameter cork borer. The sand was vortexed for 30 s at high speed in 10 ml of 0.05% Tween 80 in phosphate buffer. The suspension was diluted and spread on to a selective medium. The number of CFU was then determined after incubation in the dark at 25°C for 5–6 days.

3. A minimum of 30 virgin females and 20 virgin males of a non-diapausing laboratory strain of *Melanoplus sanguinipes* were placed into cages, 40 cm square. The cages had holes in the bottom so that the containers of the inoculated soil could be inserted with the surface of the sand being level with the cage bottom. The cages were maintained at a 25/20°C day/night temperature regime with a 16:8 h (light:dark) photoperiod.

4. Adults copulated and the pronota of the first seven females per cage to oviposit were marked with paint. The duration of each oviposition period was recorded, and upon completion, these adults were sacrificed, and the extent of abdominal infestation with conidia was quantified by excising the abdomens, sealing the cut end with molten Parafilm, and washing in 5 ml of buffer in 20 ml vials, vigorously shaken at 300 rpm for 2 h on a rotary shaker. The spore suspension was diluted, plated on selective medium and the number of CFU determined.

5. The remaining adults were maintained within the cages on a diet of bran and wheat leaves. At the end of 7 days, containers of sand were replaced with freshly inoculated sand for a further 7 days. At the time of removal, populations of conidia within the sand were assessed as described previously. Dead insects were removed daily and cadavers were surface sterilized in 0.5% sodium hypochlorite with 0.1% Tween 80 followed by two rinses in sterile water. The presence of hyphal bodies in the haemolymph or outgrowth of *B. bassiana* in cadavers held under moist conditions was recorded.

6. Each egg container was incubated at 25/20°C day/night with a 16:8 h (light:dark) photoperiod and nymphal emergence was recorded daily. At the time of first nymphal emergence, densities of viable conidia were enumerated as previously described. Ten newly emerged nymphs per replicate were collected and anaesthetized in CO_2. Nymphs were homogenized and the CFU were determined on a selective medium.

7. Remaining nymphs were maintained in cages on a diet of wheat seedlings for a period of 10 days. Dead nymphs were removed daily and placed on moistened filter paper within Petri plates. After 10 days, egg pods were sifted from the sand and the number of unhatched eggs per female was determined.

8. All experiments were repeated and analysed as completely randomized designs. Extensive mortality attributed to the fungus occurred in ovipositing females, associated males and in emergent nymphs.

Inoculation through contact with contaminated substrates

Insects are sometimes inoculated by allowing them to walk over the surface of a substrate, such as filter paper, which was pretreated with a known concentration of inoculum. Although such methods are an improvement over allowing insects to walk over sporulating cultures, there are still difficulties in ensuring that each insect receives precise and equitable doses. For instance, when using this method, it would be expected that the more mobile insects would acquire more propagules than the less active individuals. Nevertheless, this inoculation method may have utility in some situations.

Aphids with M. anisopliae

Butt *et al.* (1994) used spore-impregnated filter paper to assay *M. anisopliae* against *Lipaphis erysimi* and *Myzus persicae*. Similar methods have also been used to assay pathogens against thrips (Butt, unpublished) and corn earworm, *Heliothis zea* larvae (Champlin *et al.*, 1981).

1. Conidia from 8–12-day-old sporulating cultures of two *M. anisopliae* isolates were harvested in a 0.03% solution of Tween 80 and diluted to the desired concentrations.

2. *Myzus persicae* and *Lipaphis erysimi* were placed for 15 s on filter paper impregnated with conidia by vacuum filtration of a 10 ml conidial suspension of 1×10^7 or 1×10^{10} conidia ml^{-1}. Aphids were then transferred to a healthy Chinese cabbage leaf in a ventilated perspex box ($5.5 \times 11.5 \times 17.5$ cm) lined with moist tissue paper. Control insects were treated similarly with 0.03% Tween 80. The aphids were incubated at 23°C in a 16:8 h (light:dark) photoperiod and humid conditions were maintained for the first 24 h by placing the boxes between wet paper towels.

3. Mortality of both *M. persicae* and *L. erysimi* was 100% within 4 days post-inoculation at 1×10^7 or 1×10^{10} conidia ml^{-1} with little or no control mortality (0–3%). The earliest deaths were recorded on the first day after inoculation and sporulation occurred 1 to 2 days after death. Young, healthy aphids which contacted mycosed insects also succumbed to the *M. anisopliae* isolates.

Bait

Inoculum can be incorporated directly into the diet and presented to the insects as a bait. Although this method is most often used with fungi that infect through the gut (e.g. *Ascosphaera aggregata*), it can also be used as a method for inoculation of fungi that invade through the external integument. While feeding, insects contaminate their mouthparts and body with the pathogen propagules.

Leaf-cutting bees with Ascosphaera aggregata

Ascosphaera aggregata is one of the few species of entomogenous fungi which infects the host through the gut. Consequently, a bioassay method has been developed whereby the inoculum is introduced on an artificial diet to study the susceptibility of different ages of larvae of leaf-cutting bees, *Megachile rotundata*, to this fungus (Vandenberg, 1992). In a later study, a similar bioassay technique was used to demonstrate that larval susceptibility was much reduced when larvae were fed a natural diet (Goettel *et al.*, 1993; Fig. 4.5). This study demonstrates the importance in choice of diet if results are used to predict events under natural conditions.

1. A pollen/agar-based diet was prepared and dispensed asceptically into wells of sterile flexible microtitre plates. Sections of 16 wells were cut and placed in 60 × 15-mm sterile plastic Petri dishes. Eggs were obtained from field-collected bee cells and were transferred to the sterile diet.
2. Ascospores were obtained by scraping field-collected cadavers and stored at −20°C. Inoculum was prepared by suspending spores in sterile buffer and grinding between two microscope slides to break up the spore balls.

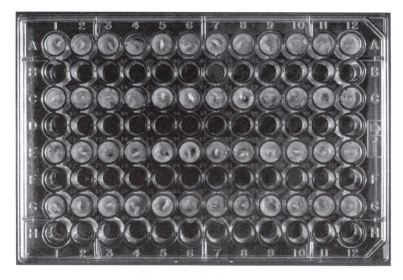

Fig. 4.5. Ninety-six well microtitre plate used for bioassay with leaf-cutting bees. Artificial media are separated with empty wells to help prevent possible cross-contamination. Photo by courtesy of Grant Duke.

Appropriate dilutions were made and concentrations determined using a haemocytometer.

3. Bees were inoculated within 1 h of inoculum preparation by applying 2 µl of the spore suspension to the diet surface adjacent to the mouthparts of the newly emerged larvae. Larvae were inoculated at 1, 2, 3 or 5 days of age. A total of 15 assays were carried out.

4. Larvae were checked daily for mortality. Unhatched eggs and larvae which died within the first 24 h were not included in the analyses. Uncertain diagnoses were verified by microscopy or fungal isolation into pure culture.

5. A dose–mortality relationship was found. There was an increase in LD_{50} with increasing age. The estimated LD_{50} values ranged between 120 and 1698 spores per bee, depending on age of larvae at time of inoculation.

Grasshoppers with B. bassiana

A leaf surface treatment bioassay has been used successfully to inoculate numerous fungi against several insect hosts (Ignoffo *et al.*, 1983 and references therein; Inyang *et al.*, 1998). Inglis *et al.* (1996a) used an oil-bait bioassay method to compare the virulence of several isolates of *B. bassiana* against the grasshopper, *M. sanguinipes*. In subsequent studies, this method was used to demonstrate the effect of bait substrate and formulation on virulence of this fungus (Inglis *et al.*, 1996c). It was demonstrated that the efficacy of this method depends on the extent to which nymphs become surface-contaminated with conidia during ingestion.

1. Conidia of several isolates of *B. bassiana* were obtained from cultures grown in the dark at 25°C on potato dextrose agar (PDA) for 7–10 days. Conidial viability was assessed on PDA amended with 0.005% Benlate (Dupont), 0.04% penicillin and 0.1% streptomycin after 24 h incubation at 25°C. Conidia were scraped from the surface of the PDA and suspended in sunflower oil. Conidial densities were determined using a haemocytometer and adjusted as necessary to obtain final concentrations of 1×10^5, 3.2×10^4, 1×10^4, 3.2×10^3 and 1×10^3 viable conidia.

2. Nymphs hatched from eggs laid by field-collected adults were reared on a diet of bran and wheat leaves. Third-instar nymphs were individually placed in sterile 20-ml vials stoppered with a sterile polyurethane foam plug, and starved for 12 h.

3. Five-microlitre aliquots of conidial suspensions were pipetted on to 5-mm diameter lettuce discs. A control consisted of oil applied to the discs alone. The inoculated discs were then pierced in the centre with a pin and suspended approximately 2 cm into the vial from the foam plug, and presented to the starving nymphs (Fig. 4.6). Nymphs were held at 25°C under incandescent and fluorescent lights for 12 h. Nymphs that underwent ecdysis or did not consume the disc after this period were excluded from the experiment.

4. Groups of 12 to 15 nymphs per treatment were transferred to 21 × 28 × 15 cm Plexiglass containers equipped with a perforated metal floor to reduce contact with frass (Fig. 4.7). Cages were incubated at a 25/20°C day/night and 16:8 h (light:dark) photoperiod regime and the nymphs were maintained on a diet of wheat leaves. Alternatively, in some assays, grasshoppers were kept singly in plastic cups (Inglis *et al.*, 1996b; Fig. 4.8).

5. The experiment was arranged as a randomized complete block design with four blocks conducted in time. The total number of nymphs per isolate–dose combination ranged from 46 to 61 nymphs. Nymphs that died and subsequently produced hyphal growth of *B. bassiana* on moistened filter paper were considered to have died of mycosis.

6. The oil-bait bioassay method facilitated the rapid inoculation of grasshopper nymphs. Within 1 h, 350 nymphs could be easily inoculated using this method. A dose–mortality relationship was demonstrated and substantial differences in virulence between isolates were found.

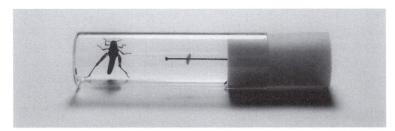

Fig. 4.6. Inoculation method for grasshoppers. An inoculated leaf disc is pinned to the inside of a foam plug and presented to a starved nymph within a shell vial. Photo by courtesy of Doug Inglis.

Fig. 4.7. A Plexiglass bioassay chamber used to incubate groups of inoculated grasshoppers. Photo by courtesy of Doug Inglis.

Fig. 4.8. A plastic container used to incubate single grasshopper nymphs. Photo by courtesy of Doug Inglis.

Inoculation using forcibly discharged conidia

Most fungi in the Entomophthorales produce forcibly-discharged conidia. These conidia are usually relatively short-lived and it is often not possible to harvest and enumerate them before using as inoculum. Consequently, many bioassays with these fungi use methods to inoculate the host directly from sporulating cultures or cadavers. In using such methods, much attention must be paid to the specific conditions that are required to induce spore discharge (Papierok and Hajek, 1997). Depending on the species involved, spores can be obtained from cultures maintained on agar medium, sporulating cadavers or hydrated marcescent mycelium.

Potato leafhopper with Zoophthora radicans

Wraight *et al.* (1990) used the forcibly discharged conidia of *Zoophthora radicans* from cultures and infected cadavers to inoculate the potato leafhopper, *Empoasca fabae*.

1. Dry mycelium of *Z. radicans* was prepared according to McCabe and Soper (1985), milled and sieved to retain particles between 1 and 0.5 mm. The mycelial particles were spread evenly on to water agar in Petri dishes and incubated at 21–22°C for approximately 12 h to obtain abundant sporulation.
2. Bioassay chambers consisted of 35-mm diameter plastic Petri dishes. Most of the upper surface of the lid was excised, leaving a narrow strip across the

centre for attachment with a small screw to a flat Plexiglass base. A cowpea leaf was then sandwiched between the base and modified lid. Each leaf was misted with water and each chamber was covered with a matching, unmodified lid.

3. Newly moulted, 5th-instar nymphs of *E. fabae* from a laboratory colony on cowpea were anaesthetized with CO_2 and randomly collected in groups of five. Individuals were placed dorsal side up on the wet leaf surface in a bioassay chamber. The chamber lid was then replaced with a lid containing the sporulating fungus. The leafhoppers were continuously exposed to the sporulating culture for 7 min. During this exposure period, the culture was continuously rotated.

4. After inoculation, each group of insects was transferred to clean chambers with fresh leaves. Each chamber was sealed in plastic bags and incubated at 20–22°C and 90–100% RH.

5. The LD_{50} was estimated at 4.1 spores per leafhopper.

Aquatic insects

Bioassay of aquatic insects is usually accomplished by introducing the inoculum directly into the water. However, use of high concentrations of inoculum in static aqueous systems may have adverse effects on water quality. Therefore, at times it is necessary to replenish the water depending on host species. Also, continuous exposure to propagules of some fungi such as *Tolypocladium cylindrosporum* may not be ideal, because the effective dose may vary according to length of exposure, as hosts continually reingest excreted conidia that remain viable (Goettel, 1987). This problem can be overcome by using a limited exposure time (Nadeau and Boisvert, 1994).

Mosquitoes with Culicinomyces clavisporus

Sweeney (1983) used a static bioassay method to determine the time–mortality responses of mosquito larvae inoculated with *Culicinomyces clavisporus*.

1. *C. clavisporus* was cultured in a broth of 1.25% corn steep liquor, 0.2% glucose and 0.1% yeast extract for 7 days at 20–24°C. Conidia were separated from the mycelium by filtration through fine gauze, then pelleted by centrifugation followed by two washes with sterile water. Conidia were counted using a haemocytometer and adjusted to the desired concentration.

2. Within 4–6 h of emergence, 5th-instar larvae of *Anopheles hilli* were placed in groups of 40 into plastic trays (18 × 12 × 5 cm) with 200 ml of water. Conidia were added on the following day.

3. The trays were incubated at 25°C and larvae were fed daily with powdered animal food pellets. Dead larvae were removed daily and the experiments were terminated after 12 days.

4. Nine separate experiments were performed with six to eight concentrations of conidia in each experiment. Five separate trays were dosed at each concentration and five trays were kept as a control.

5. A dose–mortality relationship was found and time to death decreased with increasing dose.

Mosquitoes with Coelomomyces

Toohey *et al.* (1982) used a bioassay to determine the intermediate copepod host in Fiji for a *Coelomomyces* sp. The fungus is a pathogen of mosquitoes which requires an alternate host to complete its development.

1. Cultures of five species of copepods and three species of mosquito were reared in the laboratory in rain water in transparent or opaque cups (6 × 10 cm).

2. Field- and laboratory-reared *Coelomomyces*-infected mosquito larvae which had been dead for less than 24 h were used as the inoculum.

3. Inoculum and 150–200 copepods of various ages were placed into each bioassay cup containing 200 ml of boiled treehole water. Ten to 12 days later, 20 first-instar *Aedes* larvae were placed in each cup. Cups were examined three times a week and dead larvae, pupae and adults were removed and examined microscopically for signs of infection. If infection was not apparent, a second group of larvae were added. Crushed mouse chow was added periodically for food.

4. Controls consisted of a set of three cups, one with only copepods, a second only with inoculum and the third of both copepods and larvae. There were at least five replicates for each copepod species tested and a total of 20 controls for all the species tested.

5. Only one species of copepod, *Elaphoidella taroi*, was found to be the intermediate host.

Novel bioassay methods

As stated previously, bioassays must be adapted to suit the host, pathogen and objectives of the bioassay. At times, the approach taken is very novel and sometimes even controversial. Novel approaches must balance efficacy and usefulness of the results.

Silverleaf whitefly with P. fumosoroseus, V. lecanii *and* B. bassiana

Landa *et al.* (1994) used a novel approach to bioassay entomopathogenic
fungi against the whitefly, *Bemisia argentifolii*. The bioassay is based on
rapid characterization of the growth rate and development of the fungi on
whitefly nymphs. It can be used in determining effects of environmental factors, adjuvants and pesticides on development of these fungi in whiteflies.
This method could be adapted for use with many other small insects.

1. Isolates of *P. fumosoroseus, V. lecanii* and *B. bassiana* were cultured on
PDA at 25°C in constant light for 7–10 days. Conidia were harvested by rinsing the cultures with 0.05% aqueous Tween 80. The suspension was mixed
using a vortex mixer and conidia were enumerated using a haemocytometer and adjusted to a final concentration of 1.0×10^7 conidia ml^{-1}.
2. Test materials (adjuvants, pesticides) were then added to the conidial suspensions. Drops of the test suspension were then placed on sterile microscope slides, 30 drops in three rows per slide. Laboratory-reared, early
4th-instar nymphs were singly placed in the centre of each drop and a total
of 25 nymphs were placed on each slide. The remaining five drops were
used as controls.
3. The slides were dried in a laminar airflow hood, placed in plastic Petri
dishes with a sterile wet filter paper on the bottom and incubated for 7 days
at 25°C under constant light. Each fungus was assayed using ten slides.
4. The influence of the bioassay protocol on the development of nymphs
was assessed. Early 4th-instar nymphs were incubated on microscope slides
in the diluted Tween 80 only for 7 days. The number of emerged adults was
assessed daily.
5. A rating system, named the Fungus Growth Development Index (FGDI),
was used to assess the degree of fungal development on the insect host.
Ratings were made at either daily intervals or at 24, 72 and 120 h. An FGDI
of 0.5 represented the first sign of viability of conidia, 1.5 for colonization
of the host and 2.5 for initial sporulation on the cadaver.

Semi-field and field-based assays

Field trials are essential to demonstrate that fungal isolates identified as virulent in laboratory trials are efficacious in the field. Initial trials may be done
using small cages enclosed in gauze or potted plants enclosed in a nylon
sleeve which allow for easy monitoring of insect pests. Trials may be done
in 'walk in' cages containing potted plants to ease the collection of insects
and assess the efficacy of the pathogen under field conditions. However,
most small-scale trials are done in randomized plots (3 m $\times$ 3 m) alongside,

or occasionally within, a growing crop. Although it may be more difficult to find the target insect there are various ways of assessing the impact of the pathogen. For example, insects may be collected randomly within each plot and incubated in humid chambers favouring fungal development. This could reveal more about targeting of the pathogen and its potential impact on the pest population. Alternatively, incubating healthy insects with plant parts collected from trial sites can reveal a considerable amount about the persistence (distribution on the plant and viability) of inoculum under field conditions. Assessing plant damage, or the number of larvae infesting leaves or flowers in control and treated plots is another way of assessing the impact of the pathogen. Fewer larvae would be expected to be found in plots where the pathogen was deployed. Field trials may not only be conducted on growing field crops but also rooted cuttings (Dorschner *et al.*, 1991).

Field trials against subterranean pests are technically more difficult for several reasons. First, the soil is a natural reservoir of many insect-pathogenic fungi so it would not be surprising to find target pests in control plots killed by fungi related to introduced pathogens. Second, targeting of the pathogen is not easy. Most often, inoculum is applied as a drench or ploughed into the soil using specialized equipment, but some workers have even used helicopters to treat large areas of pasture (e.g. Keller *et al.*, 1989).

Field bioassay of B. bassiana *against grasshoppers*

Inglis *et al.* (1997b) used a field cage bioassay to study the influence of environmental conditions on mycosis of grasshoppers caused by *B. bassiana*.

1. Conidia of *B. bassiana*, obtained from Mycotech Corp., were suspended in 1.5% (w/v) oil emulsion amended with 4% clay and applied to 12 ha of rangeland at a rate of 112 l ha^{-1}. Grasshoppers were collected in sweep nets immediately after treatment and placed in cages (41 × 61 × 48 cm) (Fig. 4.9), 100 hoppers per cage. Treatments consisted of cages: (i) placed in a glasshouse located at the laboratory, (ii) exposed to full sunlight, (iii) shaded from sunlight by a black plastic screen, and (iv) protected from UVB radiation by a UVB-absorbing plastic film (Fig. 4.9). Field cages were arranged as a randomized complete block with four sub-blocks, each containing three cage treatments per sub-block.

2. Grasshoppers were maintained on a diet of wheat seedlings and rangeland grasses. Cadavers were removed daily and assessed for mycosis by placing on moist filter paper.

3. Higher prevalence and more rapid development of the disease were observed in grasshoppers kept in shaded cages than in cages exposed to full sunlight or protected from UVB radiation.

Fig. 4.9. Field cages used to study the effects of solar radiation and shade on virulence of *B. bassiana* against grasshoppers under field conditions. Photo by courtesy of Doug Inglis.

Honey bee mediated infection of pollen beetle (Meligethes aeneus) by M. anisopliae

Butt *et al.* (1998) evaluated dissemination of fungal inoculum by honey bees against pollen beetles in oilseed rape (= canola) using field-caged insects. This method has also been shown to control seed weevil (*Ceutorhynchus assimilis*) and has the potential to control most floral pests including thrips (T.M. Butt, unpublished observations).

1. Trials were carried out in winter oilseed rape between late April and late May, and in spring oilseed rape between mid-June and late July.
2. Nine insect-proof cages (2.7 × 2.7 × 1.8 m high) were erected over the flowering crop infested with adult pollen beetles. Small colonies of honey bees were placed in the corner of each of six of the cages; each consisted of about six British Standard combs of bees and brood housed in a single British Standard Modified National hive body. Three of the hives had modified entrances containing an inoculum dispenser similar to that used by Peng *et al.* (1992). This consisted of a Perspex tray to contain the inoculum, through which the bees walked on leaving the hive. Bees returned to the hive *via* an entrance below the dispenser to prevent inoculum being brought into the hive. Inoculum was replenished at 48-h intervals. The three treatments (bees without inoculum, bees with inoculum, and no bees or inoculum) were randomized.
3. Ninety pollen beetles were collected from each cage at intervals of 3–6 days in winter rape and 7 days in spring rape, and were placed in groups of 30 in ventilated Perspex boxes (5.5 × 11.5 × 17.5 cm) lined with moist

tissue paper and incubated at 23°C and 16:8 h (light:dark) photoperiod. Three freshly cut inflorescences of rape were placed in each box as food. Mortality was recorded daily for 14 days. Dead beetles were removed and placed in a Petri dish lined with moist filter paper to encourage external conidiation of the fungus.

4. The first mortalities due to *M. anisopliae* were 3–5 days and 2–6 days after the sample was taken in winter and spring rape, respectively. The final mortalities for samples 1 and 2 were approximately 60% on winter rape and 99% and 69%, respectively, on spring rape. These results suggest that honey bees are effective in delivering conidia of *M. anisopliae* to flowers of oilseed rape and in the subsequent control of pollen beetles.

Checklist of Bioassay Preconditions and Requirements

There are several aspects which need to be checked to ensure effective bioassays with fungal pathogens.

1. It is important to ensure that the pathogen:
- has not lost virulence during culturing,
- inoculum is viable and percentage germination is determined,
- application method is satisfactory.

2. The target insect must be:
- healthy,
- not overcrowded or stressed,
- isolated if carnivorous or cannibalistic.

3. The bioassay chamber must:
- allow survival of control insects,
- not contain harmful substances, such as formaldehyde in food.

4. All bioassays should have:
- large enough sample size and enough replicates per treatment to make the results meaningful,
- the assays repeated at least once,
- field plots which are randomized,
- internal environments in the field cages which approximate to the external environment,
- sampling procedures which reflect the field fitness of pathogens.

Concluding Remarks

Bioassays are central to the successful development of fungi as microbial control agents. Although useful in providing valuable information on the insect–pathogen–environment interactions, the validity of bioassay results depend on the bioassay design, execution, analysis and interpretation of results. The ultimate challenge is to develop bioassays that can be used to predict field efficacy. It is therefore imperative that pertinent environmental parameters be incorporated into bioassay designs. For instance, knowledge of an LD_{50} or LT_{50} obtained from comparative laboratory assays of numerous isolates under static conditions provides minimal useful information as far as predicting the potential efficacy of a strain under field conditions is concerned.

Bioassay designs must be constantly improved to provide more meaningful information. The advent of increasingly sophisticated equipment such as incubators, environmental monitoring and inoculum application devices has allowed for the development of more complex bioassay designs which provide more pertinent results. Computerized statistical analyses have made it possible to model environmental parameters and process data with greater ease. As our understanding of the pertinent parameters important in fungal epizootiology increases, bioassays must be adapted so that they will provide information applicable for prediction of efficacy under field conditions.

We have provided some of the important parameters that need to be considered in the development and execution of a bioassay with an entomopathogenic fungus. We have also provided numerous examples of bioassays to illustrate the many methods and bioassay designs that have been used with an array of fungal and target species combinations. It is hoped this provides the reader with adequate information that should stimulate and facilitate the design of novel and pertinent bioassays which will provide useful information for the understanding of fungal biology, host–pathogen interactions, epizootiology and ultimately aid in the development of these microorganisms as microbial control agents of pest insects.

References

Adamek, L. (1965) Submerse cultivation of the fungus *Metarhizium anisopliae* (Metsch). *Folia Microbiologia (Praha)* 10, 255–267.

Aregger, E. (1992) Conidia production of the fungus *Beauveria brongniartii* on barley and quality evaluation during storage at 2°C. *Journal of Invertebrate Pathology*, 59, 2–10.

Auld, B.A. (1992) Mass production, formulation and application of fungi as biocontrol agents. In: Lomer, C.J. and Prior, C. (eds) *Biological Control of Locusts and Grasshoppers.* CAB International, Wallingford, UK, pp. 219–229.

Baath, E. (1991) Tolerance of copper by entomogenous fungi and the use of copper-amended media for isolation of entomogenous fungi from soil. *Mycological Research* 95, 1140–1142.

Bailey, L.A. and Rath, A.C. (1994) Production of *Metarhizium anisopliae* spores using nutrient impregnated membranes and its economic analysis. *Biocontrol Science and Technology* 4, 297–307.

Baker, C.A. and Henis, J.M.S. (1990) Commercial production and formulation of microbial biocontrol agents. In: Alan, R. (ed.) *New Directions in Biological Control: Alternatives for Suppressing Agricultural Pests and Diseases.* Liss, New York, pp. 333–344.

Ball, B.V., Pye, B.J., Carreck, N.L., Moore, D. and Bateman, R.P. (1994) Laboratory testing of a mycopesticide on non-target organisms: The effects of an oil formulation of *Metarhizium flavoviride* applied to *Apis mellifera. Biocontrol Science and Technology* 4, 289–296.

Bartlett, M.C. and Jaronski, S.T. (1988) Mass production of entomogenous fungi for biological control of insects. In: Burge, M.N. (ed.) *Fungi in Biological Control Systems.* Manchester University Press, Manchester, UK, pp. 61–85.

Beilharz, V.C., Parberry, D.G. and Swart, H.J. (1982) Dodine: A selective agent for certain soil fungi. *Transactions of the British Mycological Society* 79, 507–511.

Bell, J.V. and Hamalle, R.J. (1971) Comparative mortalities between field-collected and laboratory-reared wireworm larvae. *Journal of Invertebrate Pathology* 18, 150–151.

Bidochka, M.J., Miranpuri, G.S. and Khachatourians, G.G. (1993) Pathogenicity of *Beauveria bassiana* (Balsamo) Vuillemin toward lygus bug (Hem., Miridae). *Journal of Applied Entomology* 115, 313–317.

Blanchere, H., Calvez, J., Ferron, P., Corrieu, G. and Peringer, P. (1973) Etude de la formulation et de la conservation d'une preparation entomopathogene a base de blastospores de *Beauveria tenella* (DELACR) Siemaszko. *Annales de Zoologie et Ecologie Animale* 5, 69–79.

Boucias, D.G., Bradford, D.L. and Barfield, S. (1984) Susceptibility of velvetbean caterpillar and soybean looper (Lepidoptera: Noctuidae) to *Nomuraea rileyi*: Effects of pathotype, dosage, temperature, and host age. *Journal of Economic Entomology* 77, 247–253.

Bradley, C.A., Black, W.E., Kearns, R. and Wood, P. (1992) Role of production technology in mycoinsecticide development. In: Leatham, G.F. (ed.) *Frontiers in Industrial Mycology.* Chapman & Hall, New York, pp. 160–173.

Butt, T.M., Beckett, A. and Wilding, N. (1990) A histological study of the invasive and developmental processes of the aphid pathogen *Erynia neoaphidis* (Zygomycotina: Entomophthorales) in the pea aphid *Acyrthosiphon pisum. Canadian Journal of Botany* 68, 2153–2163.

Butt, T.M., Barrisever, M., Drummond, J., Schuler, T.H., Tillemans, F.T. and Wilding, N. (1992) Pathogenicity of the entomogenous, hyphomycete fungus, *Metarhizium anisopliae* against the chrysomelid beetles *Psylliodes chrysocephala* and *Phaedon cochleariae. Biocontrol Science and Technology* 2, 325–332.

Butt, T.M., Ibrahim, L., Ball, B.V. and Clark, S.J. (1994) Pathogenicity of the entomogenous fungi *Metarhizium anisopliae* and *Beauveria bassiana* against crucifer pests and the honey bee. *Biocontrol Science and Technology* 4, 207–214.

Butt, T.M., Ibrahim, L., Clark, S.J. and Beckett, A. (1995) The germination behaviour of *Metarhizium anisopliae* on the surface of aphid and flea beetle cuticles. *Mycological Research* 99, 945–950.

Butt, T.M., Carreck, N.L., Ibrahim, L. and Williams, I.H. (1998) Honey bee mediated infection of pollen beetle (*Meligethes* spp.) by the insect-pathogenic fungus, *Metarhizium anisopliae. Biocontrol Science and Technology* 8, 533–538.

Carruthers, R.I., Larkin, T.S., Firstencel, H. and Feng, Z. (1992) Influence of thermal ecology on the mycosis of a rangeland grasshopper. *Ecology* 73, 190–204.

Champlin, F.R., Cheung, P.R.K., Pekrul, S., Smith, R.J., Burton, R.L. and Grula, E.A. (1981) Virulence of *Beauveria bassiana* mutants for the pecan weevil. *Journal of Economic Entomology*, 74, 617–621.

Chew, F.S. (1988) Biological effects of glucosinolates. In: Cutler, H.G. (ed.) *Biologically Active Natural Products: Potential Use in Agriculture*. American Chemical Society, Washington, DC, pp. 155–181.

Daoust, R.A. and Roberts, D.W. (1983) Studies on the prolonged storage of *Metarhizium anisopliae* conidia: effect of growth substrate on conidial survival and virulence against mosquitoes. *Journal of Invertebrate Pathology* 41, 161–170.

Delmas, J.C. (1973) Influence du lieu de contamination tegumentaire sur la developpement de la mycose a *Beauveria tennella* (Delacr.) Siemaszo (Fungi Imperfecti) chez les larves du coleoptere *Melolontha melolontha* L. *Comptes Rendus de l'Academie des Sciences, Paris* 277, 433–435.

Dorschner, K.W., Feng, M.-G. and Baird, C.R. (1991) Virulence of an aphid-derived isolate of *Beauveria bassiana* (Fungi: Hyphomycetes) to the hop aphid *Phorodon humuli* (Homoptera: Aphididae). *Environmental Entomology* 20, 690–693.

Fargues, J. and Rodriguez-Rueda, D. (1979) Sensibilité des oeufs des Noctuides *Mamestra brassica* et *Spodoptera littoralis* aux Hyphomycetes *Paecilomyces fumoso-roseus* et *Nomuraea rileyi*. *Comptes Rendus de l'Academie des Sciences, Paris* 290, 65–68.

Fargues, J.F. and Robert, P.H. (1983) Effect of passaging through scarabeid hosts on the virulence and host specificity of two strains of the entomopathogenic hyphomycete *Metarhizium anisopliae*. *Canadian Journal of Microbiology* 29, 576–583.

Fargues, J., Delmas, J.C., Augé, J. and Lebrun, R.A. (1991) Fecundity and egg fertility in the adult Colorado beetle (*Leptinotarsa decemlineata*) surviving larval infection by the fungus *Beauveria bassiana*. *Entomologia Experimentalis et Applicata* 61, 45–51.

Fargues, J., Goettel, M.S., Smits, N., Ouedraogo, A., Vidal, C., Lacey, L.A., Lomer, C.J. and Rougier, M. (1996) Variability in susceptibility to simulated sunlight of conidia among isolates of entomopathogenic Hyphomycetes. *Mycopathologia* 135, 171–181.

Fargues, J., Goettel, M.S., Smits, N., Ouedraogo, A. and Rougier, M. (1997a) Effect of temperature on vegetative growth of *Beauveria bassiana* isolates from different origins. *Mycologia* 89, 383–392.

Fargues, J., Ouedraogo, A., Goettel, M.S. and Lomer, C.J. (1997b) Effects of temperature, humidity and inoculation method on susceptibility of *Schistocerca gregaria* to *Metarhizium flavoviride*. *Biocontrol Science and Technology* 7, 345–356.

Feng, M.G., Poprawski, T.J. and Khachatourians, G.G. (1994) Production, formulation and application of the entomopathogenic fungus *Beauveria bassiana* for insect control: Current status. *Biocontrol Science and Technology* 4, 3–34.

Feng, Z., Carruthers, R.I., Roberts, D.W. and Robson, D.S. (1985) Age-specific dose–mortality effects of *Beauveria bassiana* on the European corn borer, *Ostrinia nubilalis*. *Journal of Invertebrate Pathology* 46, 259–264.

Ferron, P. (1978) Biological control of insect pests by entomogenous fungi. *Annual Review of Entomology* 23, 409–442.

Ferron, P., Fargues, J. and Riba, G. (1991) Fungi as microbial insecticides against pests. In: Arora, D.K., Ajello, L. and Mukerji, K.G. (eds) *Handbook of Applied Mycology*. Marcel Dekker, New York, pp. 665–706.

Fransen, J.J., Winkelman, K. and van Lenteren, C. (1987) The differential mortality at various life stages of the greenhouse whitefly, *Trialeurodes vaporariorum* (Homoptera: Aleyrodidae) by infection with the fungus *Aschersonia aleyrodis* (Deuteromycotina: Coelomycetes). *Journal of Invertebrate Pathology* 50, 158–165.

Goettel, M.S. (1984) A simple method for mass culturing entomopathogenic hyphomycete fungi. *Journal of Microbiological Methods* 3, 15–20.

Goettel, M.S. (1987) Studies on the bioassay of the entomopathogenic fungus *Tolypocladium cylindrosporum* in mosquitoes. *Journal of the American Mosquito Control Association* 3, 561–567.

Goettel, M.S. (1988) Pathogenesis of the Hyphomycete *Tolypocladium cylindrosporum* in the mosquito *Aedes aegypti. Journal of Invertebrate Pathology* 51, 259–274.

Goettel, M.S. and Inglis, D.G. (1997) Fungi: Hyphomycetes. In: Lacey, L.A. (ed.) *Manual of Techniques in Insect Pathology*. Academic Press, London, pp. 213–249.

Goettel, M.S. and Jaronski, S.T. (1997) Safety and registration of microbial agents for control of grasshoppers and locusts. In: Goettel, M.S. and Johnson, D.L. (eds) *Microbial Control of Grasshoppers and Locusts. Memoirs of the Entomological Society of Canada* 171, pp. 83–99.

Goettel, M.S. and Roberts, D.W. (1992) Mass production, formulation and field application of entomopathogenic fungi. In: Lomer, C.J. and Prior, C. (eds) *Biological Control of Locusts and Grasshoppers*. CAB International, Wallingford, UK, pp. 230–238.

Goettel, M.S., St Leger, R.J., Rizzo, N., Staples, R.C. and Roberts, D.W. (1989) Ultrastructural localization of cuticle-degrading protease produced by the entomopathogenic fungus *Metarhizium anisopliae* during penetration of host (*Manduca sexta*) cuticle. *Journal of General Microbiology* 135, 2233–2239.

Goettel, M.S., Vandenberg, J.D., Duke, G.M. and Schaalje, G.B. (1993) Susceptibility to chalkbrood of alfalfa leafcutter bees, *Megachile rotundata*, reared on natural and artificial provisions. *Journal of Invertebrate Pathology* 61, 58–61.

Hajek, A.E. and St Leger, R.J. (1994) Interactions between fungal pathogens and insect host. *Annual Review of Entomology* 39, 293–322.

Hajek, A.E. and Wheeler, M.M. (1994) Application of techniques for quantification of soil-borne entomophthoralean resting spores. *Journal of Invertebrate Pathology* 64, 71–73.

Hajek, A.E., Butler, L., Walsh, S.R.A., Silver, J.C., Hain, F.P., Hastings, F.L., Odell, T.M. and Smitley, D.R. (1996) Host range of the gypsy moth (Lepidoptera: Lymantriidae) pathogen *Entomophaga maimaiga* (Zygomycetes: Entomophthorales) in the field versus laboratory. *Environmental Entomology* 25, 709–721.

Hall, I.M., Dulmage, H.T. and Arakawa, K.Y. (1972) Laboratory tests with entomogenous bacteria and the fungus *Beauveria bassiana* against the little house fly species *Fannia canicularis* and *F. femoralis. Experimental Entomology* 1, 105–108.

Hall, R.A. (1976) A bioassay of the pathogenicity of *Verticillium lecanii* conidiospores on the aphid, *Macrosiphoniella sanborni. Journal of Invertebrate Pathology* 27, 41–48.

Hall, R.A. (1980) Effect of repeated subculturing on agar and passaging through an insect host on pathogenicity and growth rate of *Verticillium lecanii*. *Journal of Invertebrate Pathology* 36, 216–222.

Hall, R.A. (1984) Epizootic potential for aphids of different isolates of the fungus, *Verticillium lecanii*. *Entomophaga* 29, 311–321.

Hallsworth, J.E. and Magan, N. (1994a) Effects of KCl on accumulation of acyclic sugar alcohols and trehalose in conidia of three entomopathogens. *Letters in Applied Microbiology* 18, 8–11.

Hallsworth, J.E. and Magan, N. (1994b) Effect of carbohydrate type and concentration on polyhydroxy alcohol and trehalose content of conidia of three entomopathogens. *Microbiology* 140, 2705–2713.

Hallsworth, J.E. and Magan, N. (1994c) Improved biological control by changing polyols/trehalose in conidia of entomopathogens. *British Crop Protection Conference – Pests and Diseases 1994*, 8D, 1091–1096.

Hallsworth, J.E. and Magan, N. (1995) Manipulation of intracellular glycerol and erythritol to enhance germination of conidia of entomopathogens at low water availability. *Microbiology* 141, 1109–1115.

Hare, J.D. and Andreadis, T.G. (1983) Variation in the susceptibility of *Leptinotarsa decemlineata* (Coleoptera: Chrysomelidae) when reared on different host plants to the fungal pathogen, *Beauveria bassiana* in the field and laboratory. *Environmental Entomology* 12, 1892–1897.

Hartmann, G.C. and Wasti, S.S. (1974) Infection of the gypsy moth, *Porthetria dispar* with the entomogenous fungus *Conidiobolus coronatus*. *Entomophaga* 19, 353–360.

Hedgecock, S., Moore, D., Higgins, P.M. and Prior, C. (1995) Influence of moisture content on temperature tolerence and storage of *Metarhizium flavoviride* conidia in an oil formulation. *Biocontrol Science and Technology* 5, 371–377.

Humber, R.A. (1997) Fungi: Preservation of cultures. In: Lacey, L.A. (ed.) *Manual of Techniques in Insect Pathology*. Academic Press, London, pp. 269–279.

Ignoffo, C.M. (1981) The fungus *Nomuraea rileyi* as a microbial insecticide. In: Burges, H.D. (ed.) *Microbial Control of Pests and Plant Diseases: 1970–1980*. Academic Press, London, pp. 513–538.

Ignoffo, C.M., Garcia, C., Kroha, M., Samsinakova, A. and Kalalova, S. (1983) A leaf surface treatment bioassay for determining the activity of conidia of *Beauveria bassiana* against *Leptinotarsa decemlineata*. *Journal of Invertebrate Pathology* 41, 385–386.

Inglis, G.D., Goettel, M.S., and Johnson, D.L. (1993) Persistence of the entomopathogenic fungus, *Beauveria bassiana* on phylloplanes of crested wheatgrass and alfalfa. *Biological Control* 3, 258–270.

Inglis, G.D., Feniuk, R.P., Goettel, M.S. and Johnson, D.L. (1995a) Mortality of grasshoppers exposed to *Beauveria bassiana* during oviposition and nymphal emergence. *Journal of Invertebrate Pathology* 65, 139–146.

Inglis, G.D., Goettel, M.S. and Johnson, D.L. (1995b) Influence of ultraviolet light protectants on persistence of the entomopathogenic fungus, *Beauveria bassiana*. *Biological Control* 5, 581–590.

Inglis, G.D., Johnson, D.L. and Goettel, M.S. (1995c) Effects of simulated rain on the persistence of *Beauveria bassiana* conidia on leaves of alfalfa and wheat. *Biocontrol Science and Technology* 5, 365–369.

Inglis, G.D., Johnson, D.L. and Goettel, M.S. (1996a) An oil-bait bioassay method used to test the efficacy of *Beauveria bassiana* against grasshoppers. *Journal of Invertebrate Pathology* 67, 312–315.

Inglis, G.D., Johnson, D.L. and Goettel, M.S. (1996b) Effects of temperature and ther-moregulation on mycosis by *Beauveria bassiana* in grasshoppers. *Biological Control* 7, 131–139.

Inglis, G.D., Johnson, D.L. and Goettel, M.S. (1996c) Effect of bait substrate and for-mulation on infection of grasshopper nymphs by *Beauveria bassiana*. *Biocontrol Science and Technology* 6, 35–50.

Inglis, G.D., Johnson, D.L. and Goettel, M.S. (1997a) Field and laboratory evaluation of two conidial batches of *Beauveria bassiana* (Balsamo) Vuillemin against grasshoppers. *Canadian Entomologist* 129, 171–186.

Inglis, G.D., Johnson, D.L. and Goettel, M.S. (1997b) Effects of temperature and sun-light on mycosis (*Beauveria bassiana*) (Hyphomycetes: Sympodulosporae) of grasshoppers under field conditions. *Environmental Entomology* 26, 400–409.

Inglis, G.D., Johnson, D.L., Cheng, K.-J. and Goettel, M.S. (1997c) Use of pathogen combinations to overcome the constraints of temperature on entomopathogenic Hyphomycetes against grasshoppers. *Biological Control* 8, 143–152.

Inglis, G.D., Johnson, D.L., Kawchuk, L.M. and Goettel, M.S. (1998) Effect of soil tex-ture and soil sterilization on susceptibility of ovipositing grasshoppers to *Beauveria bassiana*. *Journal of Invertebrate Pathology* 71, 73–81.

Inyang, E., Butt, T.M., Doughty, K.J., Todd, A.D. and Archer, S. (1999) The effects of isothiocyanates on the growth of the entomopathogenic fungus *Metarhizium anisopliae* and its infection of the mustard beetle. *Mycological Research* 103, 974–980.

Jenkins, N.E. and Goettel, M.S. (1997) Methods for mass production of microbial control agents of grasshoppers and locusts. In: Goettel, M.S. and Johnson, D.L. (eds) *Microbial Control of Grasshoppers and Locusts. Memoirs of the Entomological Society of Canada* 171, pp. 37–48.

Jenkins, N.E. and Prior, C. (1993) Growth and formation of true conidia by *M. flavoviride* in a simple liquid medium. *Mycological Research* 97, 1489–1494.

Jenkins, N.E. and Thomas, M.B. (1996) Effect of formulation and application method on the efficacy of aerial and submerged conidia of *Metarhizium flavoviride* for locust and grasshopper control. *Pesticide Science* 46, 299–306.

Keller, S., Keller, E., Schweizer, C., Auden, J.A.L. and Smith, A. (1989) Two large field trials to control the cockchafer (*Melolontha melolontha* L.) with the fungus *Beauveria brongniartii* (Sacc.) Petch. In: McFarlane, N.R. (ed.) *Progress and Prospects in Insect Control*. BCPC Monograph No. 43, pp. 183–190.

Kerry, B.R., Kirkwood, I.A., deLeij, F.A.A., Barba, J., Leijdens, M.B. and Brookes, P.C. (1993) Growth and survival of *Verticillium chlamydosporium* Goddard, a para-site of nematodes, in soil. *Biocontrol Science and Technology* 3, 355–365.

Kerwin, J.L. and Petersen, E.E. (1997) Fungi: Oomycetes and Chytridiomycetes. In: Lacey, L.A. (ed.) *Manual of Techniques in Insect Pathology*. Academic Press, London, pp. 251–268.

Kerwin, J.L., Duddles, N.D. and Washino, R.K. (1991) Effects of exogenous phos-pholipids on lipid composition and sporulation by three strains of *Lagenidium giganteum*. *Journal of Invertebrate Pathology* 58, 408–414.

Ko, W.H., Chase, L.L. and Kunimoto, R.K. (1973) A microsyringe method for deter-mining concentration of fungal propagules. *Phytopathology* 63, 1206–1207.

Krueger, S.R., Villani, M.G., Martins, A.S. and Roberts, D.W. (1992) Efficacy of soil applications of *Metarhizium anisopliae* (Metsch) Sorokin conidia, and standard and lyophilized mycelial particles against scarab grubs. *Journal of Invertebrate Pathology* 59, 54–60.

Lacey, L.A. and Brooks, W.M. (1997) Initial handling and diagnosis of diseased insects. In: Lacey, L.A. (ed.) *Manual of Techniques in Insect Pathology.* Academic Press, London, pp. 1–15.

Lacey, L.A., Martins, A. and Ribero, C. (1994) The pathogenicity of *Metarhizium anisopliae* and *Beauveria bassiana* for adults of the Japanese beetle, *Popillia japonica* (Coleoptera: Scarabaeidae). *European Journal of Entomology* 91, 313–319.

Lai, P.Y., Tamashiro, M. and Fuji, J.K. (1982) Pathogenicity of six strains of ento-mogenous fungi to *Coptotermes formosanus. Journal of Invertebrate Pathology* 39, 1–5.

Landa, Z., Osbourne, L., Lopez, F. and Eyal, J. (1994) A bioassay for determining pathogenicity of entomogenous fungi on whiteflies. *Biological Control* 4, 341–350.

Latge, J.-P., and Moletta, R. (1988) Biotechnology. In: Samson, R.A., Evans, H.C. and Latge, J.-P. (eds) *Atlas of Entomopathogenic Fungi.* Springer-Verlag, Berlin, pp. 152–164.

Li, Z., Butt, T.M., Beckett, A. and Wilding, N. (1993) The structure of dry mycelia of the entomophthoralean fungi *Zoophthora radicans* and *Erynia neoaphidis* fol-lowing different preparatory treatments. *Mycological Research* 97, 1315–1323.

Lord, J.C. and Roberts, D.W. (1986) The effects of culture medium quality and host passage on zoosporogenesis, oosporogenesis, and infectivity of *Lagenidium giganteum* (Oomycetes; Lagenidiales). *Journal of Invertebrate Pathology* 48, 355–361.

McCabe, D. and Soper, R.S. (1985) Preparation of an entomopathogenic fungal insect control agent. US Patent No. 4530834.

McClatchie, G.V., Moore, D., Bateman, R.P. and Prior, C. (1994) Effects of tempera-ture on the viability of the conidia of *Metarhizium flavoviride* in oil formula-tions. *Mycological Research* 98, 749–756.

McCoy, C.W. (1990) Entomogenous fungi as microbial pesticides. In: Baker, R.R. (ed.) *New Directions in Biological Control: Alternatives for Suppressing Agricultural Pests and Diseases.* Alan R. Liss, New York, pp. 139–159.

McCoy, C.W., Beavers, G.M. and Tarrant, C.A. (1985) Susceptibility of *Artipus flori-danus* to different isolates of *Beauveria bassiana. Florida Entomologist* 3, 402–409.

McCoy, C.W, Samson, R.A. and Boucias, D.G. (1988) Entomogenous fungi. In: Ignoffo, C. and Mandava, N.B. (eds) *CRC Handbook of Natural Pesticides, Vol. 5: Microbial Insecticides, Part A: Entomogenous Protozoa and Fungi.* CRC Press, Boca Raton, Florida, pp. 151–234.

Mesquita, A.L.M., Lacey, L.A., Mercadier, G. and LeClant, F. (1996) Entomopathogenic activity of a whitefly-derived isolate of *Paecilomyces fumosoroseus* (Deuteromycotina: Hyphomycetes) against the Russian wheat aphid, *Diuraphis noxia* (Hemiptera: Sternorrhyncha Aphididae) with the description of an effec-tive bioassay method. *European Journal of Entomology* 93, 69–75.

Milner, R.J. and Soper, R.S. (1981) Bioassay of *Entomophthora* against the spotted alfalfa aphid *Therioaphis trifolii* f. *maculata. Journal of Invertebrate Pathology* 37, 168–173.

Mitchell, D.J., Kannwischer-Mitchell, M.E. and Dickson, D.W. (1987) A semi-selective medium for the isolation of *Paecilomyces lilacinus* from soil. *Journal of Nematology* 19, 255–256.

Moore, D. and Caudwell, R.W. (1997) Formulation of entomopathogens for the control of grasshoppers and locusts. In: Goettel, M.S. and Johnson, D.L. (eds) *Microbial Control of Grasshoppers and Locusts. Memoirs of the Entomological Society of Canada* 171, pp. 49–67.

Moore, D., Bateman, R.P., Carey, M. and Prior, C. (1995) Long term storage of *Metarhizium flavoviride* conidia in oil formulations for the control of locusts and grasshoppers. *Biocontrol Science and Technology* 5, 193–199.

Moore, D., Douro-Kpindou, O.K., Jenkins, N.E. and Lomer, C.J. (1996) Effects of moisture content and temperature on storage of *Metarhizium flavoviride* conidia. *Biocontrol Science and Technology* 6, 51–61.

Morrow, B.J., Boucias, D.G. and Heath, M.A. (1989) Loss of virulence in an isolate of an entomopathogenic fungus, *Nomuraea rileyi*, after serial *in vitro* passage. *Journal of Economic Entomology* 82, 404–407.

Nadeau, M.P. and Boisvert, J.L. (1994) Larvicidal activity of the entomopathogenic fungus *Tolypocladium cylindrosporum* (Deuteromycotina: Hyphomycetes) on the mosquito *Aedes triseriatus* and the black fly *Simulium vittatum* (Diptera: Simulidae). *Journal of the American Mosquito Control Association* 10, 487–491.

Nagaich, B.B. (1973) *Verticillium* species pathogenic on aphids. *Indian Journal of Phytopathology* 26, 163–165.

Oger, R. and Latteur, G. (1985) Description et précision d'une nouvelle méthode d'estimation de la virulence d'une Entomophthorale pathogène de pucerons. *Parasitica* 41, 135–150.

Papierok, B. (1978) Obtention *in vivo* des azygospores d'*Entomophthora thaxterianan* Petch, champignon pathogene de pucerons (Homopteres, Aphididae). *Comptes Rendus de l'Academie des Sciences Paris* 286D, 1503–1506.

Papierok, B. (1982) Entomophthorales: Virulence and bioassay design. In: *Invertebrate Pathology and Microbial Control, Proceedings, IIIrd International Colloquium on Invertebrate Pathology*, University of Sussex, Brighton, UK, pp. 176–181.

Papierok, B. and Wilding, N. (1981) Etude du comportement de plusieurs souches de *Conidiobolus obscurus* (Zygmoycétes Entomophthoraceae) vis-à-vis des pucerons *Acyrthosiphon pisum* et *Sitobion avenae* (Hom. Aphididae). *Entomophaga* 26, 241–249 (in French).

Papierok, B. and Hajek, A.E. (1997) Fungi: Entomophthorales. In: Lacey, L.A. (ed.) *Manual of Techniques in Insect Pathology*. Academic Press, London, pp. 187–212.

Peng, G., Sutton, J.C. and Kevan, P.G. (1992) Effectiveness of honey bees for applying the biocontrol agent *Gliocladium roseum* to strawberry flowers to suppress *Botrytis cinerea. Canadian Journal of Plant Pathology* 14, 117–129.

Pereira, R.M. and Roberts, D.W. (1990) Dry mycelium preparations of entomopathogenic fungi, *Metarhizium anisopliae* and *Beauveria bassiana. Journal of Invertebrate Pathology* 56, 39–46.

Prenerová, E. (1994) Pathogenicity of *Paecilomyces farinosus* toward *Cephalcia abietis* eonymphs (Insecta, Hymenoptera): enhancement of bioactivity by *in vivo* passaging. *Journal of Invertebrate Pathology* 64, 62–64.

Prior, C., Jollands, P. and Le Patourel, G. (1988) Infectivity of oil and water formulations of *Beauveria bassiana* (Deuteromycotina: Hyphomycetes) to the cocoa weevil pest *Pantorhytes plutus* (Coleoptera: Curculionidae). *Journal of Invertebrate Pathology* 52, 66–72.

Ramoska, W.A. and Todd, T. (1985) Variation in efficacy and viability of *Beauveria bassiana* in the chinch bug (Hemiptera: Lygaeidae) as a result of feeding activity on selected host plants. *Environmental Entomology* 14, 146–148.

Roberts, D.W. (1989) World picture of biological control of insects by fungi. *Memórias do Instituto Oswaldo Cruz* Rio de Janeiro Numero Especial. Supl. III. 84, 168pp.

Roberts, D.W. and Hajek, A.E. (1992) Entomopathogenic fungi as bioinsecticides. In: Leatham, G.F. (ed.) *Frontiers in Industrial Mycology.* Chapman Hall, New York, pp. 144–159.

Roberts, D.W., Dunn, H.M., Ramsey, G., Sweeney, A.W. and Dunn, N.W. (1987) A procedure for preservation of the mosquito pathogen *Culicinomyces clavisporus. Applied Microbiology and Biotechnology* 26, 186–188.

Rombach, M.C., Aguda, R.M. and Roberts, D.W. (1988) Production of *Beauveria bassiana* (Deuteromycotina; Hyphomycetes) in different liquid media and subsequent conidiation of dry mycelium. *Entomophaga* 33, 315–324.

Samsinakova, A. (1966) Growth and sporulation of submerged cultures of the fungus *Beauveria bassiana* in various media. *Journal of Invertebrate Pathology* 8, 395–400.

Samsinakova, A. and Kalalova, S. (1983) The influence of a single spore isolate and repeated subculturing on the pathogenicity of conidia of the entomophagous fungus *Beauveria bassiana. Journal of Invertebrate Pathology* 42, 156–161.

Samsinakova, A., Kalalova, S., Vlcek, V. and Kybal, J. (1981) Mass production of *Beauveria bassiana* for regulation of *Leptinotarsa decemlineata* populations. *Journal of Invertebrate Pathology* 38, 169–174.

Schabel, H.G. (1976) Oral infection of *Hylobius pales* by *Metarhizium anisopliae. Journal of Invertebrate Pathology* 27, 377–383.

Schreiter, G., Butt, T.M., Beckett, A., Moritz, G. and Vestergaard, S. (1994) Invasion and development of *Verticillium lecanii* in the Western Flower Thrips, *Frankliniella occidentalis. Mycological Research* 98, 1025–1034.

Sneh, B. (1991) Isolation of *Metarhizium anisopliae* from insects on an improved selective medium based on wheat germ. *Journal of Invertebrate Pathology* 58, 269–273.

St Leger, R., Butt, T.M., Goettel, M.S., Staples, R. and Roberts, D.W. (1989a) Production *in vitro* of appressoria by the entomopathogenic fungus *Metarhizium anisopliae. Experimental Mycology* 13, 274–288.

St Leger, R., Butt, T.M., Staples, R. and Roberts, D.W. (1989b) Synthesis of proteins including a cuticle-degrading protease during differentiation of the entomopathogenic fungus *Metarhizium anisopliae. Experimental Mycology* 13, 253–262.

St Leger, R., Goettel, M., Roberts, D.W. and Staples, R.C. (1991) Prepenetration events during infection of host cuticle by *Metarhizium anisopliae. Journal of Invertebrate Pathology* 58, 168–179.

Sweeney, A.W. (1981) Prospects for the use of *Culicinomyces* fungi for biocontrol of mosquitoes. In: Laird, M. (ed.) *Biocontrol of Medical and Veterinary Pests.* Praeger, New York, pp. 105–121.

Sweeney, A.W. (1983) The time mortality response of mosquito larvae infected with the fungus *Culicinomyces. Journal of Invertebrate Pathology* 42, 162–166.

Tanada, Y. and Kaya, H.K. (1993) *Insect Pathology.* Academic Press, London.

Toohey, M.K., Prakash, G., Goettel, M.S. and Pillai, J.S. (1982) *Elaphoidella taroi*: the intermediate copepod host in Fiji for the mosquito pathogenic fungus *Coelomomyces. Journal of Invertebrate Pathology* 40, 378–382.

Vandenberg, J.D. (1990) Safety of four entomopathogens for caged adult honey bees (Hymenoptera: Apidae). *Journal of Economic Entomology* 83, 756–759.

Vandenberg, J.D. (1992) Bioassay of the chalkbrood fungus *Ascosphaera aggregata* on larvae of the alfalfa leafcutting bee, *Megachile rotundata*. *Journal of Invertebrate Pathology* 60, 159–163.

van Winkelhoff, A.J. and McCoy, C.W. (1984) Conidiation of *Hirsutella thompsonii* var. *synnematosa* in submerged culture. *Journal of Invertebrate Pathology* 43, 59–68.

Veen, K.H. and Ferron, P. (1966) A selective medium for isolation of *Beauveria bassiana* and *Metarhizium anisopliae*. *Journal of Invertebrate Pathology* 8, 268–269.

Vestergaard, S., Butt, T.M., Gillespie, A.T., Schreiter, G. and Eilenberg, J. (1995) Pathogenicity of the hyphomycete fungi *Verticillium lecanii* and *Metarhizium anisopliae* to the western flower thrips, *Frankliniella occidentalis*. *Biocontrol Science and Technology* 5, 185–192.

Vey, A. and Fargues, J. (1977) Histological and ultrastructural studies of *Beauveria bassiana* infection in *Leptinotarsa decemlineata* larvae during ecdysis. *Journal of Invertebrate Pathology* 30, 207–215.

Vidal, C., Lacey, L.A. and Fargues, J. (1997) Pathogenicity of *Paecilomyces fumosoroseus* (Deuteromycotina: Hyphomycetes) against *Bemisia argentifolii* (Homoptera: Aleyrodidae) with a description of a bioassay method. *Journal of Economic Entomology* 90, 765–772.

Wasti, S.S. and Hartmann, G.C. (1975) Experimental parasitization of larvae of the gypsy moth, *Porthetria dispar* (L.) with the entomogenous fungus, *Beauveria bassiana* (Balsamo) Vuill. *Parasitology* 70, 341–346.

Wilding, N. (1981) Pest control by Entomophthorales. In: Burges, H.D. (ed.) *Microbial Control of Pests and Plant Diseases, 1970–1980*. Academic Press, London, pp. 539–554.

Wraight, S.P., Butt, T.M., Galaini-Wraight, S., Allee, L., Soper, R.S. and Roberts, D.W. (1990) Germination and infection processes of the entomophthoralean fungus *Erynia radicans* on the potato leafhopper, *Empoasca fabae*. *Journal of Invertebrate Pathology* 56, 157–174.

Zimmermann, G. (1982) Effect of high temperatures and artificial sunlight on the viability of conidia of *Metarhizium anisopliae*. *Journal of Invertebrate Pathology* 40, 36–40.

Zimmermann, G. (1986) The 'Galleria bait method' for detection of entomopathogenic fungi in soil. *Journal of Applied Entomology* 102, 213–215.

Appendix 4.1: Selective Media for Isolation of Entomogenous Fungi

Veen's agar medium (1 l) (Veen and Ferron, 1966)

35 g Mycological agar (Difco) or 10 g Oxoid neutralized soya peptone, 10 g dextrose, 15 g No. 1. agar (or Bacto-agar), 1 g chloramphenicol (store 4°C), and 0.5 g cycloheximide (= Actidione; store 4°C). Add 1 l distilled water, stir, and cover. Autoclave for 10–15 min at 18–20 psi. Cool to *c.* 52°C and pour plates in laminar flow cabinet.

Oatmeal dodine agar (Beilharz *et al.*, 1982)

1. Antibiotic stock solution: add 4 g penicillin G (Sigma) and 10g streptomycin sulphate (Sigma) to 40 ml sterile distilled water under sterile conditions. Store at 4°C.
2. Crystal violet stock solution: add 0.1 g crystal violet (Sigma) to 200 ml distilled water. Store in the dark.
3. Add 17.5 g oatmeal agar (Difco) and 2.5 g agar (Fisons) slowly to 0.5 l distilled water while stirring vigorously and heat to boil.
4. Add 0.5 ml of the fungicide dodine (*N*-dodecylguanidine monoacetate; Cyprex 65WP, American Cyanamid Co.) and 5 ml crystal violet stock solution to the medium.
5. Autoclave for 20 min at 15 psi.
6. Allow medium to cool to 50–55°C and add 2 ml of antibiotic stock solution under sterile conditions.
7. Swirl flask well to ensure thorough mixing of compounds and pour while warm. There should be enough media for twenty 9-cm diameter Petri dishes.

Selective agar medium (1 l) (Kerry *et al.*, 1993)

37.5 mg carbendazim, 37.5 mg thiabendazole, 75 mg rose bengal, 17.5 g NaCl, 50 mg each of streptomycin sulphate, aureomycin and chloramphenиcol, 3 ml Triton X-100, and 17 g corn meal agar (Oxoid) in 1 l distilled water. This medium is appropriate for selecting some *Paecilomyces* spp. and *Verticillium* spp. from soil.

Paecilomyces lilicanus medium (Mitchell *et al.*, 1987)

To prepare 1 l of medium, mix the following: 39 g PDA, 10–30 g NaCl, 1 g Tergitol, 500 mg pentachloronitrobenzene, 500 mg benomyl, 100 mg streptomycin sulphate, and 50 mg chlorotetracycline hydrochloride.

Wheat germ based selective agar medium (1 l) (Sneh, 1991)

1. Prepare an aqueous extract of wheat germ – mix 30 g wheat germ in 1 l water, autoclave for 10 min and filter through four layers of cheesecloth.
2. Mix wheat germ extract (1 l) with 0.25 g chloramphenicol (heat stable) + 0.8 mg benlate (50% benomyl), 0.3 g dodine (65% *n*-dodecyl-guanidine acetate), 10 mg crystal violet and 15 g agar.
3. Autoclave and pour into plates.

Copper-based selective agar medium (1 l) (Baath, 1991)

2% malt extract (Oxoid), 1.5% Agar (Difco) amended with 2–4 mg $CuSO_4 \cdot 5H_2O$ per litre. *Cordyceps militaris* and *Paecilomyces farinosus* are tolerant of high Cu levels (400 mg l^{-1}), followed by *Metarhizium anisopliae* and *Beauveria bassiana*. Most other soil-borne fungi including nematophagous species of *Verticillium* were less tolerant.

Appendix 4.2: General Culture Media

Medium	Ingredients	g l^{-1}
Straw agar medium	Supernatant of boiled straw	40
	Agar (Difco)	8
	Aureomycin	0.05
	Streptomycin	0.05
	Chloramphenicol	0.05
Soya peptone medium	Soya peptone	10
	K_2HPO_4	0.3
	$MgSO_4 \cdot 7H_2O$	0.3
	NaCl	0.15
	$CaCl_2 \cdot 6H_2O$	0.3
	$MnSO_4 \cdot 6H_2O$	0.008
	$CuSO_4 \cdot 5H_2O$	0.0002
	$FeSO_4 \cdot 7H_2O$	0.002
Minimum medium	K_2HPO_4	0.3
	$MgSO_4 \cdot 7H_2O$	0.3
	NaCl	0.15
	$CaCl_2 \cdot 6H_2O$	0.3
	$MnSO_4 \cdot 6H_2O$	0.008
	$CuSO_4 \cdot 5H_2O$	0.0002
	$FeSO_4 \cdot 7H_2O$	0.002
	Agar	20.0
MC medium	Potassium phosphate dibasic	36
	Sodium phosphate heptahydrate	1.1
	Magnesium sulphate heptahydrate	0.6
	Potassium chloride	1
	Glucose	10
	Ammonium nitrate	0.7
	Yeast extract	5
	Agar	20
Sabouraud dextrose agar (SDA)	Mycopeptone	10
	Dextrose	40
	Agar	15
Oatmeal agar (OA)	Oatmeal	30
	Agar	20
Potato dextrose agar (PDA)	PDA (Oxoid)	39
Malt extract agar (MEA)	Malt extract	30
	Mycological peptone	5
	Agar (technical grade)	15

Continued

Medium	Ingredients	g l^{-1}
Sabouraud dextrose agar with yeast (SDAY)	Dextrose	40
	Neopeptone	10
	Yeast extract	10
	Agar	15
V8 Juice	V8	200
	CaCO$_3$	3
	Agar	20
Blastospore-producing medium	Corn steep liquor	20
	Sucrose	30
	KH$_2$PO$_4$	2.26
	Na$_2$HPO$_4$·12H$_2$O	3.8
	MgSO$_4$·7H$_2$O	0.123
	FeSO$_4$·7H$_2$O	0.023
	ZnSO$_4$	0.020
	K$_2$SO$_4$	0.174
	CaCl$_2$·2H$_2$O	0.147
PYG with supplements	Peptone	1.25
	Glucose	3.0
	Yeast extract	1.25
	Agar	20
	Vegetable oil (e.g. soybean, maize)	1–2 ml
	Sterol (e.g. cholesterol, ergosterol)	0.01–0.1
	Lecithin	0.05–0.1
	CaCl$_2$·2H$_2$O	0.07
Blastospore-producing medium	Glucose	25
	Soluble starch	25
	Corn steep	20
	NaCl	5
	CaCO$_3$	5

Note: most solid media can be used as liquid media by excluding the agar. Conversely, adding agar can convert a liquid medium to a solid medium. The pH of most media ranges between 5 and 9 with most workers adjusting to pH 7 with 1 M NaOH or HCl.

Bioassays of Microsporidia

J.V. Maddox,[1] W.M. Brooks[2] and L.F. Solter[1]

[1]Center for Economic Entomology, Illinois Natural History Survey and Illinois Agricultural Experiment Station, Champaign, Illinois, USA; [2]Department of Entomology, North Carolina State University, Raleigh, North Carolina, USA

Introduction

Despite their acknowledged deficiencies as short-term microbial control agents, the entomophilic protozoa played a prominent role in the development of the field of insect pathology. The famous microscopist and the 'father of protozoology', Antony van Leeuwenhoek, observed small animalcules, probably trypanosomatid flagellates, in the midgut contents of a house fly *c.* 1680. Gregarines were reported by Dufour in 1828 in the digestive tracts of an earwig and several beetles; subsequently both Burnett (1851) and Leidy (1856) observed flagellates in the house fly, *Musca domestica.* Leidy also described a flagellate from two scarab beetles and a ciliate from the oriental cockroach. Along with earlier work by Bassi (1835, 1836) on the muscardine disease of *Bombyx mori*, insect pathology as a science was established by Pasteur through his classical study of the pebrine disease of the silkworm in the 1860s. In his famous memoir *Etudes sur la Maladie des Vers á Soie* published in 1870, Pasteur demonstrated that the peculiar microscopic corpuscles seen by earlier students of silkworm diseases were the cause of pebrine disease. By retaining eggs produced only by uninfected adults, Pasteur provided a method for ensuring a disease-free stock of silkworms which eventually saved sericulture as an industry in France and other countries of the world. Although the true nature of the parasite as a protozoan was unknown to Pasteur or to Naegeli (1857) who named the parasite *Nosema bombycis*, observations by Balbiani (1882) and Stempell (1909) established the identity of the parasite as a protozoan in the order Microsporida. Pasteur (1874) also made one of the first definite suggestions that microorganisms might be useful as control agents of destructive insects when he suggested that 'les corpuscles' of pebrine be used against the grape phylloxera, a pest threatening grape production at the time in France.

Many protozoa were described from insects and other invertebrates in the early 1900s. Some, such as the microsporidium *Ameson pulvis* from the green crab, *Carcinus maenas*, were even suggested as possible biological control agents (Pérez, 1905). The first apparent attempt to use a protozoan as a microbial control agent was conducted by Taylor and King (1937) in the United States. They applied faeces of grasshoppers containing cysts of the amoeba *Malameba locustae* mixed with bran and molasses along roads and fences. Infected grasshoppers were found after a few weeks but control was not demonstrated. Probably because of the chronic nature of most protozoan infections and the often erratic results of attempts to use other entomogenous control agents, no serious efforts were made until the 1950s to use protozoa as microbial control agents. Limited attempts were made by Hall (1954) and Zimmack *et al.* (1954) in the USA and by Weiser and Veber (1955, 1957) in Czechoslovakia, but the first extensive tests to utilize protozoa as microbial control agents were carried out against the cotton boll weevil, *Anthonomus grandis*, in the southern USA (McLaughlin, 1966, 1967; Daum *et al.*, 1967; McLaughlin *et al.*, 1968, 1969). More recently, the successful production and use of the microsporidium *Nosema locustae* against grasshoppers on rangelands (Henry, 1971; Henry *et al.*, 1973; Henry and Oma, 1974; Henry and Onsager, 1982) led to the registration of this protozoan in 1980 by the US Environmental Protection Agency. This species remains the only protozoan which has been produced commercially as a microbial insecticide. Extensive reviews on protozoa as microbial control agents have been presented by McLaughlin (1971), Pramer and Al-Rabiai (1973), Brooks (1980, 1988), Henry (1981), Canning (1982) and Wilson (1982).

Unlike most of the other groups of entomopathogens, the entomogenous protozoa are generally recognized for their low virulence, producing chronic infections of a debilitative nature. Few are fast-acting and suitable for use as short-term, microbial pesticides. Most are better suited for use either as classical biological control agents or in inoculative augmentation programmes against insect pests with high economic injury levels.

This limited potential as short-term control agents is explained in part by the mode of action of protozoa as entomopathogens. Some protozoa develop extracellularly where they may compete for nutrients, especially if the host is stressed as when reared on a suboptimal diet (Harry, 1967; Dunkel and Boush, 1969). The majority of the more harmful entomophilic protozoa develop intracellularly in specific tissues or organs, although some may produce infections of a systemic nature. Infection is usually initiated in a host insect by the peroral ingestion of a spore or cyst but some species may also be transmitted vertically from parent to progeny *via* the transovarial route.

Protozoa are transmitted horizontally from one host to another by resistant spores or cysts. The spores of microsporodia are oval, dense, refractive forms that generally range from 3–10 µm in length. Under phase-contrast microscopy, viable spores appear highly refractive and quickly settle to the bottom of a fresh mount preparation (Fig. 5.1A). They have a similar oval appearance when viewed with a scanning electron microscope (Fig. 5.1D). Fixed with methanol and stained with Giemsa (Vavra and Maddox, 1976; Undeen and Vavra, 1997; Undeen, 1998), microsporidian spores have a unique appearance (Fig. 5.1B) and can be distinguished from other microorganisms. Internally, microsporidian spores have a complex ultrastructure with a coiled polar tube, one or two nuclei, and an extrusion apparatus (Vavra, 1976) as seen in transmission electron micrographs (Fig. 5.1C). Upon ingestion by a susceptible host, the polar tube extrudes and injects the sporoplasm into host cells (usually midgut cells), initiating the infection.

Once a spore or cyst is ingested, usually by the larval or nymphal stage of a host, it germinates in the host's gut to initiate infection. Protozoan species, such as microsporidia, that develop intracellularly, gradually replace the normal cellular constituents with various life-cycle stages of the parasite and eventually cause cell death. As infection proceeds from cell to cell, the functional capacity of organs and tissues is progressively impaired. A typical *Nosema*-type microsporidian life cycle is represented in Fig. 5.2. The life cycles of other genera of microsporidia are often much more complex.

Infected insects often show various cytopathological effects such as nuclear and cellular hypertrophy, the formation of neoplastic-like xenomas, and extensive alteration to cytoplasmic organelles such as chromosomes, endoplasmic reticulum, mitochondria, ribosome bodies, protein granules and vacuoles. There is little information to suggest how such changes are induced. Toxin production by protozoa has not been demonstrated (Weiser, 1969) and adjacent, non-infected cells usually exhibit normal nuclear and cytoplasmic architecture. Proteolytic enzymes resulting in cellular lysis may be produced by protozoa but this has never been confirmed. Dysfunctional organs or tissues caused by cellular replacement and destruction by protozoan stages give rise to diseases of a debilitative nature usually expressed by signs and symptoms such as irregular growth, sluggishness, loss of appetite, larval or pupal death, malformed adults, or adults with reduced vigour, fecundity and longevity (Brooks, 1988). Consequently, bioassays with protozoa must be carefully designed to measure a specific response, more often than not a response other than percentage mortality.

Relatively little work has been published on bioassays with protozoan pathogens of insects, probably because of their lack of use as microbial insecticides. Only cursory attention has been paid to protozoa in most reviews dealing with bioassays (Burges and Thomson, 1971), but Vavra and Maddox (1976) provided a general discussion and review of bioassay approaches and techniques used with various species of pathogenic protozoa as of the mid-1970s. More recently, general approaches to protozoan bioassays have been included in reviews of protozoan research techniques prepared by Undeen and Vavra (1997) and Undeen (1998). Most of the

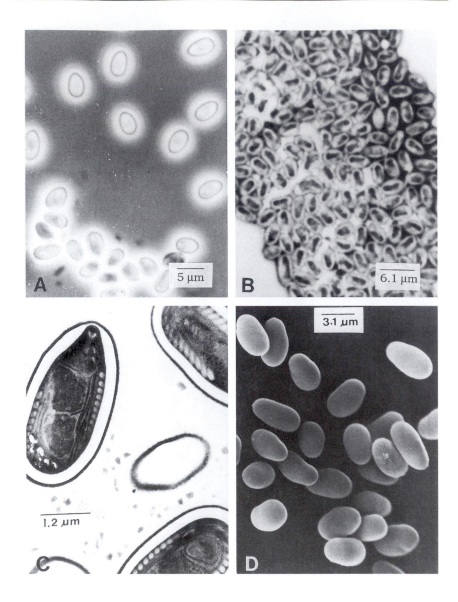

Fig. 5.1. Spores of the microsporidium, *Nosema algerae.* (A) Phase-contrast microscopy. Mature spores are highly refringent. (B) Spores stained with Giemsa and viewed under bright field microscopy show characteristic dark 'core' of microsporidian spores stained with Giemsa. (C) Transmission electron photomicrograph of mature spores. Spores contain two nuclei and a coiled polar tube. (D) Scanning electron photomicrograph. Spores of most microsporidia from terrestrial insects have few surface features. Photos by courtesy of CRC Press.

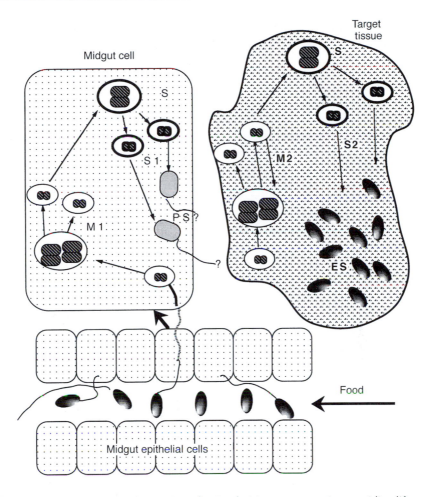

Fig. 5.2. Diagrammatic representation of a simple *Nosema*-type microsporidian life cycle in an insect host. The infection is initiated when the host ingests infective, environmentally resistant spores, usually with food. The spores pass through the crop, enter the midgut and germinate in response to proper pH and ionic conditions. The polar tube of the microsporidian spore penetrates the peritrophic membrane (not shown) and injects the sporoplasm (most of the internal content of the spore) into the midgut cells. In the cytoplasm of the midgut cells, the sporoplasm divides at least once. This is the first merogonic cycle (**M1**) of the life cycle. The first sporogonic cycle (**S1**) begins after the formation of a sporont (**S**). In the first sporogony each sporont divides and forms two primary spores (**PS**). Primary spores germinate in the midgut cells, presumably a mechanism for spreading the infection to target tissues. The exact mechanism(s) by which the infection moves from infected midgut cells to target tissues is not known (**?**). The second merogonic cycle (**M2**) in target tissues is usually extensive. Meronts divide many times and often fill the cytoplasm of the infected tissue. The second sporogony is similar to the first sporogony except that environmentally resistant spores (**ES**) are formed. The environmental spores are released into the environment either in silk from infected silk glands, in faeces and/or from the cadaver after the infected host dies. The environmental spores are responsible for horizontal transmission of the microsporidium.

published information on protozoan bioassays concerns microsporidia and this chapter deals exclusively with microsporidia.

Why Bioassay Microsporidia?

Microsporidia, like other entomogenous Protozoa, have limited potential for development as microbial insecticides, so why should estimation of the relative infectivity be important? Quite simply, many questions about the epizootiology of microsporidia require some knowledge about the number of microsporidian spores of a given species necessary to infect a host or kill a host. Host range information is equally important, not only for developing epizootiological theory, but also for determining the environmental impact and safety of non-indigenous microsporidia proposed for introduction as classical biological control agents.

Although microsporidia are not being developed as microbial insecticides, they are important natural control agents for many species of insects. Laboratory and field experiments have been conducted on microsporidia for the purpose of evaluating their role in the population dynamics of host insect species. Many of these experiments, often dealing with effects such as fecundity, vertical transmission, and host development, involve some type of bioassay. More traditional bioassays are also needed to assess the viability of the population of spores used in experiments. It is often a serious mistake to assume that the population of microsporidian spores used in an experiment is 100% viable. A sound knowledge of bioassay techniques is necessary if we are to evaluate accurately the questions we pose in experiments involving microsporidia.

The phylum Microspora is an incredibly heterogeneous group of parasites with a seemingly endless number of life histories and host–parasite relationships. Over 120 genera and 1000 species of microsporidia have been described. We cannot possibly address all of the fascinating possibilities and exceptions represented by the phylum Microspora, but we hope to cover the most important challenges to be faced when bioassaying spores of microsporidia.

Methods (other than bioassay) for estimating viability of spores

The following methods, even if effective, cannot substitute for bioassays because they do not estimate the susceptibility of a host population or different host species to infection by microsporidia. Nevertheless, information on spore viability or the proportion of a population of spores that is alive would complement any bioassay because it would ensure more accurate evaluation of the results. The three methods discussed briefly below have not been used successfully by many microsporidiologists, but with some refinement they may be developed as complementary tools for bioassays.

Vital stains

Several workers have attempted to develop a stain capable of distinguishing between dead and living spores. These attempts, mostly unpublished, have been unsuccessful for several reasons. The sporoplasm inside the spore might be alive yet the apparatus necessary for extruding the polar filament and injecting the sporoplasm into a host cell may be non-functional (Undeen, 1978; Undeen and Avery, 1988). Even if a vital stain could distinguish between a living and dead sporoplasm, no infection will occur unless the extrusion apparatus is functional. A vital stain used on spores can only indicate if the sporoplasms are alive; this may not indicate that the spores are able to infect hosts. Maddox (personal observation) was able to determine whether spores of the microsporidium *Octosporea muscaedomesticae* were infective by using a silver protargol stain, but only if the dead spores were killed by extreme heat. Spores that died from prolonged storage did not appear different from living spores when stained with silver protargol. A vital stain capable of distinguishing between living and dead spores would be very useful, but it is unlikely that such a stain will be available in the near future.

Spore extrusion

As with vital stains, spore extrusion alone is not an indication of whether a spore can infect a host. Spores with dead or damaged sporoplasms may extrude their filaments and not be able to infect a susceptible host (Undeen *et al.*, 1984; Undeen and Van der Meer, 1994). A correlation between spore extrusion and infectivity of spores has been shown for *Nosema algerae*.

Spore density and sugar content

Undeen and Solter (1996) have shown that the differences in spore density are a characteristic of living and dead spores. The lower density of dead spores is reflected in a reduction of total sugars. Dead spores banded higher in a Ludox density gradient than viable spores of the same species (Undeen and Avery, 1983). Additionally, mature spores are more refringent than immature spores and can be measured using spectrophotometry (Undeen and Solter, 1997).

Production of microsporidian spores

Since most protozoan pathogens of insects are obligate pathogens, few can be cultured and mass produced in liquid or on solid media. In most cases, spores of the protozoan are produced in living cells of a habitual host or an alternate host species. A number of species of microsporidia have been produced in tissue culture systems (Jaronski, 1984; Brooks, 1988), but limited yields of spores and high media costs limit the usefulness of such systems for the mass production of protozoan infective stages. Some species of entomophilic flagellates and ciliates can also be cultured on liquid or solid media (Brooks, 1988), but their limited potential as microbial control agents does not warrant a detailed review of cultivation efforts with these groups. Here, emphasis will be placed on the propagation of obligatory pathogenic species of microsporidia with an infective stage (environmental spore) that must be ingested by a susceptible host to initiate infection.

Microsporidia can be harvested and isolated from naturally infected hosts collected from the field, but these hosts may be infected with other pathogens and may not contain sufficient mature spores to conduct bioassays. Spores for bioassays are, therefore, usually obtained from hosts in advanced stages of infection after these hosts are fed spores in the laboratory as neonatal or early stage larvae. Inoculated insects are held under favourable environmental conditions to permit maximum growth of both the microsporidium and the host. Experimentation may be necessary to determine inoculative dosage, host age, rearing temperature, and developmental or incubation period to maximize spore production prior to harvest. While some microsporidia can be propagated in a non-target host species, most are produced in their habitual or usual host. Some species, especially those infecting aquatic hosts, have complicated life cycles which may involve an obligatory intermediate host species (Sweeny *et al.*, 1985). Propagation of such species is very difficult and may involve culture of the intermediate host itself.

Spores are harvested from heavily infected insects by homogenization of infected tissues or the entire host in a tissue grinder or blender. Most of the larger cellular debris can be removed by straining the homogenate through several layers of cheesecloth, cotton batting or a fine mesh screen. The filtrate is then centrifuged to form a pellet of spores; relatively clean spore suspensions can be obtained with two to three repeated washings and centrifugation using sterile distilled water. Such preparations are often immediately suitable for use in conducting bioassays, but purer spore suspensions can be obtained by using the triangulation method of Cole (1970) or by density-gradient centrifugation techniques using Ludox® (Undeen and Alger, 1971; Kelly and Knell, 1979; Undeen and Avery, 1983) or Percol (Jouvenaz, 1981; Iwano and Kurtti, 1995). These and other techniques that can be used to purify spore suspensions are fully discussed by Undeen (1997). Although not required for conducting bioassays, pure sterile spore suspensions can be stored at temperatures around 5°C for much longer periods of time than spore suspensions with organic debris and microbial growth. High levels of spore viability can also be maintained by freezing infected cadavers prior to spore harvest or the spore suspension itself at temperatures of −10 to −12°C (see review by Brooks, 1988). The most reliable method for storing spores for extended periods of time is storage in liquid nitrogen. Microsporidia from most terrestrial hosts can be stored indefinitely in liquid nitrogen (Maddox and Solter, 1996).

Only one protozoan, the microsporidium *Nosema locustae*, has been produced on a commercial scale as a microbial insecticide for use against grasshoppers (Henry and Oma, 1981). The propagation host, *Melanoplus bivittatus*, was reared on both live plant substrate and artificial diet; thus a description of the mass production methods used to produce this microsporidium will serve to illustrate the specifics associated with the propagation of a protozoan for use in laboratory bioassays or in a field testing programme. The propagation host species was chosen from an array of susceptible grasshopper species because of its relatively large size and its tolerance to the formation of large numbers of spores prior to host death. According to Henry and Oma (1981), grasshopper nymphs were reared to the fifth instar on a diet of lettuce, seedlings of balbo rye and wheat bran in large screened cages. The grasshoppers were fed lettuce sprayed with spores (*c.* 10^6 spores per cage) for 2 consecutive days and again 2 days later. Two to three weeks post-inoculation, the grasshoppers were transferred to small vials to reduce cannibalism and fed on an agar-based diet containing either wheat bran or crushed dog food. Infected cadavers were stored at −10°C prior to harvesting the spores. Improvements in the production technique eventually resulted in a spore yield of about 3×10^9 spores per grasshopper. Subsequently, Henry (1985) obtained significantly higher spore yields using grasshopper species with greater survival potential in the laboratory.

Because of the difficulties in rearing and infecting mosquitoes and the relatively small yield of spores per mosquito larva, Undeen and Maddox (1973) studied the potential of using surrogate or alternate hosts for propagation of the microsporidium *Nosema algerae*. Larvae of the corn earworm, *Helicoverpa zea*, were susceptible to intrahaemocoelic injection of spores, each larva yielding as many spores as 2000 mosquito larvae. Neonate *H. zea* larvae are also susceptible to *N. algerae* spores fed *per os* (Anthony *et al.*, 1978).

Microsporidia in insects that can be reared on an artificial diet can be more easily produced by simply feeding spores applied to the surface of the diet in plastic cups. For example, Fuxa and Brooks (1979) were able to obtain a yield of 1.7×10^{10} spores per larva when third-instar *H. zea* were exposed to 6.6 spores mm^{-2} of *Vairimorpha necatrix* on artificial diet surface and reared for 15 days at 26.6°C. Many other species of microsporidia have also been propagated in a variety of insect hosts using these unique approaches (Brooks, 1980, 1988).

Spore concentrations

Counting spores

The number of spores per unit of media (water, food, cadaver, etc.) used to inoculate the selected host must be accurately and precisely determined to conduct a meaningful bioassay. Even when the bioassay technique involves the use of spore-filled cadavers or contaminated food, spores must be counted under a compound microscope using some type of bacterial counter. We are not aware of any acceptable method for counting microsporidian spores other than suspending spores in water and determining the concentration in the water suspension using a bacterial counter or haemocytometer (Burges and Thomson, 1971). Several models of bacterial counters are available but we prefer the Petroff–Hausser® counter (Hausser Scientific Partnership, Horsham, Pennsylvania, USA) because it is relatively thin and allows the use of phase-contrast microscopy for counting spores. We will describe the use of the Petroff–Hausser® counter, but readers should refer to the user's manual for their particular bacterial counter for detailed instructions on the use and calculation of the concentration of spores in a water suspension.

Before counting spores, the bacterial counter must be 'loaded' with the suspension containing microsporidian spores. Incorrect counts result unless three things are carefully considered: (i) the spore suspension must be free of host tissues and other debris, (ii) the spore suspension must be thoroughly agitated before each count, and (iii) the counter must be completely filled, but not to overflowing. If the spore suspension contains extraneous debris, especially larger particles, spores will not readily pass under the coverslip of the counter and the debris may filter out many of the spores in the suspension, producing a serious counting error. Microsporidian spores are very dense and settle quickly to the bottom of a water suspension. The suspension of spores must be agitated before loading the counter to avoid a counting error. To load the counter, a sample is removed from the spore suspension with a Pasteur pipette and a small droplet of suspension is allowed to flow under the coverslip. A rubber bulb or a thumb over the end of the pipette can be used to release the suspension slowly. The counter should be loaded as soon as possible after the sample is removed from the suspension to avoid settling of spores in the pipette. The area between the coverslip and the bottom of the counter should be completely filled with the suspension, but if the counter is overfilled and the suspension flows over the edges of the counting bridge, the counter should be cleaned and the procedure started again. Loading the counter takes practice, but once mastered, it is not a difficult procedure. It is helpful to practise under the low power of a dissecting microscope a few times to observe flow of the suspension under the coverslip and to determine when the counter is overfilled.

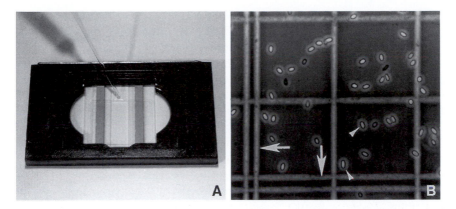

Fig. 5.3. (A) A bacteria counter or haemocytometer is used to determine the concentration of microsporidian spores in a water suspension. The model illustrated is a Petroff–Hausser® bacteria counter. (B) The counting chamber of the Petroff–Hausser® is divided into 25 large squares (separated by double lines, large arrows). Each large square contains 16 small squares (divided by single lines). Four of the small squares are illustrated. Spores in one preselected small square within each large square are counted. Only the brightly refringent spores (small arrows) should be counted; dark spores are not viable.

Once the counter (Fig. 5.3A) is loaded properly, counting spores is a relatively simple procedure. Only refractive spores should be counted. Dark, non-refractive spores (Fig. 5.3B) are immature sporoblasts or spores that have extruded their polar tubes. Such spores are not infectious and, if included in the counts, will produce inaccurate bioassay results. The Petroff–Hausser® counter is divided into 25 large squares divided by double lines and within each large square are 16 small squares divided by single lines. Ideally each small square should contain two to ten spores (Fig. 5.3B). If there is less than one spore per small square the operator should count the number per large square. A specific small square in each of the 25 large squares (such as the upper left or upper right, etc.) should be counted each time the counter is filled. The counter should be refilled and counts made at least three times. The average number of spores per small square multiplied by 2×10^4 is the number of spores per microlitre. The owner's manual for each bacterial counter will give details on how to count spores and how to calculate the number of spores in the suspension.

Making dilutions

An initial suspension of microsporidian spores, prepared as above, must be serially diluted in water for spore counts and for preparation of a range of doses needed to conduct a bioassay. The standard procedure for serial dilutions should be used. One log (1:9) or 0.5 log (1:4) dilutions are most often used. In order to ensure accurate dilutions, the suspensions of spores must be agitated thoroughly during the dilution procedure. Otherwise, the spores will settle to the bottom and the spore concentrations will not be accurate.

Determining the range of spore dosages

The number of spores fed to the test insects depends on the microsporidian species, the age of the host, and the purpose of the bioassay. When the purpose of the bioassay is to determine dosage–mortality relationships, relatively high dosages of spores are required, but when the purpose is to determine dosage–infectivity relationships, much lower dosages are appropriate. If the infectivity or mortality caused by the microsporidium being tested is unknown, it is usually wise to feed at least 6 log (1:9) dilutions to ensure that the spore concentrations will span the range of responses necessary to design a more precise bioassay.

Feeding Spores to Host Insects

Spores can be fed to host insects in a variety of ways. We have divided the techniques into methods for feeding an absolute number of spores and methods for exposure to a relative concentration of spores, where it is not possible to determine exactly how many spores each individual host insect eats.

Absolute methods

Droplet method

The droplet method of Hughes *et al.* (1986) is an effective method of feeding known spore dosages to neonate larvae of phytophagous chewing insects, primarily Lepidoptera and Coleoptera. The method is based on the premise that when unfed neonate larvae encounter a drop of liquid they will always drink the same amount. A simple device such as a rubber or cork stopper into which are driven finishing nails or blunt pins can be used to distribute small droplets of spore suspension mixed with an indicator dye on to wax paper or contact paper (Fig. 5.4A). Neonate larvae are allowed to wander on the waxed paper and drink the droplets (Fig. 5.4B). Larvae have consumed the calibrated amount of fluid when the midgut is coloured to its entire length. Fresh droplets must be prepared at least every 5 min because the spores will settle to the bottom of the droplet, affecting the number of spores ingested by a larva. Each host species must be 'calibrated', a time-

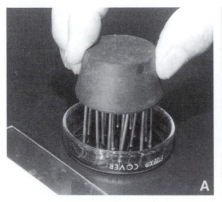 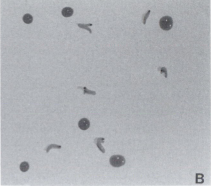

Fig. 5.4. The droplet method for feeding insect pathogens, developed by Hughes *et al.* (1986). This method allows precise control of the quantity of a spore suspension ingested by unfed neonate insect larvae. (A) Droplets are applied by dipping the illustrated device into a well agitated spore/dye suspension and touching the nails to a waterproof surface such as contact paper. This device was constructed by tapping finish nails into a large rubber stopper. (B) Droplets and neonate larvae. Larvae are observed under a field microscope while feeding and are removed to untreated diet when the gut is full as indicated by the dye in the suspension. The average amount of suspension ingested is calibrated for each host species using fluorescence spectrophotometry prior to performing bioassay.

consuming process (Van Beek and Hughes, 1986) but once the average amount of liquid ingested by a larva is determined for an insect species, it is not necessary to repeat the calibration procedure. This method is very useful in situations when many bioassays will be run against a single insect host species or when large numbers of insects are needed. It is not a convenient method for bioassaying a single microsporidian species against many insect species, because each insect species must be calibrated.

Loop method

The loop method (Vavra and Maddox, 1976) involves feeding a known quantity of a spore suspension to individual insects. It is one of the most accurate methods of administering a spore dosage to a host. A wire loop, typically a bacteriological inoculating loop, is used. The loop must be filled with the spore suspension in the same manner each time. If the loop is removed from the spore suspension with the loop parallel to the spore suspension it will hold a different amount of fluid than if it is removed from the suspension with the loop perpendicular to the spore suspension (Fig. 5.5A). Calibrated loops may be purchased, but it is easy to calibrate a bacteriological loop using a capillary pipette. The pipette is placed against the liquid in the loop and the liquid is pulled into the pipette (Fig. 5.5B, C). If the liquid does not fill the entire pipette, the distance of the fluid in the pipette can be measured in millimetres and compared with the length in millimetres of the pipette to the calibration line. A simple ratio will give the volume. For example, if the 25 µm indicated volume on a microlitre pipette is 30 mm in length, and liquid from the loop occupies 10 mm in the pipette, the loop holds 8.33 µm. ($x = 25/30 \times 10$). Feeding insect hosts from a loop is usually a relatively simple procedure. The loop holding the spore suspension is held so that the liquid touches the insect's mouthparts. The insect should drink the liquid within about 10 s (Fig. 5.5D, E). If the insect does not drink the liquid within 15 s, that individual should be discarded. Some insects, such as flies, can be induced to drink by adding a small amount of sugar to the suspension. Starving insects for several hours before feeding may induce a stronger feeding response.

The loop method is appropriate for larger insects such as late-instar lepidopteran larvae and adult house flies, but has limited use for feeding small insects such as neonate lepidopteran larvae. It has the advantage of allowing the investigator to feed a known dose of spores to an individual host at a specific time. When the object of the bioassay is to determine generation time of the microsporidium or the timing of the occurrence of life-cycle events, this method should be seriously considered. Some insect species do not readily drink from a loop and it is very difficult to use this method when small insects are involved. Because of the time required to feed individual insects, this method may not be appropriate when large numbers of insects are needed in the bioassay.

Blunt syringe (gavage) method

A blunt syringe has been used for administering known dosages of pathogens as well as other biologically active substances (Martouret, 1962; Fournie *et al.*, 1990). A blunt hypodermic syringe needle is gently pushed down the oesophagus and into the crop of the host insect (Fig. 5.5F). Some experimental hosts, such as species of actively feeding, late-instar lepidoptera, will readily swallow the needle, while other host species require forceful entry through the mouthparts. This method shares the advantage of delivering a known dose to an individual host at a specific time with the loop method. We believe it has some serious shortcomings which must be considered if it is to be used as a bioassay technique. First, it has the potential of delivering microsporidian spores directly into the crop, possibly reducing the time spores are exposed to the contents of the crop. Extrusion of spores and the subsequent infection is dependent on exposure of spores to the required pH and anions and includes a lag period from time of exposure to germination (Undeen, 1990). The crop may be involved in most cases. Thus the bioassay may not be directly comparable with bioassays

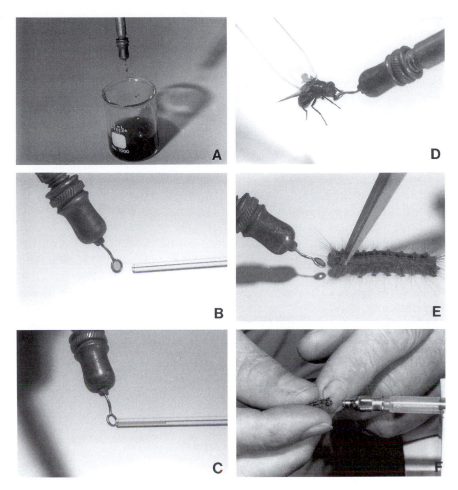

Fig. 5.5. (A–C) Calibration of a bacterial loop. The loop must be calibrated before it can be used for feeding known quantities of microsporidian spore suspensions to test insects. (A) The loop should be filled by dipping into the suspension at the same angle for each sample. In this example, the loop is withdrawn perpendicular to the surface of the suspension. (B) A micropipette is used to measure a series of five samples. (C) The volume of fluid held by the loop is determined by measuring the volume of fluid in the pipette. Using the loop (D and E) and the gavage method (F) for feeding insects. (D) Large flies (e.g. *Phormia* sp.) are held by a vacuum tube made from a Pasteur pipette and allowed to feed from the loop. (E) Lepidopteran larvae that do not have a strong response to touch may be held gently behind the head capsule with forceps. (F) The gavage method of feeding has been used to deliver precise amounts of spore suspension directly to the crop, but has several drawbacks. Spore suspensions are dyed for photographic contrast only.

where spores are fed in a more natural manner. Another serious source of error in using this method is that spores settle in the barrel of the syringe and, over time, the syringe delivers an increasingly lower number of spores. Angus (1964) developed a method for agitating particulates in a syringe by placing a small metal ball in the barrel. A rotating magnet under the syringe moved the metal ball and agitated the suspension in the syringe. Finally, there is always the possibility that the gut may be punctured by the blunt needle and spores placed directly in the haemocoel. We believe that the blunt needle syringe method should be used only if the other feeding techniques will not work with a particular host.

Leaf disc and diet plug methods

The principle of the leaf disc method (Henry, 1967; Maddox, 1968) is that an insect host will consume the entire piece of leaf on which a known number of microsporidian spores has been placed (Fig. 5.6A, B, C) or the entire piece of diet into which spores have been incorporated. Experimental hosts that do not ingest the leaf disc or diet in entirety after a prescribed time period are discarded from the bioassay.

Relative methods

When the objective of a bioassay is to compare the infectivity or the mortality caused by different species, isolates or populations of microsporidia, relative bioassay methods are appropriate and often easier to use than the absolute methods described above.

Diet surface feeding

One of the simplest methods for bioassaying microsporidia is to place known concentrations of spores on the surface of artificial diet on which the insects feed. The spore suspension must be uniformly distributed over the diet and allowed to dry before experimental insects are introduced. The amount of liquid suspension placed on the diet surface depends on the size of the diet container. Maddox and Solter (1996) routinely placed 40 µl of a microsporidian spore suspension on the surface of the diet in cups with a diameter of 30 mm. If the insects in the bioassay are not cannibalistic, several individuals can be placed in the cup after the spore suspension is applied. This procedure works best when the insects used in the bioassay feed on the diet surface and do not bore or tunnel into the diet. Tunnelling species often consume highly variable numbers of spores before they tunnel into the diet. The procedure for setting up this type of bioassay is as fol-

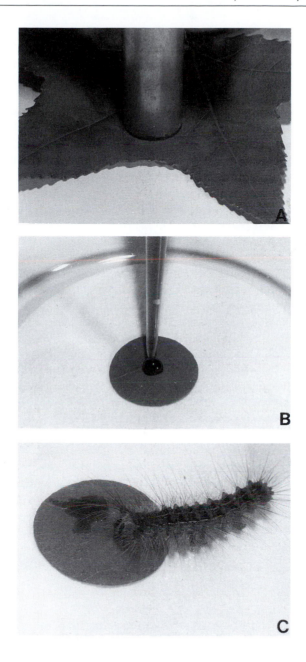

Fig. 5.6. The leaf disc method for feeding known spore dosages. (A) A leaf disc cut with a cork borer. (B) A known quantity of a spore suspension applied with a pipette to the surface of the leaf disc. (C) The disc is fed to a single insect. Only individuals that consume the entire leaf disc are included in the bioassay. Spore suspension is dyed for photographic contrast only.

lows: first dilute the spore suspensions so that the range of dilutions spans the spore concentrations necessary to obtain the information that the bioassay is designed to produce. Remove 40 μl from each tube containing a suspension and pipette on to the diet surface. Pipette tips should be changed for each suspension or pipetting should begin with the lowest spore concentration and proceed to the highest. Spread the suspension over the surface of the diet using the blunt end of either a glass rod or Pasteur pipette. Test insects are placed on the diet and allowed to feed for 24 h. Test insects ideally should be in the same stadium if immature stages are used. Immatures nearing a moult should not be included in the bioassay because feeding often ceases immediately before moulting and larvae may not feed again for several hours after moulting. Insects are removed from the treated diet after 24 h and placed on fresh untreated diet. If the insects are not removed from the treated diet, mortality and infectivity are unlikely to be affected by continued exposure to spores after the initial infection occurs. Leaving larvae on treated diet until they are examined for the presence of infection can, however, result in false-positive determinations. This occurs if ingested spores are present in the gut lumen and are mistaken for spores produced by an active infection. After insects are removed from the treated diet, they may be held individually in untreated diet cups or, if groups of insects were fed spores on the treated diet, they may be held in the untreated diet cups as groups. Again, the preference depends on the purpose of the bioassay. When determination of infectivity is the only purpose of the bioassay, insects may be kept together if they are examined for infection before there is a possibility that an infected host can produce spores and cause patent infections in other uninfected hosts.

The diet surface feeding method is a suitable choice when large numbers of insects are required in the bioassay and when knowledge of the absolute number of spores ingested by an insect is not a requirement for the bioassay. It is also an excellent method for bioassays that are conducted to determine the relative infectivity of a population of microsporidian spores or to compare the infectivity of several species of microsporidia to a single host. This method works best when young insect larvae are used.

Foliage dipping

When the host insects used in a bioassay cannot be reared on artificial diet, it is possible to dip foliage into spore suspensions and feed test insects on the treated foliage (Solter *et al.*, 1993). The spore suspensions are prepared as described for the diet surface treatments but, rather than putting the suspension on the diet surface, leaves or some part of the food plant of the insect are dipped into the suspension and the test insects allowed to feed on the treated foliage for a period of time. This method is usually not as precise as the diet surface method because there is some variability in adherence of microsporidian spores to different foliage plants. Investigators must use caution when comparing data from bioassays conducted at different times because the plants used in the bioassays may vary in quality throughout the season, resulting in changes in adherence of spores to the foliage and feeding behaviour of the host. The same cautions concerning selection of test insects mentioned in the diet surface feeding methods apply for foliage dipping. The foliage dipping method, like the diet surface method, is suitable for determining the relative infectivity of a population of microsporidian spores or comparing the infectivity of several species of microsporidia to a single host. It should be used when the host does not feed readily on artificial diet.

Special considerations

Microsporidia are represented by more species with complex life cycles and a range of hosts from many environments than are most other groups of insect pathogens. Questions about intermediate or alternate hosts and insects living in aquatic environments may require novel research approaches for bioassays. Epizootiological questions can also be addressed under laboratory conditions, again requiring the use of unconventional bioassays.

Aquatic insects

Most bioassays on aquatic insects have involved mosquito larvae and it is difficult, if not impossible, to feed absolute spore dosages to individual larvae. Only relative methods have been used for bioassays of mosquito larvae (Canning and Hulls, 1970; Undeen, 1998). Spore dilutions are made as described earlier. Spores are placed in a Petri dish and the spore concentration is expressed as number of spores per mm^2 of the bottom surface of the Petri dish. Three variables greatly influence the outcome of the bioassay: larval age, depth of the water in the Petri dish, and physical characteristics of the bottom of the dish (Vavra and Maddox, 1976). When comparing different populations of microsporidian spores, different microsporidian species or different host species, these variables must be kept constant.

Mixed populations of healthy and diseased hosts

Epizootiological questions often involve the variables that affect the transmission of microsporidia from one host to another (Anderson and May, 1981; Canning, 1982; Onstad *et al.*, 1990). These questions can often be addressed by placing known numbers or ratios of infected and healthy hosts in a common arena for a period of time which is at least as long as the generation time of the microsporidium. Hosts are then examined for the presence of infection.

Bioassays of this type are especially valuable for extended host specificity studies. Initial host specificity studies usually involve one of the absolute or relative methods described above in which spores are fed directly to the experimental host (Undeen and Maddox, 1973). Non-target hosts that prove susceptible to the direct feeding bioassays may be the subjects of bioassays involving a mixture of healthy and infected hosts held in a common arena for a period of time. One of the best examples of this type of bioassay involves microsporidia from European gypsy moths and native North American forest Lepidoptera, as well as microsporidia from native North American Lepidoptera and the gypsy moth. Insect hosts, which are infected when fed spores directly, may not develop infections when exposed to infected conspecific individuals (Solter *et al.*, 1997; Solter and Maddox, 1998). In order to address these questions, bioassays may range from simply combining healthy and infected hosts in the same diet cup to placing healthy and infected hosts on cut or sleeved host-plant branches.

Host specificity bioassays

Studies of the effects of entomopathogens on non-target hosts have become increasingly important as these pathogens are considered for use as classical biological control agents. Before introduction of a non-indigenous microsporidium is seriously considered, it is important to evaluate susceptability of representative non-target hosts to the microsporidium. In bioassays of this type, the investigator does not know a priori if the experimental host will become infected with the microsporidium. The purpose of the bioassay is to determine if the non-target host is susceptible to infection and, if so, what range of spore dosages are infective.

Intermediate/alternate hosts

Many species of microsporidia from hosts that live in aquatic environments have complex life cycles involving two host species and several types of microsporidian spores. The best known examples are microsporidia in the genus *Amblyospora* (Andreadis, 1985; Sweeny *et al.*, 1985). These microsporidia infect both mosquitoes and copepods. Infected adult female mosquitoes transmit the infection transovarially to their progeny. Infected larval mosquitoes die and release large numbers of spores, which infect only copepods. Copepods, in turn, produce spores that only infect mosquito larvae, and these infected larvae develop into infected adult mosquitoes. Different spore types are produced in each host: adult mosquitoes, larval mosquitoes infected transovarially and copepods. It is very likely that many, if not most, of the microsporidia from aquatic invertebrates have similar life cycles involving multiple hosts and several spore types. Few bioassays have been conducted with this group of microsporidia and it is beyond the scope of this chapter to deal extensively with it. Nevertheless, if complex life cycles are not considered when contemplating bioassays on microsporidia from aquatic invertebrate hosts, misleading results will be produced.

Reading Bioassays and Evaluating Results

Because microsporidia produce so many sublethal effects, the objective of many microsporidian bioassays is not simply to determine mortality or infectivity. It is often important to learn how infections affect the rate of development, feeding rate, mobility, mating behaviour, fecundity and ovipositional behaviour. In addition, microsporidia may kill their hosts by depleting resources in infected cells or by physically damaging the midgut tissues in the initial phases of the infection. In the latter situation the host dies, not from development of microsporidia in the host tissues, but from physical damage to the gut and associated bacterial septicaemia.

Mortality

Mortality from midgut damage/bacterial septicaemia

Insects that die from physical damage to the midgut rather than from microsporidian development in the host tissues become flaccid and discoloured and may contain many bacteria in the blood (Maddox *et al.*, 1981). The host tissues may contain a few microsporidian spores but these are usually primary spores and do not appear as refractive under phase-contrast microscopy as do the environmental spores. When bacterial septicaemia and/or initial invasive activity of the microsporidia is the cause of death, insects usually die within 2 to 4 days following ingestion of spores. Some species of microsporidia, such as *Vairimorpha necatrix*, cause bacterial septicaemia at relatively low spore doses, while other microsporidia, such as *Endoreticulatus schubergi*, seldom cause bacterial septicaemia, even at high spore doses. As a general rule, if the test insect dies within 4 days following ingestion of spores, suspect bacterial septicaemia.

Mortality from microsporidiosis

When microsporidian development in host tissues is the cause of death, insects usually die over a long time period and seldom die sooner than 5 days following the ingestion of spores. A variety of external symptoms and signs may develop, based on the particular microsporidium involved. Nevertheless, all insects must be examined microscopically to confirm the presence of large numbers of microsporidian spores in the target tissues. Insects, dead from microsporidiosis, will usually contain massive numbers of spores. Bacteria may also be present in the haemocoel and, when present, may contribute to the death of the insect, but are not the exclusive cause of death as is the case with bacterial septicaemia. When a wide range of spore dosages is used in a single bioassay, insects may die from bacterial septicaemia at the highest dosages and from microsporidiosis at the lower dosages, with a mixture of bacterial septicaemia and microsporidiosis at the midrange dosages.

Determining dose–response relationships

There are many probit or computer program methods that may be used for constructing regression lines and calculating the LD_x (spore dose required to cause x% mortality) and LT_x (time required for x% of mortality to occur at a specific spore dose) values for different species of microsporidia (Robertson and Preisler, 1992). LD_x values may be expressed as overall mortality, but it is often useful to distinguish between death from bacterial septicaemia and death from microsporidiosis as indicated above. Therefore, for a single species of microsporidia one may obtain both LD_x and LT_x values for death caused by bacterial septicaemia and LD_x and LT_x values for death from microsporidiosis.

Although the LT_x is a very important concept for many species of microsporidia, LT bioassays for microsporidia have been rather superficial and have not utilized either of the two sampling designs suggested by Robertson and Preisler (1992). Future research on microsporidia should consider use of more appropriate sampling designs.

Delayed mortality

Simple bioassays may evaluate mortality at a single predetermined time after treatment but, for time–mortality (LT) studies, mortality must be evaluated over the entire time period during which mortality occurs. Because many species of microsporidia do not kill their hosts immediately and often cause death in later developmental stages, the time element of the study is very important in order to capture the true mortality effects caused by many species of microsporidia. We will give specific examples in a later section. Here, we will only deal with general concepts.

Many species of microsporidia cause mortality over an extended period after initial infection of the host. For example, if spores are fed to the test insects in the third larval stadium, mortality may occur in all subsequent larval instars and in the pupal stage. In addition, the life span of infected adults may be shortened. In situations such as these, mortality should be recorded as often as possible; daily recording of mortality is ideal. In addition to the date of mortality, the stage of the host in which death occurred should also be recorded. It is then possible to calculate not only LD and LT values, but also a demographic summary of the stages when death occurred.

Microsporidia that are transovarially transmitted from infected females to their progeny may cause high levels of mortality in the egg stage before hatch and/or after hatch in larval stages. For many species of microsporidia this is an important source of mortality, but it has seldom been addressed using carefully designed bioassays.

Some microsporidian species are more virulent than others. When no virulence data have been reported for the species of microsporidia under consideration, we suggest that preliminary bioassays be conducted using a wide range of spore dosages and relatively small numbers of hosts. This greatly reduces the possibility of using a range of spore dosages that produce either 0 or 100% responses throughout the entire range of doses.

Sublethal effects

The microsporidia, probably more than any other group of important insect pathogens, cause many sublethal effects in infected hosts. Rate of development (Solter *et al.*, 1990), feeding rate (Maddox, 1966), number and fertility of eggs laid (Siegel *et al.*, 1986), efficiency of movement (Kramer, 1965), mating efficiency (Gaugler and Brooks, 1975) and behavioural characteristics (Kramer, 1965; Gaugler and Brooks, 1975) may all be influenced by microsporidian infections. Varying combinations of these phenomena have been observed for many species of microsporidia but few studies have been conducted to evaluate the effect of spore dosages on the intensity of these phenomena. Gathering information about sublethal effects often requires a subset of experiments following the administration of the range of spore dosages over which the phenomena are to be observed. The rate of development, feeding rate, or number of eggs laid is relatively simple to determine, but effects on mating efficiency, mobility and other behavioural characteristics are much more difficult to evaluate.

Infectivity

ID_x *(spore dose required to cause x% infectivity)*

Many questions about microsporidian biology are related to infectivity. This may be in addition to or in place of questions related to mortality. The techniques for determining the ID_x differ from the techniques for determining the LD_x; rather than observing mortality, each host is examined microscopically for the presence of infection.

The most reliable indicator of infection is the presence of spores in the target tissues of the experimental host. This requires that each host be dissected and the most obvious or dominant target tissue examined under a compound microscope for the presence of microsporidian spores. There are three major possibilities of error when conducting infectivity bioassays.

The first possibility of error is examining the experimental insects too soon after spores are fed. Microsporidia require a period of time after spores are ingested by a susceptible host until new spores are formed in the target tissues. This is called the generation time and it is influenced by the spore dose (lower spore dosages have longer generation times), temperature, and the specific microsporidium for which the bioassay is being conducted. Generation times may vary greatly between microsporidian species. The investigator must be sure that experimental hosts are not examined before the generation time is complete at all dosages to avoid an underestimate of the true prevalence of infection and overestimation of the ID_x.

A second possibility of error is contamination of the preparation examined under the compound microscope for the presence of spores. If spores observed in this preparation are from any source other than from infected tissues of the experimental insect, a false-positive diagnosis will result. All equipment used for dissecting the experimental insects (forceps, scissors, pins, trays, etc.) must be scrupulously cleaned after every dissection. A second source of spores other than spores formed in infected hosts is the inoculum itself. If larvae are fed spores using any of the methods where insects are exposed to spores for a long period of time (diet surface method, diet incorporation method) and the diet is not changed, it is possible that uninfected insects may contain spores from the inoculum in the midgut contents. This may cause an incorrect positive diagnosis. When in doubt about the source of spores in the preparation examined under the compound microscope, Giemsa stains should be made. Vegetative forms (meronts, sporonts and/or sporoblasts) will confirm that an active infection occurred. Vegetative forms are only produced in the cells of a living host and do not survive outside a host cell; therefore they cannot be present in the inoculum. Removing the peritrophic membrane and gut contents before examining midgut tissues is helpful.

The third possibility of error is the horizontal transmission from infected to healthy insects in the treated group. This can only occur if treated insects are kept in a common container or if changing the food for treated larvae allows contamination from infected to healthy individuals. It is possible to keep treated insects in a common container if all individuals are examined for infection before there is a possibility that an individual insect, horizontally infected by a treated infected insect, could produce sufficient spores to be scored as infected. Nevertheless, it is always preferable to keep treated insects in individual containers unless conditions of the bioassay experiment and availability of supplies make this an impossibility.

Vertical transmission

We have previously discussed the use of bioassays for evaluating mortality in the progeny of infected females. Bioassays can also be used to evaluate infection in the progeny of infected female insects. Procedures are similar to those given for evaluating mortality except that the progeny are individually examined for the presence of infection instead of mortality.

Specific examples

Vairimorpha necatrix: LD_x; LT_x, bacterial septicaemia and microsporidiosis

The microsporidium *Vairimorpha necatrix* is a highly virulent microsporidium that primarily infects the fat body of lepidopteran larvae such as the armyworm, *Pseudaletia unipuncta*, and the corn earworm, *Heliocoverpa zea*. When larvae are fed high doses, damage to the midgut epithelium may allow opportunistic gut bacteria to enter the haemocoel. Lethal bacterial septicaemia occurs before the microsporidium produces spores in the target tissues. Even when bacterial septicaemia does not occur, most infected larvae die from microsporidiosis before pupation. The following example using *V. necatrix* illustrates a typical LD/LT bioassay. It also illustrates the partitioning of mortality caused by bacterial septicaemia and the mortality caused by microsporidiosis as well as the separation of mortality by larval instar.

Fresh spores for bioassays are produced in the host species or in an appropriate laboratory host and then are cleaned and counted for use in bioassays. If the investigator is willing to accept bioassay results expressed as spores per microlitre of suspension (i.e. concentration; IC_x, LC_x) one of the relative methods (diet surface, foliage dipping) may be used to infect larvae. If, however, results must be expressed as spores per individual host (i.e. dose; ID_x, LD_x) one of the absolute methods (loop, leaf disc) must be used.

Both the ID_{50} and LD_{50} for *Vairimorpha necatrix* are very low, less than 100 spores per larva, for most host species (Maddox *et al.*, 1981). Therefore, in order to obtain a range of responses from 0 to 100%, at least three spore dosages should fall between 0 and 100 spores per larva. Spores are counted and serial dilutions are prepared as described earlier. A total of five spore dosages should be used. Spores are fed to host larvae using the loop method and larvae are held individually in 30-ml cups containing diet. At least 15 larvae should be used per spore dose for each treatment and each treatment replicated three times. Each replicate should be set up independently, preferably on different days with new spore dilutions to avoid pseudoreplication (Robertson and Preistler, 1992). The larvae are observed once a day and the date of death, spore concentration and larval instar of the host at death recorded. All dead larvae are examined micro-scopically for microsporidian infections and cause of death. These data will provide LD_x, LT_x and ID_x information. Because most infected larvae even-tually die, the LD_x and the ID_x are similar if not identical for most isolates of *V. necatrix*. At these low dosages most larvae will die from microsporidio-sis and not bacterial septicaemia.

In order to evaluate the bacterial septicaemia effect, much higher dosages (10^4–10^6 spores per larva) must be used. The bioassays should be set up as for the lower dosages described above. If mortality is plotted against days in a log/log graph (for procedure see Ignoffo, 1964), the LT_{50} can be estimated for each spore dose. Maddox *et al.* (1981) reported an LT_{50} value of 3.7 and 5.7 days for dosages of 4.5×10^6 and 2×10^6 spores, respectively, while the next consecutive lower dosages of 5×10^5 and below had LT_{50} values greater than 11 days. Most of the larvae dying before 4 days died from bacterial septicaemia, while those dying later died from microsporidiosis.

Nosema pyrausta: *sublethal effects, vertical transmission, LC_x and IC_x*

The microsporidium *Nosema pyrausta* is a pathogen of the European corn borer (ECB), *Ostrinia nubilalis*. It is transmitted both by ingestion of spores (horizontal transmission) and by transovarial transmission (vertical trans-mission). Moderate infections of *N. pyrausta*, initiated in mid- or late-instar host larvae, do not generally prevent pupation and eclosion of adults. Longevity and fecundity of infected adults may be reduced but mating and oviposition can occur. *N. pyrausta* causes many sublethal effects which may be evaluated as part of the bioassay. Mating efficiency, fecundity and mobil-ity of both larval and adult stages can be quantitatively evaluated as part of the bioassay. Mortality as a result of *N. pyrausta* infections occurs through-out the life cycle of the corn borer but the most significant mortality occurs in the progeny of infected females. The following is an example of a bioas-say used to evaluate the effect of spore dose and larval age at the time of ingesting spores on mortality throughout the life cycle of the ECB and espe-cially the infection and mortality of transovarially infected offspring.

N. pyrausta spores are produced in ECB larvae and the spores are har-vested, cleaned and counted. A relative method of feeding spores is chosen for the bioassay, such as spreading spores on the surface of meridic diet. Several spore concentrations are fed to selected instars, for example, 40 µl of 10^4, 10^5 and 10^6 spores µl^{-1} suspensions are spread on meridic diet in 30-ml cups and, for each suspension, first to fifth instars are treated. Numbers of treated individuals surviving to pupation are recorded and the pupae separated by sex. Each female pupa is placed in an individual oviposition cage with two uninfected male pupae. Eclosing female adults are allowed to mate and oviposit. The female is dissected to confirm infection, and egg masses of approximately the same size from infected females are randomly chosen (e.g. three egg masses per female) and placed individually in diet cups until hatch. The number of larvae hatching from the egg masses is recorded and, ideally, the larvae are reared individually in diet cups. Long-term rearing in collective cups may result in horizontal transmission. Larval mortality is recorded and dead larvae are dissected to determine presence of infection. Remaining larvae are dissected at a preselected time and examined for presence of infection.

Because there are a large number of variables in this bioassay (age of larvae, spore dose, female oviposition, transovarial transmission) it is difficult to collect enough data in this bioassay for developing an appropriate statistical analysis of the dose–response relationships. Nevertheless, it is important to conduct bioassays of this type for many species of microsporidia for which the most important effect on the host is either sublethal or expressed in the transovarially infected individuals of the next generation.

Summary

Statistically valid bioassays have been conducted on very few species of microsporidia. This is primarily because microsporidia have little potential as microbial insecticides. In addition, microsporidia cause many sublethal and delayed effects and it is often very time consuming and difficult to design bioassays that properly evaluate these effects. Nevertheless, microsporidia are important naturally occurring control agents and many questions about their role in the population dynamics of the host require information about both the lethal and sublethal effects we have discussed in this chapter. We believe that well designed microsporidian bioassays will be invaluable in answering some of these questions.

References

Anderson, R.M. and May, R.M. (1981) The population dynamics of microparasites and their invertebrate hosts. *Philosophical Transactions of the Royal Society of London, Ser B.*, 291, 451–524.

Angus, T.A. (1964) A magnetic stirring device for syringes. *Journal of Insect Pathology* 6, 125–126.

Andreadis, T.G. (1985) Experimental transmission of a microsporidian pathogen from mosquitoes to an alternate copepod host. *Proceedings of the National Academy of Sciences of the USA* 82, 5574–5577.

Anthony, D.W., Savage, K.E., Hazard, E.I., Avery, S.W., Boston, M.D. and Oldacre,

S.W. (1978) Field tests with *Nosema algerae* Vavra and Undeen (Microspora, Nosematidae) against *Anopheles albimanus* Wiedemann in Panama. *Miscellaneous Publications of the Entomological Society of America* 11, 17–28.

Balbiani, E.G. (1882) Sur les microsporidies ou psorospermies des Articulés. *Comptes Rendus de l'Academie des Sciences Paris, Ser. D*, 95, 1168–1171.

Bassi, A. (1835) *Del Mal del Segno Calcinaccio o Moscardino Malattia che Affligge i Bachi da Seta e sul Modo di Liberarne le Bigattaie anche le piú Infestate. Parte I. Teoria.* Orcesi, Lodi, 67 pp.

Bassi, A. (1836) *Del Mal del Segno Calcinaccio o Moscardino Malattia che Affligge i Bachi da Seta e sul Modo di Liberarne le Bigattaie anche le piú Infestate. Parte II. Pratica.* Orcesi, Lodi, 58 pp.

Brooks, W.M. (1980) Production and efficacy of protozoa. *Biotechnology and Bioengineering* 22, 1415–1440.

Brooks, W.M. (1988) Entomogenous protozoa. In: Ignoffo, C. (ed.) *Handbook of Natural Pesticides, Microbial Insecticides*, Part A, *Entomogenous Protozoa and Fungi*, Vol. 5. CRC Press, Boca Raton, Florida, pp. 1–149.

Burges, H.D. and Thomson, E.M. (1971) Standardization and assay of microbial insecticides. In: Burges, H.D. and Hussey, N.W. (eds) *Microbial Control of Insects and Mites.* Academic Press, New York, pp. 591–622.

Burnett, W.J. (1851) The organic relations of some Infusoria, including investigations concerning the structure and nature of the genus *Bode* (Ehr.). *Proceedings of the Boston Society of Natural History* 4, 124–125.

Canning, E.U. (1982) An evaluation of protozoal characteristics in relation to biological control of pests. *Parasitology* 84, 119–149.

Canning, E.U. and Hulls, R. (1970) A microsporidian infection of *Anopheles gambia* Gibbs from Tanzania; interpretation of its mode of transmission. *Journal of Protozoology* 17, 531–539.

Cole, R.J. (1970) The application of the 'triangulation' method to the purification of *Nosema* spores from insect tissues. *Journal of Invertebrate Pathology* 15, 193–195.

Daum, R.J., McLaughlin, R.E. and Hardee, D.D. (1967) Development of the bait principle for boll weevil control: cottonseed oil, a source of attractants and feeding stimulants for the boll weevil. *Journal of Economic Entomology* 60, 321–325.

Dufour, L. (1828) Sur la Grégarine, nouveau genre de ver qui vit en troupeau dans les intestins de divers insectes. *Annales de Sciences Naturelles* 1, 366–368.

Dunkel, F.V. and Boush, G.M. (1969) Effect of starvation on the black carpet beetle, *Attagenus megatoma*, infected with the eugregarine *Pyxinia frenzeli. Journal of Invertebrate Pathology* 14, 49–52.

Fournie, J.W., Foss, S.S., Courtney, L.A. and Undeen, A.H. (1990) Testing of insect microsporidia (Microspora: Nosematidae) in nontarget aquatic species. *Diseases of Aquatic Organisms* 8, 137–144.

Fuxa, J.R. and Brooks, W.M. (1979) Mass production and storage of *Vairimorpha necatrix* (Protozoa: Microsporida). *Journal of Invertebrate Pathology* 33, 86–94.

Gaugler, R.R. and Brooks, W.M. (1975) Sublethal effects of infection by *Nosema heliothidis* in the corn earworm, *Heliothis zea. Journal of Invertebrate Pathology* 26, 57–63.

Hall, I.M. (1954) Studies of microorganisms pathogenic to the sod webworm. *Hilgardia* 22, 535–565.

Harry, O.G. (1967) The effect of a eugregarine, *Gregarina polymorpha*

(Hammerschmidt), on the larva of *Tenebrio molitor* (L.). *Journal of Protozoology* 14, 539–547.

Henry, J.E. (1967) *Nosema acridophagus* sp. n., a microsporidian isolated from grasshoppers. *Journal of Invertebrate Pathology* 9, 331–341.

Henry, J.E. (1971) Experimental application of *Nosema locustae* for control of grasshoppers. *Journal of Invertebrate Pathology* 18, 389–394.

Henry, J.E. (1981) Natural and applied control of insects by protozoa. *Annual Review of Entomology* 26, 49–73.

Henry, J.E. (1985) Effect of grasshopper species, cage density, light intensity, and method of inoculation on mass production of *Nosema locustae* (Microsporida: Nosematidae). *Journal of Economic Entomology* 78, 1245–1250.

Henry, J.E. and Oma, E.A. (1974) Effect of prolonged storage of spores on field applications of *Nosema locustae* (Microsporida: Nosematidae) against grasshoppers. *Journal of Invertebrate Pathology* 23, 371–377.

Henry, J.E. and Oma, E.A. (1981) Pest control by *Nosema locustae*, a pathogen of grasshoppers and crickets. In: Burges, H.D. (ed.) *Microbial Control of Pests and Plant Diseases, 1970–1980*. Academic Press, New York, pp. 573–586.

Henry, J.E. and Onsager, J.A. (1982) Large-scale test of control of grasshoppers on rangeland with *Nosema locustae*. *Journal of Economic Entomology* 75, 31–35.

Henry, J.E., Tiahrt, K. and Oma, E.A. (1973) Importance of timing, spore concentration, and levels of spore carrier in applications of *Nosema locustae* (Microsporida: Nosematidae) for control of grasshoppers. *Journal of Invertebrate Pathology* 21, 263–272.

Hughes, P.R., Van Beek, N.A.M. and Wood, H.A. (1986) A modified droplet method for rapid assay of *Bacillus thuringiensis* and baculoviruses in noctuid larvae. *Journal of Invertebrate Pathology* 48, 187–192.

Ignoffo, C.M. (1964) Bioassay technique and pathogenicity of a nuclear polyhedrosis of the cabbage looper, *Trichoplusia ni* (Hübner). *Journal of Insect Pathology* 6, 237–245.

Iwano, H. and Kurtti, T.J. (1995) Identification and isolation of dimorphic spores from *Nosema furnacalis* (Microspora: Nosematidae). *Journal of Invertebrate Pathology* 65, 230–236.

Jaronski, S.T. (1984) Microsporida in cell culture. *Advances in Cell Culture* 3, 183–229.

Jouvenaz, D.P. (1981) Percoll: an effective medium for cleaning microsporidian spores. *Journal of Invertebrate Pathology* 37, 319.

Kelly, J.F. and Knell, J.D. (1979) A simple method of cleaning microsporidian spores. *Journal of Invertebrate Pathology* 31, 280–288.

Kramer, J.P. (1965) Effects of an octosporeosis on the locomotor activity of adult *Phormia regina* (Meigen) (Dipt. Calliphoridae). *Entomophaga* 10, 339–342.

Leidy, J. (1856) A synopsis of Entozoa and some of their ecto-congeners observed by the author. *Proceedings of the Academy of Natural Sciences, Philadelphia* 8, 42–58.

Maddox, J.V. (1966) Studies on a microsporidiosis of the armyworm, *Pseudaletia unipuncta* (Haworth). PhD thesis, University of Illinois, Urbana, Illinois.

Maddox, J.V. (1968) Generation time of the microsporidium *Nosema necatrix* in larvae of the armyworm, *Pseudaletia unipuncta*. *Journal of Invertebrate Pathology* 11, 90–96.

Maddox, J.V. and Solter, L.F. (1996) Long term storage of microsporidian spores in liquid nitrogen. *Journal of Eukaryotic Microbiology* 43, 221–225.

Maddox, J.V., Brooks, W.M. and Fuxa, J.R. (1981) *Vairimorpha necatrix*, a pathogen of agricultural pests: potential for pest control. In: Burges, H.D. (ed.) *Microbial Control of Pests and Plant Diseases, 1970–1980.* Academic Press, New York, pp. 587–594.

Martouret, D. (1962) Etude pathologiques sur la mode d'action de *Bacillus thuringiensis. VII International Congress of Entomology*, Vienna, 1960, Vol. 2, pp. 849–855.

McLaughlin, R.E. (1966) Infection of the boll weevil with *Mattesia grandis* induced by a feeding stimulant. *Journal of Economic Entomology* 59, 909–911.

McLaughlin, R.E. (1967) Development of the bait principle for boll-weevil control. II. Field-cage tests with a feeding stimulant and the protozoan *Mattesia grandis. Journal of Invertebrate Pathology* 9, 70–77.

McLaughlin, R.E. (1971) Use of protozoa for microbial control of insects. In: Burges, H.D. and Hussey, N.W. (eds) *Microbial Control of Insects and Mites.* Academic Press, New York, pp. 151–172.

McLaughlin, R.E., Daum, R.J. and Bell, M.R. (1968) Development of the bait principle for boll weevil control. III. Field-cage tests with a feeding stimulant and the protozoan *Mattesia grandis* (Neogregarinida) and a microsporidian. *Journal of Invertebrate Pathology* 12, 168–174.

McLaughlin, R.E., Cleveland, T.C., Daum, R.J. and Bell, M.R. (1969) Development of the bait principle for boll weevil control. IV. Field tests with a bait containing a feeding stimulant and the sporozoans *Glugea gasti* and *Mattesia grandis. Journal of Invertebrate Pathology* 13, 429–441.

Naegeli, C. (1857) Über die neue Krankheit der Seidenraupe und verwandte Organismen. *Botanisches Zentralblatt* 15, 760–761.

Onstad, D.W., Maddox, J.V., Cox, D.J. and Kornkven, E.A. (1990) Spatial and temporal dynamics of animals and the host-density threshold in epizootiology. *Journal of Invertebrate Pathology* 55, 76–84.

Pasteur, L. (1870) *Études sur la Maladie des Vers a Soie*, Vols 1 and 2. Gauthier-Villars, Paris.

Pasteur, L. (1874) (On the use of fungi against *Phylloxera*. Discussion). *Comptes Rendus de l'Academie des Sciences, Paris* 79, 1233–1234 (in French).

Pérez, C. (1905) Microsporidies parasites des crabes d'Arcachon. *Travaux des Laboratories de la Sociéte Scientifique et Station Zoologique d'Arcachon* 8, 15–36.

Pramer, D. and Al-Rabiai, S. (1973) Regulation of insect populations by protozoa and nematodes. *Annals of the New York Academy of Science* 217, 85–92.

Robertson, J.L. and Preisler, H.K. (1992) *Pesticide Bioassays with Arthropods.* CRC Press, Boca Raton, Florida, 127 pp.

Siegel, J.P., Maddox, J.V. and Ruesink, W.G. (1986) Lethal and sublethal effects of *Nosema pyrausta* on the European corn borer, *Ostrinia nubilalis*, in Central Illinois. *Journal of Invertebrate Pathology* 48, 167–173.

Solter, L.F. and Maddox, J.V. (1998) Physiological host specificity of microsporidia as an indicator of ecological host specificity. *Journal of Invertebrate Pathology* 71, 207–216.

Solter, L.F., Onstad, D.W. and Maddox, J.V. (1990) Timing of disease influenced processes in the life cycle of *Ostrinia nubilalis* infected with *Nosema pyrausta. Journal of Invertebrate Pathology* 55, 337–341.

Solter, L.F., Roberts, S.J., Maddox, J.V. and Armbrust, E.J. (1993) A new microsporidium in alfalfa weevil populations: distribution and characterization. *Great Lakes Entomologist* 26, 151–154.

Solter, L.F., Maddox, J.V. and McManus, M.L. (1997) Host specificity of microsporidia (Protista: Microspora) from European populations of *Lymantria dispar* (Lepidoptera: Lymantriidae) to indigenous North American Lepidoptera. *Journal of Invertebrate Pathology* 69, 135–150.

Stempell, W. (1909) Über *Nosema bombycis* Nägeli. *Archiv für Protistenkunde* 16, 281–358.

Sweeney, A.W., Hazard, E.I. and Graham, M.F. (1985) Intermediate host for an *Amblyospora* sp. (Microspora) infecting the mosquito, *Culex annulirostris*. *Journal of Invertebrate Pathology* 46, 98–102.

Taylor, A.B. and King, R.L. (1937) Further studies on the parasitic amoebae found in grasshoppers. *Transactions of the American Microscopic Society* 56, 172–176.

Undeen, A.H. (1978) Spore-hatching processes in some *Nosema* species with particular reference to *Nosema algerae* Vavra and Undeen. *Miscellaneous Publications of the Entomological Society of America* 11, 29–50.

Undeen, A.H. (1990) A proposed mechanism for the germination of microsporidian (Protozoa Microspora) spores. *Journal of Theoretical Biology* 142, 223–235.

Undeen, A.H. (1997) *Microsporidia (Protozoa): a Handbook of Biology and Research Techniques. Southern Cooperative Series Bulletin*, no. 387, Oklahoma State University.

Undeen, A.H. and Alger, N.E. (1971) A density gradient method of fractionating microsporidian spores. *Journal of Invertebrate Pathology* 18, 419–420.

Undeen, A.H. and Avery, S.W. (1983) Continuous flow-density gradient centrifugation for purification of microsporidian spores. *Journal of Invertebrate Pathology* 41, 405–406.

Undeen, A.H. and Avery, S.W. (1988) Effects of anions on the germination of *Nosema algerae* (Microspora: Nosematidae) spores. *Journal of Invertebrate Pathology* 52, 84–89.

Undeen, A.H. and Maddox, J.V. (1973) The infection of non-mosquito hosts by injection with spores of the microsporidian *Nosema algerae. Journal of Invertebrate Pathology* 22, 258–265.

Undeen, A.H. and Solter, L.F. (1996) The sugar content and density of living and dead microsporidian (Protozoa: Microspora) spores. *Journal of Invertebrate Pathology* 67, 80–91.

Undeen, A.H. and Solter, L.F. (1997) Sugar acquisition during the development of microsporidian (Microspora: Nosematidae) spores. *Journal of Invertebrate Pathology* 70, 106–112.

Undeen, A.H. and Van der Meer, R.K. (1994) Conversion of intrasporal trehalose into reducing sugars during germination of *Nosema algerae* (Protista: Microspora) spores: a quantitative study. *Journal of Eukaryotic Microbiology* 41, 129–132.

Undeen, A.H. and Vavra, J. (1997) Research methods for entomopathogenic Protozoa. In: Lacey, L.A. (ed.) *Manual of Techniques in Insect Pathology.* Academic Press, London, pp. 117–149.

Undeen, A.H., Van der Meer, R.K., Smittle, B.J. and Avery, S.W. (1984) The effect of gamma radiation on *Nosema algerae* (Microspora: Nosematidae) spore viability, germination, and carbohydrates. *Journal of Protozoology* 31, 479–482.

Van Beek, N.A.M. and Hughes, P.R. (1986) Determination of fluorescence spectroscopy of the volume ingested by neonate lepidopterous larvae. *Journal of Invertebrate Pathology* 48, 249–252.

Vavra, J. (1976) Structure of the Microsporidia. In: Bulla, L.A. and Cheng, T.C. (eds)

Comparative Pathobiology, Vol. 1. *Biology of the Microsporidia*. Plenum Press, New York pp. 1–85.

Vavra, J. and Maddox, J.V. (1976) Methods in microsporidiology. In: Bulla, L.A. and Cheng, T.C. (eds) *Comparative Pathobiology*, Vol. 1. *Biology of the Microsporidia*. Plenum Press, New York, pp. 281–319.

Weiser, J. (1969) Immunity of insects to protozoa. In: Jackson, G.L., Herman, R. and Singer, I. (eds) *Immunity to Parasitic Animals*. Academic Press, New York, pp. 129–147.

Weiser, J. and Veber, J. (1955) Moznosti biologickeho boje s prastevnickem americhym (*Hyphantria cunea* Drury). II. *Ceskoslovenska Parazitology* 2, 191–199 (in Czech).

Weiser, J. and Veber, J. (1957) Die Mikrosporidie *Thelohania hyphantriae* Weiser des weissen Bärenspinners und anderer Mitglieder seiner Biocönose. *Zeitschrift fuer Angewandte Entomologie* 40, 55–70.

Wilson, G.G. (1982) Protozoans for insect control. In: Kurstak, E. (ed.) *Microbial and Viral Pesticides*. Marcel Dekker, New York, pp. 581–600.

Zimmack, H.L., Arbuthnot, K.D. and Brindley, T.A. (1954) Distribution of the European corn borer parasite *Perezia pyraustae* and its effect on the host. *Journal of Economic Entomology* 47, 641–645.

Bioassays for Entomopathogenic Nematodes

6

I. Glazer[1] and E.E. Lewis[2]

[1]*Department of Nematology, Agricultural Research Organization, Bet Dagan, Israel;* [2]*Department of Entomology, Virginia Tech, Blacksburg, Virginia, USA*

Introduction

Insect-parasitic nematodes of the families Steinernematidae and Heterorhabditidae have been known for decades (Poinar, 1990). However, nematode preparations for controlling insect pests have become commercially available only recently (Georgis, 1992; Georgis and Manweiler, 1994). These nematodes can actively locate, infect and kill a wide range of insect species. Both *Steinernema* and *Heterorhabditis* pass through four juvenile stages before maturing. Only the third-stage juvenile ('infective' or 'dauer') can survive outside the insect host and move from one insect to another. Insect mortality, due to nematode infection, is caused by a symbiotic bacterium (*Xenorhabdus* spp. for steinernematids and *Photorhabdus* spp. for heterorhabditids) which the infective juveniles (IJs) carry in their intestines and release in the insect haemolymph (Akhurst and Boemare, 1990). Invasion by the nematode occurs through natural openings (spiracles, mouth, anus) or, in some cases, directly through the cuticle of certain insects (Bedding and Molyneux, 1982; Peters and Ehlers, 1994). The bacteria cells proliferate and eventually kill the insect host (usually within 72 h).

The symbiotic relationship between the nematode and the bacterium is rather complex; the infective juvenile serves as a vector by which the bacterium is transferred from one insect to another. After invading the haemolymph, the nematode secretes proteinaceous substances which inhibit the activity of the insect immune system (Simoes *et al.*, 1992) and affect its nervous system (Burman, 1982), thus providing the initial conditions for development of a bacterial colony. In a symbiotic manner, the bacterium renders the cadaver's interior favourable for nematode development by breaking down the haemolymph and providing a suitable diet for the nematodes. In addition, the bacterium releases substances with high antibiotic activity that protect the cadaver from the invasion of opportunistic organisms, thus allowing undisturbed development of the nematodes (Kondo and Ishibashi, 1986; Akhurst, 1990).

At temperatures ranging from 18–28°C, the life cycle is completed in 6–18 days, depending on the host insect and nematode species (Poinar, 1990; Zioni *et al.*, 1992). The invading IJs belonging to *Heterorhabditis* develop into hermaphrodites and those belonging to *Steinernema* develop into females or males. One or more generations of progeny develop within one host and reproductions continue until host-derived nutrients are depleted. At this time the nematodes become third-stage IJs (Fig. 6.1) that leave the cadaver in search of new hosts.

Many qualities render these nematodes excellent biocontrol agents: they have a broad host range, possess the ability to search actively for hosts, and present no hazard to mammals (Gaugler and Boush, 1979). Finally, the US-EPA (United States Environmental Protection Agency) has exempted the entomopathogenic steinernematids and heterorhabditids from registration and regulation requirements. They are already used commercially in high-value crops (Georgis and Manweiler, 1994).

Although entomopathogenic nematodes are listed among the important microbial control agents, no standard universal assays for evaluation of

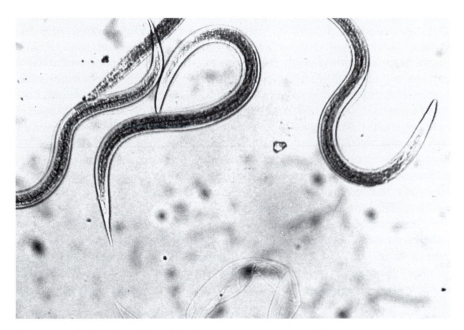

Fig. 6.1. Infective juveniles of *Steinernema carpocapsae* All strain.

nematode quality or potency exist. Lack of fundamental knowledge on the nematode–bacterium–host interaction has blocked the development of such assays. Some species such as *Steinernema carpocapsae* use an ambush strategy to find hosts (Kaya *et al.*, 1993), that is, they wait for long periods for the prey to pass their strike area. Others, like *S. glaseri*, adopt a cruising approach (Lewis *et al.*, 1992; Kaya *et al.*, 1993; Grewal *et al.* 1994a), that is, they search by moving constantly and are more effective at finding sedentary prey. Furthermore, the infection process – invasion of and establishment in the insect by the different nematode species – is not well understood. Hay and Fenlon (1995) have developed a 'zero-modified binomial model' to describe nematode invasion behaviour at different application rates. They distinguished between primary and secondary infection of the host. Their study indicated that: (i) a nematode-infected host is more susceptible to subsequent reinvasion than an unparasitized host, and (ii) a population of IJs comprises some individuals that readily infect unparasitized hosts and others that rely on prior infection before they invade. The ratio between these two types of IJs in any species/population is unknown. Due to differences in foraging strategies and invasion behaviour, the full potential of all nematode species can not be assessed by a single bioassay.

With the expansion of commercial interest in entomopathogenic nematodes, the susceptibility of many economically important insect pests has been tested in a wide range of laboratory assays. The most commonly used bioassay consists of exposure of the target insect to the infective juvenile stage of the nematode in filter paper arenas (Kaya and Hara, 1980; Morris, 1985; Miller, 1989; Morris *et al.*, 1990; Glazer, 1991, 1992). Assuming a positive correlation between nematode concentration and host mortality, probit analysis has been used to analyse data from dose–response tests and to calculate the LD_{50} (Morris *et al.*, 1990; Glazer, 1991, 1992). However, when a parasite is highly virulent, the use of probit analysis is not useful since the results of the bioassays are likely to be biased by large errors in dosage (Burges and Thomson, 1971; Huber and Hughes, 1984). Estimation of nematode virulence by LD_{50} values is questionable since a single steinernematid or heterorhabditid infective juvenile is capable of killing an insect.

Efficacy of field applications is affected by factors associated with the nematode (invasion rate and bacterial release), the bacterium (establishment and multiplication rate) and the host (behaviour and immune response), as well as with the environment (insect location, temperature, moisture, pH, soil composition and texture). A series of sand- or soil-based bioassays was developed (Molyneux, 1986; Fan and Hominick, 1991; Mannion and Jansson, 1993; Westerman, 1994) to simulate more closely the effect of these factors on nematode virulence and, thus, to obtain predictive information about nematode efficacy in the field. In these assays, either mortality of the insect host or invasion efficiency (measured as the slope resulting from the linear regression of the number of nematodes found in the insect cadaver plotted against the dose) was determined (Fan and Hominick, 1991). In a recent study Caroli *et al.* (1996) used penetration rate of a single concentration of IJs as a measure of nematode virulence.

Since no standard assay is available, we will describe in this chapter some of the most common assays and refer to their use in assessment of nematode virulence against insect pests. Virulence is defined as the 'disease-producing power' (Tanada and Kaya, 1993). The virulence assays are: penetration (Caroli *et al.*, 1996), exposure time (Glazer, 1991), sand column (Griffin and Downes, 1994), one-on-one (Miller, 1989), and host suitability (Lewis *et al.*, 1996). These bioassays rank nematode activity according to different steps in the infection process: the penetration assay reflects nematode ability to penetrate into the insect; the exposure time assay indicates the rate of penetration; the sand column assay measures nematode ability to locate and penetrate the target host; and the one-on-one assay represents the overall infection process. Host suitability is recorded by measures of host recognition behaviour.

Rearing and Handling Entomopathogenic Nematodes

Nematode sources

Steinernematid and heterorhabditid taxonomy

At present, 20 species of *Steinernema* and 11 species of *Heterorhabditis* are recognized. In addition, numerous undescribed populations of nematodes from both genera have been isolated in many countries. Poinar (1990) produced a key for identification of steinernematid and heterorhabditid nematodes.

Rearing technique

Small quantities of nematodes, needed for laboratory and greenhouse tests, can be reared on last instars of the greater wax moth, *Galleria mellonella*. This insect is easily reared in the laboratory (Woodring and Kaya, 1988). It can also be purchased in animal pet stores as it is used as fish bait. Details on the culture technique of nematodes on *G. mellonella* are described by Woodring and Kaya (1988) and Kaya and Stock (1997). Briefly, a 0.5 ml suspension of 100–200 IJs in water is applied on to filter paper in 5-cm diameter plastic Petri dishes. Five to ten *G. mellonella* larvae are transferred to each dish and the dishes are incubated. Commonly, 25°C is a sufficient temperature, but some nematodes may require a different growth temperatures (Grewal *et al.*, 1994b). After insect mortality occurs (within 24–48 h from infection), the cadavers are transferred on to 'white trap' plates for further incubation at 25°C. After 10–14 days the new IJs migrate from the

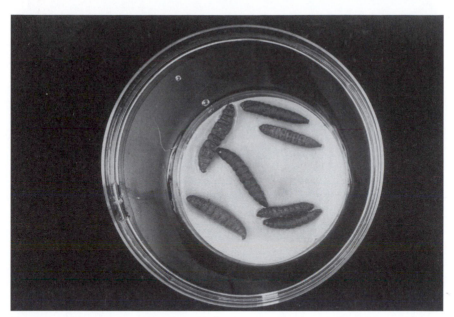

Fig. 6.2. White trap for collection of infective juveniles from infected insects (last instars of *G. mellonella*). The nematodes migrate from the filter paper into the water of the large Petri dish.

cadaver into the water surrounding the filter paper in the trap. The nematodes are then collected into tissue culture bottles and placed in storage. The storage temperature for steinernematids is 4–6°C and for heterorhabditids is 8–10°C.

Due to the large diversity among nematode populations, some of the above conditions may have to be modified for certain species/strains. For example, the most suitable host for rearing *S. scapterisci* is the mole cricket (Nguyen and Smart, 1989, 1992); *Galleria mellonella* larvae are not a suitable host for this nematode species. Incubation and storage temperatures may also need adjustment for nematode strains/species with specific adaptation to either cold or warm environments.

Preparation of nematodes for an assay

Cleaning the nematode suspension

IJs of various nematode strains/species can be obtained from different sources: from *in vivo* cultures as described above or from solid or liquid *in vitro* cultures, different storage and shipment conditions, different formulations, and so on. Some nematode suspensions may include media and storage substrate, dead non-infective stages, and some dead infective stages. We recommend using a clean suspension of viable IJs. The infective juveniles' suspension should be cleaned by physical means, for instance washing several times in water, sieving and settling in water. To avoid pre-selection for active nematodes, it is not recommended to use separation techniques based on nematode behaviour (e.g. migration into white traps or Baermann funnel).

Adjusting nematode concentration

1. Counting large numbers of nematodes is impractical, so serial dilutions are generally used. The nematode suspension should be shaken well in its container. A 50 μl aliquot is withdrawn with a micropipette and transferred to a 5-cm Petri dish. Three such samples are taken from each suspension and placed into three different Petri dishes. Fifteen millilitres of water are added to each Petri dish.

2. The nematodes in the dishes are counted under a dissecting microscope. Nematode concentration per millilitre is calculated by multiplying the average of the three 50 μl counts by 20.

3. To adjust to any concentration the following formula is used:

$$[(i/c) - 1] \times V = V_a$$

where i = initial concentration 50 μl^{-1}, c = final concentration 50 μl^{-1}, V = volume of the suspension (ml), and V_a = the amount of water (ml) to be added (if positive) or to be removed (if negative) from the suspension. If a higher concentration is needed, let the nematodes settle to the bottom of the container for at least 30 min, then remove excess water.

4. The final concentration should be checked by repeating steps 1 and 2. The final count should be within ± 10% of the needed concentration. At low concentrations (< 50 IJs ml^{-1}) estimation of nematode numbers by the procedure described above may be inaccurate due to large variation in sampling. Therefore, counts of individual nematodes of each sample are required.

Nematode storage prior to assay

All nematode suspensions should be kept at 20–25°C for 24 h prior to testing. Nematodes should be tested at the same temperature.

The Bioassay Procedures

Penetration assay

Insect preparation

Use last instar larvae of the wax moth (*G. mellonella*). Always use insects with an average weight of 0.2–0.3 g. At cool temperatures, larvae should not be stored for more than 14 days. Larvae should be at kept at 22–25°C for 48 h prior to the assay.

Arena preparation

Use 12-well plates (Corning Cell Wells, Corning, New York) or bioassay plastic trays (C-D International, Pitman, New Jersey). Place filter paper (Whatman No. 1) at the bottom of each one of ten wells (= ten replicates) for each experimental treatment.

Setting up the assay

Prepare a nematode suspension of 4000 IJs ml^{-1} (see p. 234). Shake the nematode suspension and transfer a 50 µl aliquot with a micropipette to each one of the ten filter paper-padded wells. Transfer a single larva into each well and seal the wells with their lid.

Incubation

Store the plates at 25°C for at least 40 h. Although most of the nematodes penetrate the insect within 12–24 h of incubation, an additional incubation period allows the nematodes to initiate development. Adult nematodes are observed more easily than IJs after insect dissection.

Insect dissection and nematode count

COLLECTION OF THE INFECTED LARVAE
After the appropriate incubation period collect all the dead larvae from the same treatment into 50 ml water in a 250-ml conical flask. Shake well to remove all nematodes from the surface of the insects. Remove the rinsed insects to a clean 9-cm diameter Petri dish. If needed, the larvae can be stored at this stage in the freezer, at −20°C, for several weeks prior to dissection.

ENZYMATIC DIGESTION

Cut each cadaver lengthwise into two halves and place it at the bottom of a scintillation vial. Add 3 ml of a pepsin solution. [Preparation of pepsin solution: Mix 1000 ml distilled water with 23 g NaCl. Adjust pH to 1.8–2.0 using concentrated HCl solution (note: HCl is a hazardous material. Use all necessary means of protection). Add 8 g of pepsin and mix well until it has completely dissolved. Use fresh solution (no more than 1 week old, stored at 4–6°C).] Incubate the vials in a shaking incubator (37°C, 120 rpm) for 60 min. Take the vials out, shake well by vortex for 10 s each, and return them to incubation for an additional 20 min. Add 7 ml of Tween 80 (Sigma, St Louis, Missouri) solution (0.1% in water) to each flask and shake well. The vials can then be stored at 5°C for up to 48 h before counting.

COUNTING NEMATODES

Shake the vial well and pour the contents into a 9-cm diameter Petri dish. Count the nematodes in the suspension under a dissecting microscope at a recommended magnification of 40× for infective juveniles or 10–20× for adults.

DATA COLLECTION AND ANALYSIS

Penetration should be expressed as percentage according to the following calculation:

$$\frac{N \times 100}{T} = P$$

where N = average number of nematodes counted in each cadaver, T = original average number of nematodes in the well (usually 200), and P = percentage penetration. See Table 6.1.

Table 6.1. Levels of insect mortality and P values to be expected in the penetration assay with nematode strains.

Nematode species	Strain	Insect mortality %	Range	P value	Range
H. bacteriophora	HP88	95	90–100	5.0	2.7–12.3
	NJ	95	90–100	7.3	1.5–14.0
S. feltiae	UK	95	90–100	48.4	27.8–63.5
S. riobravae		95	90–100	38.9	30.1–53.7

From Caroli *et al.*, 1996; Ricci *et al.*, 1996.

Exposure time assay

Insect preparation

See p. 235.

Arena preparation

Use 24-well plates (Corning Cell Wells, Corning, New York). Place filter paper at the bottom of each well. Use three plates for each treatment.

Setting up and conducting the assay

Prepare a nematode suspension of 8000 IJs ml^{-1} (see p. 234). Shake the nematode suspension and transfer 50 µl with the micropipette to each of the filter paper-padded wells (approximately 400 IJs per well). Add a single *G. mellonella* larva to each well and seal the wells with their lid. After 1, 2 and 3 h remove eight insects from each plate, rinse well in tap water to remove nematodes from their surface and transfer to a 5-cm diameter Petri dish padded with moist filter paper (0.5 ml tap water).

Incubation

Store the Petri dishes at 25°C for at least 40 h. Insect mortality should be recorded 44–48 h post-exposure to nematodes.

Data collection and analysis

Insect mortality is expressed as a percentage of the total number of insects tested in each replicate (usually $n = 8$). ET$_{50}$ values, which are the exposure times (in minutes) of nematodes to insects that are required to achieve 50% insect mortality, are obtained by computing the data for probit analysis in any statistical program (SAS, SX, etc.). See Table 6.2.

One-on-one assay

Insect preparation

See p. 235.

Table 6.2. Levels of ET_{50} to be expected in the exposure time assay with nematode strains.

Nematode species	Strain	ET_{50} (min)	Range
H. bacteriophora	HP88	85	65–110
	NJ	75	55–110
S. feltiae	UK	45	35–70
S. riobravae		35	20–45
S. carpocapsae	All	65	50–70

From Glazer, 1991; Ricci *et al.*, 1996.

Arena preparation

Use 24-well plates. Place filter paper at the bottom of each well. Use three plates for each treatment.

Setting up the assay

Prepare a nematode suspension of 100 IJs ml^{-1} (see p. 234). Transfer individual nematodes, under the dissecting microscope, from the suspension to each well in 25 µl distilled water. Use an additional 25 µl to flush the contents of the pipette tip into each well. Add a single insect larva to each well and seal the wells with their lid (Fig. 6.3).

Incubation

Store the plates at 25°C for at least 72 h prior to determination of insect mortality.

Data collection and analysis

Insect mortality is expressed as a percentage of the total number of insects tested in each replicate (usually $n = 24$) dead after 72 h. The one-on-one assay was routinely used at Biosys Inc. (Columbia, Maryland) for quality control of *Steinernema carpocapsae* All strain. At the laboratory of Dr Ralf (Christian Albrecht University, Kiel, Germany) a similar assay is used with two nematodes per well with *H. megidis* and *H. bacteriophora*. See Table 6.3.

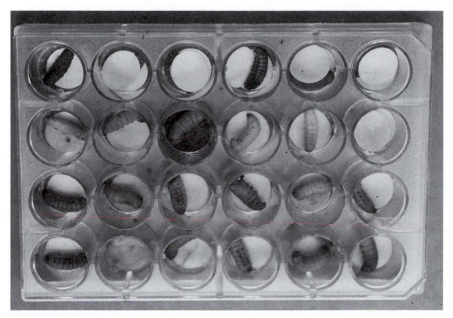

Fig. 6.3. Multiwell plate (24 wells) containing last-instar larvae of the wax moth *Galleria mellonella* used for the one-on-one infectivity bioassay.

Table 6.3. Levels of insect mortality in the one-on-one assay to be expected with nematode strains.

Nematode species	Strain	Insect mortality (%)	Range
S. feltiae	SN	35	30–70
S. riobravae		48	42–85
S. carpocapsae	All	50	45–85

Provided by Dr Grewal Parwinder, Biosys Inc., Columbia, Maryland, USA.

Sand column assay

Insect preparation

See p. 235.

Arena preparation

Use plastic vials (40 mm height × 45 mm diameter). The vials are packed with moist sand (washed sea sand with fractions retained between 400- and 250-μm diameter screens, heat-sterilized (24 h at 110°C) and moistened with 8% w/w tap water). Ensure even and uniform packing. Conduct 12 replicates per treatment.

Setting up the assay

1. Place one *G. mellonella* larva at the bottom of the plastic vial prior to adding the sand.
2. Prepare nematode suspension of 1000 IJs ml^{-1} (see p. 234).
3. Shake the nematode suspension and transfer a 100 µl aliquot with a micropipette to an indentation on the surface of the sand in each vial.

Incubation

Store the vials at 20°C for 24 h. Then remove the insects from the bottom of the vials and rinse them in tap water to remove surface nematodes. Finally, place the insects in a dry 5-cm diameter Petri dish for 24 h at 20–25°C. Those insects that die are placed on moist filter paper at 20–25°C for a further 3–4 days. After this period the cadavers are dissected and the number of nematodes is recorded.

INSECT DISSECTION AND NEMATODE COUNT
See p. 235. The sand column assay described above is used at the laboratory of Dr Christine Griffin at St Patrick's College, Maynooth, Co. Kildare, Ireland. The assay was used primarily for *Heterorhabditis* spp. (North-West European and Irish type). See Table 6.4.

Host recognition assay

The utility of host recognition behaviour as a predictive screening tool for field tests is based on two assumptions. The first is that entomopathogenic nematodes will be most effective controlling insects that they are adapted to parasitize naturally. The second is that entomopathogenic nematodes will

Table 6.4. Levels of insect mortality and *P* values to be expected in the sand column assay with nematode strains at 20°C.

Nematode species	Strain	Insect mortality (%)	*P* value
Heterorhabditis sp. NWE Group	HW79	100	54.8
Heterorhabditis sp. Irish Group	M145	100	19.9
Steinernema feltiae	OBS III	100	50.3
Steinernema carpocapsae	UK	42	1.5

From Griffin and Downes, 1991, 1994; Griffin, 1996.

respond most strongly to materials associated with insects in which they will be successful (to which they are adapted).

We present two assays of host recognition behaviour. The first is a two-step test for ambushing nematodes (*S. carpocapsae*), based upon the hierarchy of foraging events for ambushing nematodes described by Lewis *et al.* (1996). The first contact an ambushing nematode has with a prospective host is the cuticle. Once on the cuticle, they seek a portal of entry; usually the spiracles for *S. carpocapsae*. Since CO_2 emanates from the spiracles, IJs on the cuticle of an appropriate host are attracted to this material. *S. carpocapsae* IJs are not strongly attracted to host volatiles before exposure to host cuticle, or after exposure to cuticle of non-hosts (e.g. millipedes). Therefore, the proportion of nematodes attracted to a CO_2 source is an indication of the nematodes' assessment of host quality. A one-step test for cruising nematodes (*S. glaseri*) is based upon this nematode species' response to contact with host cues (Lewis *et al.*, 1992). Nematodes without exposure to host cues move in an approximately linear path. After 5 min of contact with host cuticle, they shift their foraging to 'localized search', which is slower, has a higher turning rate, and covers less area. The duration of this altered searching behaviour is indicative of these nematodes' assessment of host quality. Other tests will probably need to be developed for nematodes with foraging strategies between the extremes of cruising and ambushing.

Determination of whether the nematode to be tested ambushes

Campbell and Gaugler (1993) state that nictation behaviour, where the nematode elevates all but about 5% of the body from the substrate for extended periods, is essential to ambush foraging. This behaviour can be viewed easily on sand-covered 2% agar. Sprinkle small-particle sand sparingly over the agar, place about 200–300 nematodes in the centre of the dish and replace the cover. After 10 min, if the nematodes nictate while foraging, 30–40% will be doing so. Before working with unknowns, try this test with *S. carpocapsae* All to become familiar with the behaviour. If the nematodes nictate, the two-step process is appropriate. If not, the one-step assay is more appropriate. Many nematodes, such as *S. feltiae*, will 'nearly nictate'. That is, they will elevate 60–70% of their body from the substrate. These species will be problematic, and may require a combination of the two assays or a completely different one to yield useful information.

Two-step assay for ambushers

This test exposes IJs to insect cuticle for 30 min, then immediately tests them for their response to volatile host cues. (Prepare a series of host replicates, assay chambers, and all other necessary material in advance, because timing is crucial to proper execution of this test.)

1. The assay chambers used to assess nematode responses to volatile cues are shown in Fig. 6.4 (after Gaugler *et al.*, 1990). On a piece of glass (15 cm × 15 cm), attach pieces of Tygon tubing (1 cm o.d.) to the edges with silicon-based caulk. This will serve as the chamber base. The top is Plexiglas, with a hole cut in the centre for inoculation of nematodes, and another hole 5 cm away to accommodate a 1000-ml Eppendorf pipette tip. The top is held in place with rubber bands. Place two *G. mellonella* larvae in the pipette tip serving as a carbon dioxide source.

2. Place the candidate hosts individually in 60-mm Petri dishes lined with very wet filter papers and put them aside.

3. Prepare 2% plain agar and fill the bases of the test plates. Let the agar cool without the tops. After 1 h, seal the test plates with their tops and cover the nematode inoculation port with tape. Insert the pipette tip with *G. mellonella* into the hole provided and fix it with modelling clay so the tip is suspended 3–5 mm from the surface of the agar, taking care not to touch the agar with the tip. A gradient of volatiles will form for 1 h.

4. For a 30-min contact exposure to host cuticle, concentrate about 500–1000 nematodes by vacuum into a paste that can be picked up with a dissecting needle. Make sure the cuticle of the host is moist, and wipe the nematodes on to the cuticle and replace the cover of the Petri dish. Use one insect for each of ten replicates of the volatile test.

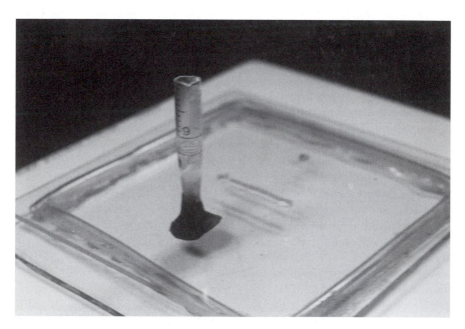

Fig. 6.4. The chamber used in the host recognition assay.

5. At the simultaneous end of the hour gradient formation and the 30-min cuticle exposure, place the host in a small tube with water and vortex it to remove nematodes. A control group that had no exposure to host cuticle should be tested. Concentrate the nematodes by vacuum, and transfer them to the gradient assay dish through the inoculation port and replace the tape.

6. After 1 h, disassemble the assay plate and collect all nematodes directly under the opening of the pipette tip and place them in a dish with water for counting (a 1-cm cork borer is convenient). Collect all other nematodes into another dish. Count the nematodes in each dish and calculate the percentage response by dividing the number of nematodes under the pipette tip by the total for each replicate.

Analysis: a high response (45–60%) indicates that the nematode is probably well-adapted to the test host. A low response (<10%) suggests a poor match between the nematode and the host. Typically, the control response for *S. carpocapsae* will be about 5%.

One-step assay for cruisers

This test exposes individual nematodes to host cuticle for 5 min, then transfers them to plain agar and records their movement.

1. Fill several Petri dishes (ten replicates suggested, but make some extra plates) with 2% agar and let them cool without the lids for 1 h. Draw a 1-cm diameter ring on the bottom of each dish.

2. Concentrate the test nematodes by vacuum and transfer them to one of the plates.

3. Place the test hosts in another Petri dish. Surround the insect with very wet filter paper (1.2 ml water per 5-cm diameter Petri dish) and make sure the cuticle of the insect is moist.

4. Pick up one nematode with a dissecting needle, and transfer it to the cuticle of the insect and replace the lid of the host dish to assure a humid environment. The exposure lasts 5 min.

5. Remove the nematode from the host cuticle with the dissecting needle, and place it in the centre of the ring. A control group should be tested with no exposure to host cuticle. Record the time it takes for the nematode to leave the ring (this is commonly called the 'giving-up time').

Analysis: the giving-up time for good matches between hosts and nematodes will be significantly longer than that for poor matches. Typical values for *S. glaseri* are 1600 s after exposure to *P. japonica* (an optimal host), *Acheata domesticus* stimulates a 700-s giving-up time and the control value is less than 100 s.

Validation

To validate the results of these tests, two measures of host suitability can be conducted. The first may be any of the pathogenicity tests mentioned earlier in this chapter that measure nematode-induced host mortality. The second measures reproductive potential (IJs produced per milligram of host tissue) for a nematode in a given host. First, record the weight of each test insect. Then, in a 24-well plate, inoculate the insects with 100 IJs. After 3–4 days, set up dead insects individually in white traps. Count all infective juveniles produced from each host. For a given series of test hosts, there should be a positive correlation between both of these measures and the behavioural response.

Concluding Remarks

1. The methods and procedures described above were developed and tested with specific nematode strains. It is likely that for other species of nematodes experimental conditions such as concentration, incubation period and incubation temperature may need to be optimized.

2. The last instar of *G. mellonella*, which was suggested here, is only a model insect with no economic importance. Due to its high susceptibility it may not be suitable for determination of small differences between nematode populations. The meal worm *Tenebrio molitor* has been suggested as an alternative model insect for determination of nematode virulence (Caroli *et al.*, 1996). Additional information is needed before this insect can be considered to replace *G. mellonella*. Other insects, which are considered as target pests for nematodes, may be tested in the different assays described above. It is recommended that *G. mellonella* larvae are always used for comparison.

3. The bioassays described above emphasize the potential of measuring quantitative behavioural responses as specific criteria for nematode virulence. Since nematodes differ in their behaviour, one common bioassay cannot be used as a universal measure of virulence for all species and strains (Grewal *et al.*, 1994a). Furthermore, different assays may be used for different purposes; to select a specific population for use against a particular insect a variety of assays which are more laborious but simulate environmental conditions (sand column assay) or the invasion by the nematode (penetration assay) should be considered. In cases where production batches of the same nematode strains are compared, simple and fast assays are needed (one-on-one and exposure time assays).

4. Determining the host range of entomopathogenic nematodes, in the host recognition assays described here, can help make decisions about what nematodes to test against a particular insect pest. Despite many years of research, very little is known about the natural host ranges of entomopathogenic nematodes. The common *Galleria*-bait technique for isolation of new strains (Bedding and Akhurst 1975) yields no information about host range, except that *G. mellonella* is included in it. The tests we describe will hopefully serve two purposes: provide background information about the host associations of newly isolated strains of entomopathogenic nematodes and assist in deciding which nematode species or strains will be worthy of field testing against particular insect pests.

In conclusion the development of standardized procedures to measure nematode virulence is a key factor in enhancing the effective utilization of entomopathogenic nematodes as biological agents. The scientific community and industry involved in the development of these microbial insecticides should put effort into the development of such procedures in the framework of international organizations such as the International Organization of Biological Control, and the Society of Invertebrate Pathology.

References

Akhurst, R.J. (1990) Antibiotic activity of *Xenorabdus* spp., bacteria symbiotically associated with insect pathogenic nematodes of the families Steinernematidae and Heterorhabditidae. *Journal of General Microbiology* 128, 3061–3065.

Akhurst, R.J. and Boemare, N.E. (1990) Biology and taxonomy of *Xenorabdus*. In: Gaugler, R. and Kaya, H.K. (eds) *Entomopathogenic Nematodes in Biological Control*. CRC Press, Boca Raton, Florida, pp. 75–90.

Bedding, R.A. and Akhurst, R.J. (1975) A simple technique for the detection of insect parasitic rhabditid nematodes in soil. *Nematologica* 21, 109–110.

Bedding, R.A. and Molyneux, A.S. (1982) Penetration of insect cuticle by infective juveniles of *Heterorhabditis* spp. (Heterorhabditidae: Nematoda). *Nematologica* 28, 354–359.

Burges, H.D. and Thomson, E.D. (1971) Standardization and assay of insecticides. In: Burges, H.D. and Hussey, N.W. (eds) *Microbial Control of Insects and Mites*. Academic Press, London, pp. 591–622.

Burman, M. (1982) *Neoaplectana carpocapsae*: toxin production by axenic insect parasitic nematode. *Nematologica* 28, 62–70.

Campbell, J.F. and Gaugler, R. (1993) Nictation behaviour and its ecological implications in the host search strategies of entomopathogenic nematodes (Heterorhabditidae and Steinernematidae). *Behaviour* 126, 155–169.

Caroli, L., Glazer, I. and Gaugler, R. (1996) Entomopathogenic nematode infectivity assay: multi variable comparison of penetration into different hosts. *Biocontrol Science and Technology* 6, 227–233.

Fan, X. and Hominick, W.M. (1991) Efficacy of *Galleria* (wax moth) baiting technique for recovering infective stages of entomopathogenic rhabditids (Steinernematidae and Heterorhabditidae) from sand and soil. *Revue de Nematologie* 14, 381–387.

Gaugler, R. and Boush, G.M. (1979) Nonsusceptibility of rats to the entomogenous nematode *Neoaplectana carpocapsae*. *Environmental Entomology* 8, 658–660.

Gaugler, R., Campbell, J.F. and McGuire, T. (1990) Selection for host finding in *Steinernema feltiae*. *Journal of Invertebrate Pathology* 54, 363–372.

Georgis, R. (1992) Present and future prospects for entomopathogenic nematodes products. *Biocontrol Science and Technology* 2, 83–99.

Georgis, R. and Manweiler, S.A. (1994) Entomopathogenic nematodes: a developing biological control technology. In: Evans, K. (ed.) *Agricultural Zoology Reviews*. Intercept, Andover, UK, pp. 63–94.

Glazer, I. (1991) Invasion rate as a measure of infestivity of sterinernematid and heterorhabditid nematodes to insects. *Journal of Invertebrate Pathology* 59, 90–94.

Glazer, I. (1992) Measures for evaluation of entomopathogenic nematode infectivity to insects. In: Gommers, F.J. and Mass, P.W.Th. (eds) *Nematology from Molecule to Ecosystem*. European Society of Nematologists, Invergowie, Dundee, pp. 195–200.

Grewal, P.S., Lewis, E.E., Gaugler, R. and Campbell, J.F. (1994a) Host finding behavior as a predictor of foraging strategy in entomopathogenic nematodes. *Parasitology* 108, 207–215.

Grewal, P.S., Selvan, S. and Gaugler, R. (1994b) Thermal adaptation of entomopathogenic nematodes: niche breadth for infection, establishment, and reproduction. *Journal of Thermal Biology* 19, 245–253.

Griffin, C.T. (1996) Effects of prior storage conditions on the infectivity of *Heterorhabditis* sp. (Nematoda: Heterorhabditidae). *Fundamental and Applied Nematology* 19, 95–102.

Griffin, C.T. and Downes, M.J. (1991) Low temperature activity in *Heterorhabditis* sp. (Nematoda: Heterorhabditidae). *Nematologica* 37, 83–91.

Griffin, C.T. and Downes, M.J. (1994) Recognition of low temperature active isolates of the entomopathogenic nematode *Heterorhabditis*. *Nematologica* 37, 83–91.

Hay, D.B. and Fenlon, J.S. (1995) A modified binomial model that describes the infection dynamics of the entomopathogenic nematode *Steinernema feltiae* (Steinernematidae: Nematoda). *Parasitology* 111, 627–633.

Huber, J. and Hughes, P.R. (1984) Quantative bioassay in insect pathology. *Bulletin of the Entomological Society of America* 30, 31–34.

Kaya, H.K. and Hara, A.H. (1980) Differential susceptibility of lepidopterous pupae to infection by the nematode *Neoaplectana carpocapsae*. *Journal of Invertebrate Pathology* 36, 389–393.

Kaya, H.K. and Stock, S.P. (1997) Techniques in insect nematology. In: Lacey, L. (ed.) *Manual of Techniques in Insect Pathology*. Academic Press, San Diego, pp. 281–324.

Kaya, H.K., Burlando, T.M. and Thurston, G.S. (1993) Two entomopathogenic nematode species with different search strategies for insect suppression. *Environmental Entomology* 22, 165–171.

Kondo, E. and Ishibashi, N. (1986) Infectivity and propagation of entomogenous nematodes, *Steinernema* spp. on the common cutworm *Spodoptera litura* (Lepidoptera: Noctuidae). *Applied Entomology and Zoology* 21, 95–108.

Lewis, E.E., Gaugler, R. and Harrison, R. (1992) Entomopathogenic nematode host finding: Response to host contact cues by cruise and ambush foragers. *Parasitology* 105, 103–107.

Lewis, E.E., Ricci, M. and Gaugler, R. (1996) Host recognition behavior reflects host suitability for the entomopathogenic nematodes, *Steinernema carpocapsae*. *Parasitology* 109, 1–7.

Mannion, C.M. and Jansson, R.K. (1993) Infectivity of five entomopathogenic nematodes to the sweet potato weevil, *Cylas formicarius* (Coleoptera: Apionidae) in three experimental arenas. *Journal of Invertebrate Pathology* 62, 29–36.

Miller, R. (1989) Novel pathogenicity assessment technique of steinernematids and heterorhabditids entomopathogenic nematodes. *Journal of Nematology* 21, 574.

Molyneux, A.S. (1986) *Heterorhabditidis* spp. and *Steinernema* (= *Neoaplectana*) spp.: Temperature, and aspects of behavior and infectivity. *Experimental Parasitology* 62, 169–180.

Morris, O.N. (1985) Susceptibility of 31 species of agricultural insect pests to entomogenous nematodes *Steinernema feltiae* and *Heterorhabditis bacteriophora*. *Canadian Entomologist* 117, 401–407.

Morris, O.N., Converse, V. and Harding, J. (1990) Virulence of entomopathogenic nematode–bacteria complexes for larvae of noctuids, a geometrid and a pyralid. *Canadian Entomologist* 122, 309–320.

Nguyen, K.B. and Smart, G.C., Jr (1989) A new steinernematid nematode from Uruguay as biological control agent of mole crickets. *Journal of Nematology* 21, 576.

Nguyen, K.B. and Smart, G.C., Jr (1992) Life cycle of *Steinernema scapterisci* Nguyen & Smart. *Journal of Nematology* 24, 160–169.

Peters, A. and Ehlers, R.U. (1994) Susceptibility of leatherjackets (*Tipula paludosa* and *T. oleracae*, Tipulidae: Nematocera) to the entomopathogenic nematode *Steinernema feltiae*. *Journal of Invertebrate Pathology* 63, 163–171.

Poinar, G.O., Jr (1990) Taxonomy and biology of Steinernematidae and Heterorhabditidae. In: Gaugler, R. and Kaya, H.K. (eds) *Entomopathogenic Nematodes in Biological Control*. CRC Press, Boca Raton, Florida, pp. 23–61.

Ricci, M., Glazer, I. and Gaugler, R. (1996) Entomopathogenic nematode infectivity assay: comparison of laboratory bioassays. *Biocontrol Science and Technology* 6, 235–245.

Simoes, N., Bettencourt, R., Laumond, C. and Boemare, N. (1992) Studies on immunoinhibitor released by *Steinernema carpocapsae*. In: *Proceedings of the European Society of Nematology 21st Meeting*, Albuferia, Portugal, pp. 123–127.

Tanada, Y. and Kaya, H.K. (1993) *Insect Pathology*. Academic Press, San Diego.

Westerman, P.R. (1994) An assay on assays. In: Burnell, A.M., Ehlers, R.U. and Masson, J.P. (eds) *Proceedings of the COST 812 Workshop on Genetics of Entomopathogenic Nematode–Bacterium Complexes*, Report EUR 15681 EN.

Woodring, J.L. and Kaya, H.K. (1988) *Steinernematid and Heterorhabditid Nematodes: a Handbook of Techniques*. Arkansas Agricultural Experiment Station, Fayetteville, Arkansas.

Zioni (Cohen-Nissan), S., Glazer, I. and Segal, D. (1992) Life cycle and reproductive potential of the entomopathogenic nematode *Heterorhabditis bacteriophora* strain HP88. *Journal of Nematology* 24, 352–358.

Statistical and Computational Analysis of Bioassay Data

7

R. Marcus[1] and D.M. Eaves[2]

[1]Department of Statistics, Agricultural Research Organization, Bet Dagan, Israel; [2]Department of Mathematics and Statistics, Simon Fraser University, Burnaby, British Columbia, Canada

Introduction

Entomopathogenic microbes and nematodes cause diverse quantitative responses in invertebrate pests. For instance, *Bacillus thuringiensis* causes fast mortality, mostly because of its toxins. With a few virus groups, mortality is somewhat delayed, as a result of an incubation period. With other microbes such as fungi, microsporidia and, sometimes, nematodes, pathogenicity is characterized by a gradual increase in mortality over time of exposure. The ultimate tools for quantifying these effects in the pests are bioassays.

In general, the quantitative response of pests in a bioassay depends on the concentration of the microbial control agent. Thus, experiments are often conducted to determine concentration–response relationships. Each experiment is designed with several groups of similar pests, each group being exposed to a different concentration of a microbial control agent. Most bioassays include untreated control groups of pests (which receive zero concentration). Each insect exposed to the material may exhibit a dichotomous response, such as mortality or survival; in practice, the mortality proportion among several pests exposed to a given concentration is recorded. The effect of the variation of concentration on pest mortality is commonly characterized by a gradual increase in mortality as the concentration increases. Parametric models may be used to describe mortality as a function of concentration. The activity of a microbial agent is traditionally measured as the median lethal concentration LC_{50} (or median lethal dose LD_{50}), which causes mortality in 50% of the pests. The efficacies of microbial materials are compared in terms of their LC_{50}s.

In bioassays recording sublethal effects, a continuous response of the insects, such as changes in body weight, is measured. The activity of a microbial control agent is expressed as the effective concentration, EC_{50}, which causes a 50% reduction in body weight or other response, as compared with an untreated control. The efficacies of microbial control agents are compared in terms of their EC_{50}s.

Mortality of pests caused by microbes and nematodes may often depend on the time of their exposure to the material. A typical experiment conducted to study the time–mortality relationship is commonly designed with a group of similar pests exposed to the material, with mortalities being recorded at distinct times of exposure during the experiment. The microbe activity is expressed in terms of the lethal time, LT_{50}, causing mortality of 50% of the pests. The activities of different materials are compared in terms of their LT_{50}s. If, in addition, several concentrations of a microbial control agent are included in the experiment, then lethal concentrations are functions of time of exposure and, similarly, lethal times are functions of concentration. The microbe activity is expressed as the median lethal concentration, $LC_{50}(t)$, causing mortality of 50% of the pests at exposure time t, or by the lethal time, $LT_{50}(x)$, causing mortality of 50% of the pests at concentration x. As time of exposure is increased, smaller and smaller concentrations will be needed to produce mortality of 50% of the pests. Thus, the median lethal concentration, $LC_{50}(t)$, decreases as time of exposure increases and, similarly, the median lethal time, $LT_{50}(x)$, decreases as concentration increases.

The bioassays developed to determine the activities of microbes and nematodes require the use of appropriate statistical and computational methods for analysing the biological data. In this chapter we describe a selection of the methods used for analysing bioassay data. Major statistical references are: Finney (1971, 1978), and Robertson and Preisler (1992); additional references are cited later, in this chapter. The statistical methods are described in a general form and can be used specifically to analyse bioassays involving microbial control agents.

This chapter is organized as follows. We start with a discussion of methods used to analyse bioassays which yield concentration–mortality data, and present an example of the use of probit analysis, applied to artificial concentration–mortality data. The evaluation of the relative efficacy of a new material in terms of a standard is considered. Some practical guidelines for conducting bioassays are provided. We then proceed to a brief discussion of bioassays which yield continuous response data. Subsequently, we describe methods for analysing bioassays which produce time–mortality data, and finally, we address the analysis of those based on time–concentration–mortality data.

Most of the results presented here were produced with procedures available in the SAS software package (Cody and Smith, 1991; SAS Institute Inc., 1990, 1996). Further available computing facilities are mentioned by Collett (1991) and Robertson and Preisler (1992).

Bioassays Yielding Concentration–Mortality Data

An example

We start with a constructed example of data obtained in an experiment conducted to study a concentration–mortality relationship, in order to illustrate basic concepts and general modelling procedures.

An experiment was conducted with groups of similar insects, that were exposed to various microbial concentrations: ten batches, each of 100 insects, were randomly assigned to five concentrations, say 0.01, 0.1, 1.0, 10.0, 100.0 (mg l^{-1}), with two batches being exposed to each concentration. After 6 days of exposure to the material, the number of dead among the insects in each of the ten batches was recorded.

Henceforth, we shall consider r batches, each of n insects, run at each of k concentrations $(x_1, \ldots, x_k)$. Let d_{ij} represent the number of dead among the n insects in batch j, which had been run at concentration x_i. The corresponding mortality proportions are defined as: $p_{ij} = d_{ij}/n$ ($i = 1, \ldots, k$; $j = 1, \ldots, r$). The pooled mortality proportion among nr insects at concentration x_i is denoted by p_i and equals $\Sigma^r_{j=1} d_{ij}/(nr)$ or $\Sigma^r_{j=1} p_{ij}/r$ ($i = 1, \ldots, k$). In our example, $k = 5$, $r = 2$ and $n = 100$. The observed mortality proportion data $\{p_{ij}\}$ and the pooled mortality proportions $\{p_i\}$ at the various concentrations $\{x_i\}$ ($i = 1, \ldots, 5$; $j = 1,2$) are presented in Table 7.1.

Basic concepts

Our first goal is to establish a concentration–mortality relationship. More specifically, we want to know the essential features of the function $\pi(x)$ of x, which represents the true or theoretical mortality proportion among a population of all insects exposed to the material at concentration x, or the probability that an insect dies after it has been exposed to concentration x. A simple plot of observed mortality proportion p_{ij}, vs. concentration x_i shows an increasing trend of mortality with concentration (Fig. 7.1). The

Table 7.1. Artificial data of mortality proportions in two batches each of 100 insects and pooled mortalities, at five concentration levels.

Conc. (mg l^{-1})	Mortalities		Pooled
0.01	0.19	0.25	0.220
0.10	0.20	0.25	0.225
1.0	0.21	0.27	0.240
10.0	0.45	0.56	0.505
100.0	0.80	0.91	0.855

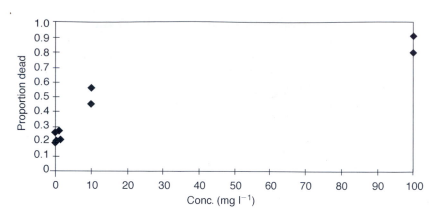

Fig. 7.1. Observed mortality proportions vs. concentration values for the data from Table 7.1.

pattern of the data suggests that the concentration scale be changed to a logarithmic scale which will result in a more orderly scatter.

We wish to complete this scatter plot by sketching or fitting a curve freehand through, or close to, the plotted points – but how should we define such a curve, and then judge its accuracy? The simplest procedure for dealing with these problems begins with our adoption of a reasonable assumption that the curve of the true mortality function $\pi(x)$ belongs to a definite designated family of curves. This is called the modelling assumption and the designated family of curves is called the model. Then, we will choose that particular curve within the model which is, in an appropriate sense, closest of all curves in the model, to the plotted points. This procedure is called fitting the model and the curve thus chosen is called the fitted or estimated curve of $\pi(x)$, written as $\hat{\pi}(x)$, so as to distinguish it from the theoretical mortality function $\pi(x)$.

All models considered use monotonically increasing curves to express $\pi(x)$, that is, $\pi(x)$ always increases as concentration x increases. The term $\pi(0)$ represents the true or theoretical mortality proportion among all insects that have not been exposed to the substance, or the untreated control insects. Note that as insects can die without receiving an entomopathogenic material on one hand, on the other hand, the microbe or nematode might not be able to kill all insects even at extremely high concentrations. Therefore, the mortality against concentration curve could reach a plateau below 100% mortality. The notation $\pi(\infty) = \lim_{x \to \infty} \pi(x)$ represents the theoretical mortality as concentration tends to infinity. Thus, $0 \leq \pi(0) < \pi(\infty) \leq 1$. Henceforth, for simplicity, we shall assume that $\pi(\infty) = 1$. Since the function $\pi(x)$ increases steadily from $\pi(0)$ to $\pi(\infty)$, it is interpretable as the cumulative representation of a probability distribution function, called the mortality distribution. The function $\pi(x)$ is also viewed as the cumulative probability distribution function of a random variable which represents the tolerance of an insect under study to the microbial agent concentration, which, if exceeded, causes the insect to die. We may also think of concentration x as a function of mortality, π. A simple notation like $x(\pi)$ suggests this, but it is customary to speak in terms of a percentage 100π rather than a mortality, π, and to express it as LC (lethal concentration) rather than x. Therefore, rather than $x(0.5)$ we write LC_{50} which is defined as the lethal concentration (or lethal dose, LD_{50}) required to kill 50% of the insect population or to have a 0.5 probability of killing a random insect.

The control-adjusted and power models

A commonly used model for describing the concentration–mortality relationship is expressed as:

$$\pi(x) = F(\alpha + \beta x) \qquad (x \geq 0) \tag{1}$$

where F represents a cumulative probability distribution function (for instance, the standard normal distribution that is commonly denoted by Φ); α and β are unknown parameters.

Let $F^{-1}(\pi)$ denote the inverse function of F. This inverse function gives a unique number, η, that corresponds to a specific cumulative mortality probability π, that is, $F(\eta) = \pi$. Thus, model (1) can also be written as:

$$F^{-1}(\pi) = \alpha + \beta x \qquad (x \geq 0) \tag{2}$$

which is a straight-line model for $F^{-1}(\pi)$, with unknown intercept α and slope β.

The commonly used probability distributions F are often named in terms of their implied inverse transformations, F^{-1}. Commonly used inverse functions are: the probit, $\Phi^{-1}(\pi) = 100\pi$ percentile of the standard normal distribution, for example, $\Phi^{-1}(0.50) = 0$, $\Phi^{-1}(0.95) = 1.64$, and $\Phi^{-1}(0.99) = 2.33$; the logit, $F^{-1}(\pi) = \log(\pi/(1 - \pi))$, and the complementary log–log (cll), $F^{-1}(\pi) = \log(-\log(1 - \pi))$. Their graphs look similar, although that of cll lacks the symmetry about $\pi = 0.5$ which the other two curves exhibit. Employing the probit or the logit function in (2) leads to the probit or logit regression model for the mortality, π.

Concentration–mortality models often relate the true mortality proportion, π, to some transformation of x, such as $\log(x)$, where log denotes the natural logarithm.[1] In this case, model (1) is expressed as:

$$\pi(x) = F(\alpha + \beta \log(x)) \qquad (x > 0) \tag{3}$$

In model (3), it is assumed that the mortality of the untreated control insects (at zero concentration) is zero. More generally, let γ denote the control mortality ($0 \leq \gamma < 1$), then model (3) is extended by including the parameter γ and can be expressed as:

$$\pi(x) = \gamma + (1 - \gamma)F(\alpha + \beta\log(x)) \qquad (x{>}0) \qquad (4)$$

Note that for a given γ, the expression

$$\pi_{adj}(x) = (\pi(x) - \gamma)/(1 - \gamma) \qquad (5)$$

represents Abbott's (1925) control-adjusted mortality, which is the theoretical mortality proportion caused by a concentration x of the microbial agent alone, that is, adjusted for the control mortality ($0 \leq \pi_{adj}(x) < 1$). Thus, model (4) can be rewritten as:

$$\pi_{adj}(x) = F(\alpha + \beta\log(x)) \qquad (x > 0) \qquad (6)$$

The inverse transformation F^{-1} produces

$$F^{-1}(\pi_{adj}) = \alpha + \beta\log(x) \qquad (7)$$

which, for a given γ, is a straight-line model for $F^{-1}(\pi_{adj})$, with unknown intercept α and slope β. We shall refer to (4), (6) or (7) as the control-adjusted model.

Another concentration–mortality model is given by

$$\pi(x) = F(\alpha + \beta.x^\lambda) \qquad (x \geq 0) \qquad (8)$$

This model relates the mortality, $\pi(x)$, to a power transformation of the concentration, x^λ ($\lambda > 0$). In this model $\pi(0) = F(\alpha)$. The inverse transformation, F^{-1}, produces

$$F^{-1}(\pi) = \alpha + \beta.x^\lambda \qquad (9)$$

which, for a given λ, is a straight-line model for $F^{-1}(\pi)$ with unknown intercept α and slope β. We shall refer to (8) or (9) as the power (λ) model.[2]

Both control-adjusted and power models are non-linear models in their parameters for describing the relationship between $\pi_{adj}(x)$ or $\pi(x)$ and concentration x. All these functions $\pi(x)$ have positively sloping S-shaped graphs. A detailed description of the control-adjusted and power models is provided in Finney (1971); see also Robertson and Preisler (1992). Although the probit and logit functions differ each from the other, their modelling results are usually quite similar. Therefore, the following analyses of concentration–mortality data are based on the probit function, $\Phi^{-1}(\pi)$.

Model fitting

Which of these possible models fits the above data best? If the control-adjusted model $\Phi^{-1}(\pi_{adj}) = \alpha + \beta\log(x)$ were appropriate, then, for some value of γ, the probit values of the observed control-adjusted mortalities, $\Phi^{-1}((p_i - \gamma)/(1 - \gamma))$, plotted against logarithm concentration values, $\log(x_i)$, ($i = 1, \dots ,k$) should resemble a straight-line relationship. Figure 7.1

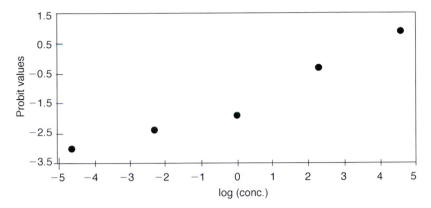

Fig. 7.2. Probit values of observed pooled control-adjusted mortalities, $\Phi^{-1}((p_i-0.219)/0.781)$, vs. logarithm concentration values, $\log(x_i)$.

clearly suggests a positive control mortality γ of around 0.2. Since p_1 (which corresponds to concentration 0.01 mg l^{-1}) is 0.22, the probit values $\Phi^{-1}((p_i - \gamma)/(1 - \gamma))$ cannot be calculated for a value of γ as large as 0.22. Therefore, the probit values are calculated for $\gamma = 0.219$ and are plotted against $\log(x_i)$ (Fig. 7.2). The scatter plot in Fig. 7.2 resembles a straight-line relationship except, perhaps, for the mortality observation at $\log(x) = 0$. Thus, the control-adjusted model appears to be a reasonable choice for modelling our concentration–mortality data. Similarly, an approximately straight-line relationship indicated by a scatter plot of probit value $\Phi^{-1}(p_i)$ against power transformation of the concentration x_i^λ, for some λ, would suggest a power (λ) model.

The maximum likelihood estimation (mle) procedure will be used for estimating the unknown model parameters. In order to use the likelihood method, certain distributional assumptions about the mortality observations are necessary. It is assumed that the observed number of dead among n insects in batch j exposed to concentration x_i follows a binomial distribution, with a theoretical mortality of π_i, that is, $d_{ij} \sim \text{Bin}(\pi_i, n)$ ($i = 1, \dots, k$; $j = 1, \dots, r$); all mortality observations d_{ij} are statistically independent. Hence, for the pooled mortality observations it follows that $\Sigma_{j=1}^{r} d_{ij} \sim \text{Bin}(\pi_i, rn)$ ($i = 1, \dots, k$). The likelihood function is an expression for the joint probability of all observations $\{d_{ij}\}$, with π_is expressed in terms of its unknown parameters. The mle method is based on maximizing the likelihood function or, equivalently, the log-likelihood function with respect to the unknown parameters. Substitution of the maximum likelihood estimators of the parameters yields the estimated curve, $\hat{\pi}(x)$, which is that particular curve within the model that makes our data the most probable data set to occur.

For binomial observations, the log likelihood function is given by

$$L(\mathbf{p};\boldsymbol{\pi}) = rn\Sigma_{i=1}^{k} [p_i\log(\pi_i) + (1-p_i)\log(1-\pi_i)]$$

where $\mathbf{p} = (p_1, \ldots, p_k)$ and $\boldsymbol{\pi} = (\pi_1, \ldots, \pi_k)$ represent the vectors of observed pooled mortality data and theoretical mortality, respectively.[3] Thus, subject to the constraints implied by the model equation, one has to maximize the function $L(\mathbf{p};\boldsymbol{\pi})$ where the π_is are expressed in terms of the parameters. We shall use the notation $L(\mathbf{p};\hat{\boldsymbol{\pi}})$ for the maximum value of $L(\mathbf{p};\boldsymbol{\pi})$ under the $\pi(x)$ model. Various computer programs are available for calculating the estimated model parameters; the computational details need not concern us, as we will focus on the steps for carrying out the fitting process and for drawing conclusions. These programs also provide the estimated variances and covariances of the parameter estimates. The square roots of the estimated variances are the standard errors (SES) of the parameter estimates, and they indicate the accuracy of the estimates. The t-value of a parameter, which is the ratio of its estimated value to its standard error, is computed and the corresponding significance value (P value) is provided; a small P value provides an evidence to reject the hypothesis that the parameter equals zero. Non-rejection of the hypothesis that the slope, β, equals zero is crucial, since in this case one would infer that the data fail to reveal any association between mortality and concentration level.

Lack-of-fit tests

The lack of fit of the $\pi(x)$ model can now be tested, by fitting a full model that allows the mortalities $\pi(x)$ for distinct concentrations x to have any values, without modelling constraints, which is certainly a model we believe to be correct. The estimated mortalities obtained by the full model are just the observed pooled mortalities at the various concentrations (in our example: $\mathbf{p} = [0.22, 0.225, 0.24, 0.505, 0.855]$). The difference between the log-likelihoods of the fitted model and of the full models, $L(\mathbf{p};\hat{\boldsymbol{\pi}}) - L(\mathbf{p};\mathbf{p})$, is the log-likelihood ratio statistic that provides a comparative measure of the two models. This difference in log-likelihoods, multiplied by -2, or LR = $2[L(\mathbf{p};\mathbf{p}) - L(\mathbf{p};\hat{\boldsymbol{\pi}})]$ is also known as the deviance difference.

A closely related measure of the model discrepancy, provided by Pearson's statistic, is defined as:

$$X^2 = \Sigma_{i=1}^k (p_i - \hat{\pi}_i)^2/\text{vâr}(p_i) \qquad (10)$$

where p_i and $\hat{\pi}_i$ denote the observed pooled mortality and the fitted mortality at concentration x_i, respectively, and $\text{vâr}(p_i) = \hat{\pi}_i(1 - \hat{\pi}_i)/(nr)$ is the estimated variance of p_i under the assumed binomial distribution. Thus, X^2 is a sum of the squared differences, $(p_i - \hat{\pi}_i)^2$, between the observed pooled and fitted mortalities, weighted by taking into account the relative reliabilities of the p_is. Under certain conditions, when the sample size is large and if the hypothesized model $\pi(x)$ is correct, each of the LR and X^2 statistics follows a χ^2 distribution with $(k - q)$ degrees of freedom (df), where k and q represent the number of parameters included in the full model and in the

Table 7.2. The control-adjusted model fitted to the data from Table 7.1.

Log likelihood value	-543.84		
Parameter	Estimate	SE	P value
Intercept α	-1.6597	0.2450	0.0001
Slope β	0.5586	0.0649	0.0001
Control mortality γ	0.2172	0.0194	

Goodness-of-fit tests			
Statistics	Value	df	P value
X^2	0.4080	2	0.8155
LR	0.4107	2	0.8144

fitted model, respectively. This distribution is denoted by $\chi^2(k - q)$ (in our example, $k = 5$, $q = 3$). Thus, the significance of each of these test statistics is judged in terms of a $\chi^2(k - q)$ distribution. Small significance levels (*P* values) indicate lack of fit of the model and throw doubt on to estimates obtained with the model.

Use of the SAS PROBIT procedure for fitting the control-adjusted model to our mortality data $\{d_{ij}\}$ produced the results presented in Table 7.2. Notice that the SAS PROBIT procedure automatically pools the mortality data over replicates at a specified concentration.

For our data, the values of the LR and X^2 lack-of-fit statistics are LR = 0.4107, and $X^2 = 0.4080$. The number of df of each of these statistics equals $5 - 3 = 2$. Both lack-of-fit statistics are not significant in terms of a $\chi^2(2)$ distribution, showing no evidence of departure of the control-adjusted model.

The estimated mortality is then given by

$$\hat{\pi}(x) = 0.2172 + 0.7828*\Phi(-1.6597 + 0.5586*\log(x)) \qquad (x > 0)$$

Although the control-adjusted model appears to fit our data satisfactorily, we will also consider the fit of the power model expressed as: $\Phi^{-1}(\pi) = \alpha + \beta x^\lambda$. Experimentation with plots of probit values, $\Phi^{-1}(p_i)$ against the power transformation of concentration, x_i^λ, for several different values of λ, suggests that λ may be around 0.5 (Fig. 7.3).

Use of the SAS PROBIT procedure for fitting the power (0.5) model to our data produced the results presented in Table 7.3.

Since we assumed that the power λ is known ($\lambda = 0.5$), only the parameters α and β were estimated. Each of the lack-of-fit tests depends on 3 df and provides a moderate indication of lack of fit for the power (0.5) model (*P* = 0.086). Fitting the power (λ) model over a range of λ values reveals that the likelihood function is maximized for $\lambda = 0.38$. If the SAS PROBIT procedure is again used, to fit the power (0.38) model to our data, the results presented in Table 7.4 are produced. Both lack-of-fit tests show

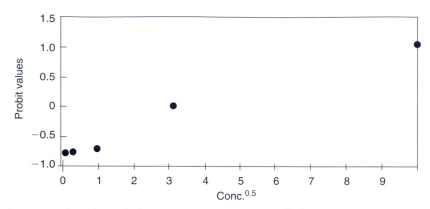

Fig. 7.3. Probit values of observed pooled mortalities, $\Phi^{-1}(p_i)$, vs. square-root concentration values, $x_i^{0.5}$.

Table 7.3. The power (0.5) model fitted to data from Table 7.1.

Log likelihood value	−546.92		
Parameter	Estimate	SE	P value
Intercept α	−0.7796	0.0817	0.0001
Slope β	0.1912	0.0193	0.0001
Goodness-of-fit tests			
Statistics	Value	df	P value
χ^2	6.5907	3	0.0862
LR	6.5747	3	0.0868

Table 7.4. The power (0.38) model fitted to data from Table 7.1.

Log likelihood value	−545.59		
Parameter	Estimate	SE	P value
Intercept	−0.9013	0.0606	0.0001
Slope	0.3447	0.0231	0.0001
Goodness-of-fit tests			
Statistics	Value	df	P value
χ^2	3.838	3	0.2794
LR	3.912	3	0.2714

non-significance ($P = 0.27$), indicating no evidence to reject the power model with $\lambda = 0.38$.

Notice that the results presented in Table 7.4 were calculated for $\lambda = 0.38$. Thus, two parameters (α and β) were estimated and each of the lack-of-fit tests was based on a $\chi^2(3)$ distribution. In general, the power λ is an unknown parameter; therefore, a power model with three parameters should be fitted to the data, and the uncertainties in the estimates $\hat{\alpha}, \hat{\beta}$, and $\hat{\lambda}$, should be taken into account in all calculations. In this case, each of the lack-of-fit tests is based on a $\chi^2(2)$ distribution rather than $\chi^2(3)$. Eaves and Marcus (1997) used an extended computational procedure for estimating the variances and covariances of $\alpha, \hat{\beta}$, and $\hat{\lambda}$.

When the lack-of-fit tests are significant, the experimenter should seek possible reasons for the model inadequacy. Residuals are usually defined by the difference of each observation from its fitted value and are used to diagnose model adequacy. A crucial feature of the binomial distribution assumption is that $\text{var}(p_{ij}) = \pi_i(1 - \pi_i)/n$ $(i = 1, \dots ,k; j = 1, \dots ,r)$. These variances have small values when π_i is around the extremes (0 or 1) and larger values when it is around 0.5. The residuals are, therefore, divided by the standard errors (SEs) of the observed mortalities, so that we are not distracted by large swings in residual magnitude that are due merely to differing mortalities. The terms $R_{ij} = (p_{ij} - \hat{\pi}_i)/[\text{vâr}(p_{ij})]^{1/2}$, where $\text{vâr}(p_{ij}) = \hat{\pi}_i(1 - \hat{\pi}_i)/n$ $(i = 1, \dots ,k; j = 1, \dots ,r)$ are called Pearson residuals and are used to diagnose model violations, as discussed by Preisler (1988) (see also Robertson and Preisler, 1992). Preisler (1988) gave a few reasons for the model to fit poorly:

1. An extreme or an outlier observation that departs too strongly from most of the data may result in a poor fit; such an observation may be detected, since its Pearson residual would be large (larger than 2 in its absolute value). Exclusion of an outlier observation from the analysis results in smaller values of the lack-of-fit statistics, and may eliminate any evidence of departure of the model.

2. The assumptions of the $\pi(x)$ model are incorrect. Patterns obtained when Pearson residuals are plotted against $\log(x_i)$ or x_i^λ (for a given value of λ) indicate that the $\hat{\pi}(x)$ curve does not fit the data. The experimenter should then try to use a more advanced model that would improve the residuals pattern. In our example, plots of Pearson residuals, R_{ij}, for each of the control-adjusted and power (0.38) models, against logarithm of the concentration, $\log(x_i)$, show no obvious trend in the discrepancies between the data and the model predictions (Fig. 7.4).

3. Sometimes, an important variable that affects mortality is omitted from the $\pi(x)$ model, resulting in a poor fit of the model. For example, insect weights may have an effect on mortality and, therefore, should be included in the model. This type of lack of fit would be revealed when a scatter plot of Pearson residuals against weights showed an obvious trend.

4. The presence of extra-binomial variability or overdispersion: this could arise when several batches of insects are run at each concentration, if there were differences in mortality probability among the several replicates at a given concentration. These differences could indicate that fluctuations of some undentified influential conditions had occurred despite the

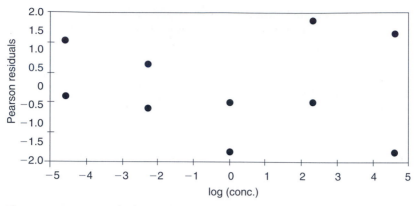

Fig. 7.4. Pearson residuals, R_{ij}, for the power (0.38) model vs. logarithm concentration values, $\log(x_i)$.

experimenters' best efforts to prevent it. In this case, the observed mortality data $\{p_{ij}\}$ exhibit extra-binomial variability or overdispersion: the variances of the pooled mortalities, $\text{var}(p_i)$, will tend to be larger than the binomial variances, $\pi_i(1 - \pi_i)/(nr)$. Thus, terms of the Pearson X^2 statistic, $(p_i - \hat{\pi}_i)/[\text{vâr}(p_i)]^{1/2}$, that are standardized by means of binomial estimated variances will have large values, resulting in large values of lack-of-fit test statistics.

A statistical method to assess whether proportion data $\{p_{ij}\}$ exhibit overdispersion across concentration levels was proposed by Cochran (1954). The method is based on a test statistic $\Sigma_i \Sigma_j \tilde{R}_{ij}^2$, where $\tilde{R}_{ij} = (p_{ij} - p_i)/[p_i(1 - p_i)/n]^{1/2}$ represents the Pearson residual for the full model. Under certain conditions, the significance of this test is judged in terms of a χ^2 distribution with $k(r - 1)$ df: low significance levels indicate that the data exhibit an extra-binomial variation. The mortality data from Table 7.1 produce $\Sigma_i \Sigma_j \tilde{R}_{ij}^2$, = 10.053 with 5 df, which carries moderate indication of extra-binomial variation ($P = 0.08$).

Statistical methods are available for fitting a concentration–mortality model, when mortality data exhibit extra-binomial variation across concentration levels. A commonly used method is based on extending the binomial distributional assumption by including an unknown dispersion parameter, ϕ, defined as $\phi = \text{var}(p_i)/[\pi_i(1 - \pi_i)/(nr)]$ for $i = 1, \dots, k$. If $\phi > 1$, the data are overdispersed, while $\phi = 1$ corresponds to binomial variation. The underdispersion case, $\phi < 1$ occurs less often in practice. When overdispersion is present, the form of the likelihood function for the mortality data $\{p_{ij}\}$ cannot be specified and the unknown model parameters are estimated by means of a quasi-likelihood theory. A quasi-likelihood function is constructed which is similar in nature to the likelihood function for estimating the unknown model parameters under the binomial distribution assumption.

Readers with an advanced statistical background are referred to the book of McCullagh and Nelder (1989) for the relevant theory. If overdispersion is present, the model parameter estimates obtained by the maximum quasi-likelihood procedure are equivalent to those obtained by the maximum likelihood method with the assumption that $\phi = 1$. However, the standard errors of all parameter estimates must be multiplied by a scale factor $\hat{\phi}^{1/2}$, where $\hat{\phi}$ denotes an estimate of ϕ. A useful criterion of overdispersion can be obtained by the use of $\hat{\phi} = \Sigma_i \Sigma_j R_{ij}^2/(kr - q)$, where the R_{ij}s represent the Pearson residuals for the fitted model. The lack of fit of the $\pi(x)$ model is judged by a scaled Pearson's statistic defined as $X^2/\hat{\phi}$, rather than by X^2. Under certain conditions, the significance of $X^2/\hat{\phi}$ is judged in terms of a χ^2 ($k - q$) distribution. For our data, the power (0.38) model yields $\hat{\phi} = 13.873/7 = 1.981$, which is larger than unity. The fitting of the power (0.38) model produces $X^2/\hat{\phi} = 3.838/1.981 = 1.937$. The significance of this ratio is judged in terms of a $\chi^2(2)$ distribution and shows not the slightest evidence against the power (0.38) model with $\phi > 1$. The SES of the parameter estimates are multiplied by $\sqrt{1.981}$ but, despite this adjustment, the *t*-values remain highly significant as in the case of $\phi = 1$ presented in Table 7.4.

If one batch ($r = 1$) were run at each concentration, it would be impossible to test for extra-binomial variation and to distinguish between true overdispersion and a systematically inadequate $\pi(x)$ model. In this case, if X^2 were judged to be significant, the experimenter should first try to identify reasons for the poor fit and to use a more advanced model that would improve the residual pattern. Alternatively, the estimated variances and covariances of the model parameter estimates could be multiplied by a heterogeneity factor, $X^2/(k - q)$, as an approximate way of taking into account an extra-binomial variation.

Estimation of mortality

The control-adjusted and the power (λ) models are both candidates for describing the concentration–mortality relationship. The respective fitted curves are:

$$\hat{\pi}(x) = \hat{\gamma} + (1 - \hat{\gamma})\Phi(\hat{\alpha} + \hat{\beta}\log(x)) \qquad (x > 0) \text{ control-adjusted model}$$
$$\hat{\pi}(x) = \Phi(\hat{\alpha} + \hat{\beta}x^\lambda) \qquad (x \geq 0) \text{ power } (\lambda) \text{ model}$$

Therefore, in our example, we are free to choose between the two models according to our convenience. One issue of convenience is whether we are more interested in the adjusted mortality $\pi_{adj}(x)$ or in the unadjusted mortality $\pi(x)$, and this may be a scientific rather than a statistical issue. We shall illustrate inferences about the adjusted mortalities only when discussing the control-adjusted model and about the (unadjusted) mortalities only when discussing the power model.

Provided that randomness was effectively achieved in the experimental procedure, the rules of probability can be used to calculate a 95% (say) confidence interval for estimating the unknown $\pi(x)$, for any given value of x. A complete analysis of the vertical accuracy of a fitted curve will be to present, on a common plot, the three curves $\hat{\pi}(x)$, $\hat{\pi}_L(x)$ and $\hat{\pi}_U(x)$, representing, respectively, the estimated curve, and the lower and upper confidence limits. The conventional interpretation of $\hat{\pi}_L(x)$ and $\hat{\pi}_U(x)$ at a given x is that they have a 95% chance of capturing the true mortality $\pi(x)$.

For a given x-value, a 95% confidence interval of $\pi_{adj}(x)$ is created by obtaining a symmetric confidence interval for $F^{-1}(\pi_{adj}(x)) = \alpha + \beta \log(x)$, of the type

$$\alpha + \hat{\beta}\log(x) \pm 1.96\text{SE}[\hat{\alpha} + \hat{\beta}\log(x)] \tag{11}$$

and transforming it to an asymmetric confidence interval by applying Φ. Thus, a confidence interval of $\pi_{adj}(x)$ for a given x is given by:

$$\Phi(\hat{\alpha} + \hat{\beta}\log(x) \pm 1.96\text{SE}[\hat{\alpha} + \hat{\beta}\log(x)]) \tag{12}$$

Table 7.5 shows the 95% lower and upper confidence limits for $\pi_{adj}(x_i)$ ($i = 1, \ldots ,5$), for the control-adjusted model. We may present 95% confidence bands on $\pi_{adj}(x)$ as a function of $\log(x)$ (Fig. 7.5). It is important to note that the graph in Fig. 7.5 is based on calculation of the upper and lower confidence limits at each value of the concentration x. Simultaneity issues of constructing 95% confidence bands on $\pi_{adj}(x)$ for all $x > 0$ are not faced here; to do that, adjustment for multiplicity would be required, which could be achieved by expanding the critical point 1.96 in eq. (11) by appropriate multipliers.

Similarly, for a given value of λ, an asymmetric confidence interval for $\pi(x) = F(\alpha + \beta x^\lambda)$ is created by:

$$\Phi(\hat{\alpha} + \hat{\beta}x^\lambda \pm 1.96\text{SE}[\hat{\alpha} + \hat{\beta}x^\lambda]) \tag{13}$$

Table 7.6 shows the 95% lower and upper confidence limits for $\pi(x_i)$ ($i = 1, \ldots ,5$) for the power (0.38) model. Figure 7.6 shows the corresponding 95% confidence bands for $\pi(x)$ as a function of $\log(x)$.

Table 7.5. Ninety-five per cent confidence limits for control-adjusted mortality $\pi_{adj}(x)$; control-adjusted model fitted to the data from Table 7.1.

	Conc. (mg l^{-1})				
	0.01	0.10	1.0	10.0	100.0
$\hat{\pi}_{adj,L}$	0.0000	0.0001	0.0162	0.2716	0.7586
$\hat{\pi}_{adj}$	0.0000	0.0016	0.0485	0.3547	0.8193
$\hat{\pi}_{adj,U}$	0.0007	0.0144	0.1191	0.4447	0.8725

$\hat{\pi}_{adj}$ is the estimated control-adjusted mortality; $\hat{\pi}_{adj,L}$ is the lower confidence limit; $\hat{\pi}_{adj,U}$ is the upper confidence limit.

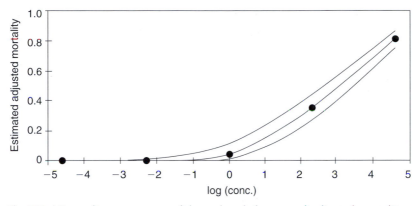

Fig. 7.5. Ninety-five per cent confidence bands for control-adjusted mortality $\pi_{adj}(x)$, as a function of concentration logarithm $\log(x)$ (control-adjusted model).

Table 7.6. Ninety-five per cent confidence limits for mortality $\pi(x)$; power (0.38) model fitted to data from Table 7.1.

	Conc. (mg l^{-1})				
	0.01	0.10	1.0	10.0	100.0
$\hat{\pi}_L$	0.1697	0.1937	0.2578	0.4348	0.8177
$\hat{\pi}$	0.2000	0.2243	0.2889	0.4703	0.8604
$\hat{\pi}_U$	0.2333	0.2575	0.3216	0.5060	0.8998

$\hat{\pi}$ is the estimated mortality; $\hat{\pi}_L$ is the lower confidence limit; $\hat{\pi}_U$ is the upper confidence limit.

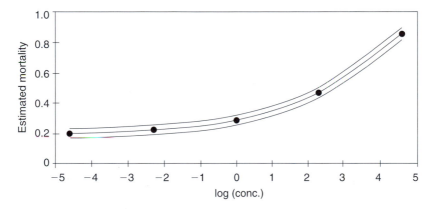

Fig. 7.6. Ninety-five per cent confidence bands for mortality $\pi(x)$ as a function of logarithm concentration $\log(x)$ (power (0.38) model).

Estimation of lethal concentrations

The efficacy of a microbial agent is traditionally measured by the LC_{50}, the concentration required to kill 50% of the pest population. Sometimes, other lethal concentrations might be of interest, for instance the lethal concentration LC_{90} required to produce a mortality of 90%. More generally, for a given value of π ($0 < \pi < 1$), let $LC_{100\pi}$ represent the lethal concentration required to achieve a 100π% mortality of the pests.

For the control-adjusted model, $LC_{100\pi}$ is obtained by solving the following equation for x:

$$\gamma + (1 - \gamma)\Phi(\alpha + \beta\log(x)) = \pi$$

The solution is:

$$\log(x) = (\Phi^{-1}(\pi_{adj}) - \alpha)/\beta \tag{14}$$

where $\pi_{adj} = (\pi - \gamma)/(1 - \gamma)$. Applying an exponential transformation to both sides of (14) produces $LC_{100\pi}$, which is expressed as:

$$LC_{100\pi} = \exp((\Phi^{-1}(\pi_{adj}) - \alpha)/\beta) \tag{15}$$

The estimated $LC_{100\pi}$, denoted by $\hat{LC}_{100\pi}$, is obtained by substituting $\hat{\alpha}$, $\hat{\beta}$ and $\hat{\gamma}$ in expression (15). [Note that $\pi_{adj} = (\pi - \hat{\gamma})/(1 - \hat{\gamma})$].

For our mortality data, $\hat{\alpha} = -1.6507$, $\hat{\beta} = 0.5586$ and $\hat{\gamma} = 0.2172$. If $\pi = 0.5$, then $\pi_{adj} = (0.50 - 0.2172)/0.7828 = 0.361$. Thus, the logarithm of $\hat{LC}_{50}$ equals $(\Phi^{-1}(0.36) + 1.6597)/0.5586 = 2.3267$, or $\hat{LC}_{50} = 10.244$ mg l^{-1}.

The SAS PROBIT procedure calculates the estimated lethal concentrations $\hat{LC}_{100\pi^*}$ required to produce a given adjusted mortality of π^*, for various values of π^* [Note that $\hat{LC}_{100\pi^*}$ causes a mortality of 100π%, with $\pi = \hat{\gamma} + (1 - \hat{\gamma})\pi^*$]; it also calculates confidence limits for $LC_{100\pi^*}$. The confidence limits are based on Fieller's theorem for assessing bounds on the error of ratios of estimates (Fieller, 1940; also see Finney, 1971). Table 7.7 gives the 95% confidence limits for $LC_{100\pi^*}$ for $\pi^* = [0.1, \ldots, 0.9]$.

Table 7.7. Ninety-five per cent confidence limits for lethal concentration $LC_{100\pi^*}$; control-adjusted model fitted to data from Table 7.1.

	π^*								
	0.1	0.2	0.3	0.4	0.5	0.6	0.7	0.8	0.9
$\hat{LC}_L$	0.79	2.14	4.35	7.84	13.3	22.1	36.3	62.2	124
$\hat{LC}_{100\pi^*}$	1.97	4.33	7.63	12.4	19.5	30.7	49.9	88.1	193
$\hat{LC}_U$	3.58	6.97	11.4	17.6	26.9	42.4	71.7	138	363

π^* = control-adjusted mortality; $\hat{LC}_{100\pi^*}$ is the estimated lethal concentration achieving adjusted mortality π^*; $\hat{LC}_L$ is the lower confidence limit; $\hat{LC}_U$ is the upper confidence limit.

Table 7.8. Ninety-five per cent confidence limits for lethal concentration $LC_{100\pi}$; power (0.38) model fitted to data from Table 7.1.

					π			
	0.2	0.3	0.4	0.5	0.6	0.7	0.8	0.9
$\hat{LC}_L$	.	0.58	3.62	9.54	18.86	33.04	55.88	100.1
$\hat{LC}_{100\pi}$	0.01	1.26	5.26	12.54	24.07	41.93	71.14	128.6
$\hat{LC}_U$	0.14	2.20	7.33	16.49	31.28	54.70	93.79	171.6

π = mortality; $\hat{LC}_{100\pi}$ is the estimated lethal concentration achieving mortality π; $\hat{LC}_L$ is the lower confidence limit; $\hat{LC}_U$ is the upper confidence limit.

We now consider a power model with known λ. The lethal concentration $LC_{100\pi}$ which produces a given mortality π is obtained by solving the equation $\Phi^{-1}(\pi) = \alpha + \beta x^\lambda$ for x. The solution is $x^\lambda = (\Phi^{-1}(\pi) - \alpha)/\beta$, or

$$LC_{100\pi} = [(\Phi^{-1}(\pi) - \alpha)/\beta]^{1/\lambda} \qquad (16)$$

For a given λ, the estimated lethal concentration, $\hat{LC}_{100\pi}$, is obtained by substituting the estimated values of $\hat{\alpha}$ and $\hat{\beta}$ in eq. (16). Table 7.8 shows the 95% confidence limits for $LC_{100\pi}$ for $\pi = [0.2, \dots, 0.9]$.

Analysis and comparison of the efficacies of two microbial agents

We may wish to analyse and compare the efficacies or killing powers of two microbes, say standard, S and test, T, applied to two independent samples of similar insects. Let $LC_{50;S}$ and $LC_{50;T}$ represent the median lethal concentrations of the standard and the test microbes, respectively, which cause 50% mortality among the insect populations. The relative efficacy of the test substance in terms of a standard substance is often measured by $LC_{50;S}/LC_{50;T}$. When this ratio is larger than one, the test substance requires a lower concentration than the standard for achieving a mortality of 50%; therefore, in this case, the killing power of the test microbe is higher than that of the standard.

It may happen that, although comparisons based on $\hat{LC}_{50;S}$ and $\hat{LC}_{50;T}$ lead to the conclusion that the test substance is more effective than the standard, comparisons based on $\hat{LC}_{90;S}$ and $\hat{LC}_{90;T}$ would lead to the conclusion that the efficacies of the two microbes were similar. Therefore, comparison of the efficacies only on the basis of their $\hat{LC}_{50}$s may miss some information. For a given value of π ($0 < \pi < 1$), let $LC_{100\pi;S}$ and $LC_{100\pi;T}$ represent the lethal concentrations of a standard and a test substance, respectively, each achieving a mortality of $100\pi\%$. In general, it is of interest to estimate the ratio $\rho_\pi = LC_{100\pi;S}/LC_{100\pi;T}$, which represents the relative efficacy of a test substance in terms of a standard, as a function of π.

We consider a constructed example of data obtained in an experiment

Table 7.9. Artificial data of mortality proportions in two batches, each of 100 insects, for five concentrations each of test and standard (T and S) substances, and a zero-concentration control.

Conc. (mg l⁻¹)	Test		Standard	
0	0.02	0.04	0.05	0.08
0.01	0.04	0.07	0.03	0.08
0.1	0.13	0.16	0.14	0.11
1	0.31	0.36	0.27	0.24
10	0.66	0.70	0.55	0.60
100	0.94	0.90	0.86	0.90

conducted with 24 batches of similar insects, each comprising 100 insects. The experiment was designed as follows: ten batches of insects were randomly exposed to a standard substance at five concentrations, with two batches at each concentration. Similarly, ten independent batches were exposed to a test substance at the same five concentrations, again with two batches at each concentration. In addition, four independent batches were used as untreated controls. After 6 days of exposure, mortality was recorded among insects in each of the 24 batches (Table 7.9).

The control-adjusted models describing the concentration–mortality relationship for each of the test and standard samples are expressed as:

Test sample: $\pi(x) = \gamma_T + (1 - \gamma_T)\Phi(\alpha_T + \beta_T \log(x))$

Standard sample: $\pi(x) = \gamma_S + (1 - \gamma_S)\Phi(\alpha_S + \beta_S \log(x))$

The results of fitting control-adjusted models separately to each of the test and the standard samples, are presented in Table 7.10. Lack-of-fit statistics are not significant for either of the two samples.

Table 7.11 presents the estimated lethal concentrations of the test and the standard substances, each achieving an adjusted mortality of π^*, and the estimated relative efficacy, $\hat{\rho}_{\pi^*}$, for $\pi^* = [0.1, \dots ,0.9]$. Table 7.11 shows that the relative efficacy, $\hat{\rho}_{\pi^*}$, is higher than 1 for all π^* and decreases from 2.60 to 1.74 as π^* increases from 0.10 to 0.90.

Indeed, it is reasonable to assume that $\gamma_T = \gamma_S = \gamma$, since the untreated control batches are taken from a pest population with a common control mortality of γ. Application of the SAS PROBIT procedure for fitting control-adjusted models that include five parameters $(\gamma, \alpha_T, \alpha_S, \beta_T, \beta_S)$ to the combined data from the two independent samples, produces the results presented in Table 7.12(a).

A little algebra reveals that the efficacy of the test substance relative to the standard is constant over all control-adjusted mortalities if, and only if, $\beta_T = \beta_S = \beta$. In this case, the relative efficacy is $\exp(\alpha_T - \alpha_S)/\beta$. The parallelism condition, $\beta_T = \beta_S = \beta$ can be tested once we fit a reduced model with four parameters $(\gamma, \alpha_T, \alpha_S, \beta)$ to the combined data from two inde-

Table 7.10. The control-adjusted models fitted to the test and standard samples data from Table 7.9.

(a) Test sample

Log likelihood value	−461.7401		
Parameter	Estimate	SE	P value
Intercept α_T	−0.4229	0.0764	0.0001
Slope β_T	0.3789	0.0275	0.0001
Control mortality γ_T	0.0354	0.0117	

Goodness-of-fit tests

Statistics	Value	df	P value
X^2	1.4884	3	0.6849
LR	1.4949	3	0.6835

(b) Standard sample

Log likelihood value	−490.3075		
Parameter	Estimate	SE	P value
Intercept α_S	−0.7847	0.1061	0.0001
Slope β_S	0.4065	0.0341	0.0001
Control mortality γ_S	0.0634	0.0117	

Goodness-of-fit tests

Statistics	Value	df	P value
X^2	1.9168	3	0.5898
LR	1.8939	3	0.5947

Table 7.11. Estimated lethal concentrations of test and standard substances and efficacy of the test substance relative to the standard; control-adjusted models are fitted to data from Table 7.9.

					π^*					
	0.1	0.2	0.3	0.4	0.5	0.6	0.7	0.8	0.9
$\hat{L}C_{100\pi^*;T}$	0.05	0.19	0.48	1.06	2.18	4.36	8.89	19.7	57.3
$\hat{L}C_{100\pi^*;S}$	0.13	0.47	1.16	2.46	4.88	9.37	18.1	37.8	100
$\hat{\rho}_{\pi^*}$	2.60	2.47	2.42	2.32	2.24	2.15	2.06	1.92	1.74

π^* = adjusted mortality; $\hat{L}C_{100\pi^*;T}$ is the estimated lethal concentration of the test substance; $\hat{L}C_{100\pi^*;S}$ is the estimated lethal concentration of the standard substance; $\hat{\rho}_{\pi^*}$ is the efficacy of the test substance relative to the standard.

pendent samples. An indication of how poorly the model with common slope, β, fits the data is the LR or deviance difference, which is twice the difference between the log-likelihood statistics of the model with parameters $(\gamma, \alpha_T, \alpha_S, \beta_T, \beta_S)$ and those of the reduced model with parameters $(\gamma, \alpha_T,$

Table 7.12. Fitting reduced control-adjusted models to the combined two-sample data from Table 7.9.

(a) Control-adjusted models with parameters (γ, α_T, α_S, β_T, β_S)

Log likelihood value	−953.312		
Parameter	Estimate	SE	P value
Intercept α_T	−0.4796	0.0746	0.0001
Intercept α_S	−0.7222	0.1043	0.0001
Slope β_T	0.3954	0.0272	
Slope β_S	0.3900	0.0381	
Control mortality γ	0.0506	0.0088	

Goodness-of-fit tests

Statistics	Value	df	P value
χ^2	3.2254	6	0.7801
LR	3.1494	6	0.7899

(b) Control-adjusted models with parameters (γ, α_T, α_S, β)

Log likelihood value	−953.323		
Parameter	Estimate	SE	P value
Intercept α_T	−0.4765	0.0710	0.0001
Intercept α_S	−0.7218	0.0821	0.0001
Slope β	0.3931	0.0272	
Control mortality γ	0.0507	0.0088	

Goodness-of-fit tests

Statistics	Value	df	P value
χ^2	3.2592	7	0.8600
LR	3.1693	7	0.8689

α_S, β). Its significance is judged in terms of a $\chi^2(1)$ distribution. The SAS PROBIT procedure for fitting these two control-adjusted models produces the results presented in Table 7.12.

The parallelism condition is not rejected, since the corresponding test statistic equals $2(953.323 - 953.312) = 0.02$, which is not significant. The estimated efficacy of the test substance relative to the standard is $\hat{\rho} = \exp((0.7218 - 0.4765)/0.3931) = 1.896$.

A similar analysis can be performed by fitting power models to data from two independent samples. Let α_S and β_S denote the parameters of a power (λ_S) model, fitted to the standard sample data. Similarly, let α_T and β_T denote the parameters of a power (λ_T) model fitted to the test sample data. The lethal concentrations of the test and the standard substances are expressed as: $LC_{100\pi;S} = [(\Phi^{-1}(\pi) - \alpha_S)/\beta_S]^{1/\lambda S}$ and $LC_{100\pi;T} = [(\Phi^{-1}(\pi) - \alpha_T)/\beta_T]^{1/\lambda T}$, respectively. If $\lambda_T = \lambda_S = \lambda$ and $\alpha_T = \alpha_S$, the relative efficacy of the test substance in terms of the standard is expressed as: $\rho = (\beta_S/\beta_T)^{1/\lambda}$. The calculations for the power models are omitted.

Experimental design

Experimental design involves the problem of choosing sample sizes and concentrations which will achieve specified precisions for the parameter estimates or confidence intervals for the parameters of interest (e.g. LC_{50} or LC_{90}) as narrow as desired. These choices must be based on previous experience with training samples. A pilot study should give a general idea of the various lethal concentrations for mortalities of interest, such as $\pi = 0.10$, 0.25, 0.50, 0.75. We may then use a plausible model and parameter estimates based on a pilot sample, for a further study.

Eaves and Marcus (1997) conducted a study of experimental design for the problem of estimating lethal concentrations;[4] they used the S-PLUS programming language to obtain confidence limits of LC_{50} and of LC_{90} for various sets of concentrations and sample sizes. The confidence limits were calculated for the control-adjusted and power models, for each of the nine possible combinations of total sample sizes, N_{tot} of 300, 900, 2700 insects, with uniform allocation of insects over each of the three sets of concentrations, [1,3,9,27,81], [20,40,80,160,320] and [1,4,16,64,256]. Thus N_{tot} = (number of different concentrations)*(number of batches tested at each concentration)*(number of insects in each batch). The first set of concentrations is centred near LC_{50}, as judged from the pilot study results presented in Table 7.8; the second list is centred near LC_{90}, and the third list spans both parameters. Thus we have nine designs classified in two ways: according to N_{tot} (three levels) and to concentration (three levels). All nine designs are intended to improve upon the pilot study, with regards to cost and estimate of precision level at some specific concentration allocation. For each combination of N_{tot} and concentrations set, the difference between the upper and lower confidence limits of LC_{50} and of LC_{90} were computed for the control-adjusted model and for the power model. The results show that, not surprisingly, the width of the confidence interval is roughly inversely proportional to the square-root of N_{tot} when the other design factors are held constant. This implies diminishing precision improvement per additional sampling unit as the sample size increases. Furthermore, targeting only one of LC_{50} and LC_{90} can result in appreciable loss of precision in the estimate of the other, with only modest precision gain for the targeted parameter.

The following are quick guidelines and recommendations for designing a quantal bioassay and for analysing concentration–mortality data.

1. Conduct a quantal bioassay with at least five reasonably widely spread substance concentrations that are supposed to provide a plausible range of mortality rates, for instance π = 0.25, 0.50, 0.75, 0.90, with three or more replicate batches at each concentration (to detect binomial overdispersion) and with a more or less uniform allocation of a total of around 750 or more insects (50 insects in each batch run at each concentration), while keeping in mind the law of diminishing returns on sample size. Thus, in practice we recommend the use of n = 50, r = 3 and k = 5.

2. Using these concentrations, estimate the upper and lower confidence limits that would be obtained for your parameter of interest (e.g. LC_{50}) by using each of several different total sample sizes, as illustrated above. Conduct a study of sufficient size to fulfil your precision requirements.

The following is a general sequence of analysis phases of quantal bioassay data.

1. Experiment with plots of response transformation ($F^{-1}(p)$ or $F^{-1}((p - \gamma)/(1 - \gamma))$, for some γ) vs. concentration transformation (x^λ or $\log(x)$) and look for a straight line, to gain an initial impression of a suitable model. Either the control-adjusted model or the power model may offer sufficient flexibility.

2. Fit a model, $\pi(x)$, to the concentration–mortality data using a statistical computer package such as SAS.

3. Test for lack-of-fit of the model. If fit is lacking, try to use a more advanced model that will improve the pattern of the Pearson residuals. If a high random variation within concentration levels is suspected, fit the model $\pi(x)$ to the overdispersed mortality proportion data.

4. Calculate estimated lethal concentrations $LC_{100\pi}$ and obtain lower and upper confidence limits for the true lethal concentrations.

When the experiment includes several substances, the above steps 1–4 should be taken for each substance.

5. When the killing efficacies of test and standard substances are compared, estimate the efficacy of the test substance relative to that of the standard. If the test substance required a lower concentration than the standard for achieving the same mortality level, its killing efficacy would be higher than that of the standard.

Finally, it should be noted that the data obtained from bioassays with microbial agents are usually considered to be valid for statistical analysis only if the control mortality does not exceed a specified value. For example, for bioassays with *Bacillus thuringiensis* it is required that control mortality is below 10%; otherwise, the bioassay data is eliminated from further analysis.

Bioassay with Continuous Response to Concentration

In bioassays recording sublethal effects, a continuous response of the insects, such as body weight, is measured. The following is a constructed example of data from a bioassay with a continuous response.

Similar insects were exposed to one of five concentrations, say, 0.01, 0.10, 1.0, 10.0 and 100.0 mg l^{-1}, of a standard microbial preparation or to a zero-concentration preparation (untreated control). Twenty-four batches, each of 50 insects were used, with four batches randomly assigned to each of the five concentrations and four batches being untreated. Weight measurements of batches of insects were recorded after 6 days of exposure. The weight observations, the means and the variances at each concentration are presented in Table 7.13a. A similar experiment designed with a test microbial preparation, at the same concentrations as the standard, yielded the data presented in Table 7.13b.

We first consider the data obtained with the standard preparation. We wish to formulate the relationship between weight and concentration, that is, to establish a function, $\mu(x)$ of x, which represents the true or theoretical weight of a batch of insects exposed to the standard preparation at concentration x.

Table 7.13. Weight measurements (g) of batches of insects, means and variances for five concentrations (mg l^{-1}) of each of the standard and the test substances and for a zero-concentration control.

(a) Standard substance

Conc.	Weight				Mean	Variance
0	35.41	37.83	44.17	50.21	41.902	44.243
0.01	35.02	37.35	46.29	48.64	41.825	44.238
0.10	30.78	38.38	40.22	45.22	38.650	35.880
1	20.24	24.56	30.16	34.41	27.342	38.692
10	10.21	15.39	18.79	21.51	16.475	23.713
100	1.23	2.37	3.81	4.59	3.00	2.238

(b) Test preparation

Conc.	Weight				Mean	Variance
0	34.21	40.39	47.58	50.60	43.195	54.220
0.01	31.79	38.81	42.18	48.02	40.150	46.106
0.10	23.38	30.62	34.17	37.53	31.425	36.725
1	14.58	17.52	20.26	28.64	20.250	36.664
10	1.02	2.52	3.31	5.79	3.160	3.976
100	1.02	0.68	0.24	0.76	0.675	0.105

Let y_{ij} denote the observed weight of the insects in batch j, exposed to the standard preparation at concentration x_i (i = 1, ... ,k; j = 1, ... ,r), where k denotes the number of distinct concentrations and r is the number of replicates at each concentration (in our example, k = 6 and r = 4). The sample mean weight at concentration x_i is denoted by $\bar{y}_i$ and equals $\sum_{j=1}^{r} y_{ij}/k$; its theoretical or true mean is μ_i. The sample variance bewteen batches at concentration x_i is denoted by s_i^2 and equals $\sum_{j=1}^{r} (y_{ij} - \bar{y}_i)^2/(k-1)$.

For a preliminary assessment of the standard sample data, the weights $\{y_{ij}\}$ were plotted against concentrations $\{x_i\}$ to provide an initial impression, which was of a trend of decreasing weight with increasing concentration, with steadily diminishing slope (Fig. 7.7).

We shall assume that the true weight, $\mu(x)$ is a monotonically decreasing function of concentration x, that is, $\mu(x)$ always decreases as concentration x increases. The term $\mu(0)$ represents the true weight of an untreated batch of insects, while $\mu(\infty)$ = $\lim_{x \to \infty} \mu(x)$ denotes the true weight as the concentration tends to infinity. Thus, $\mu(0) > \mu(\infty) \geq 0$.

Commonly used models relating a non-negative continuous response, μ, of insects exposed to a substance at concentration x are: the exponential and the hyperbolic models, described below.

1. The exponential model is expressed as:

$$\mu(x) = \exp(\alpha + \beta x) \quad (x \geq 0) \tag{17}$$

where α and β are unknown parameters that must be estimated from the data. Provided $\beta < 0$, the function $\mu(x)$ is a monotonically decreasing function of concentration x. In this model, $\mu(0) = \exp(\alpha)$ and $\mu(\infty) = 0$. The exponential model often relates a non-negative continuous response, μ, to a tranformation of x, such as: x^λ, for some known $\lambda > 0$. The model is then expressed as:

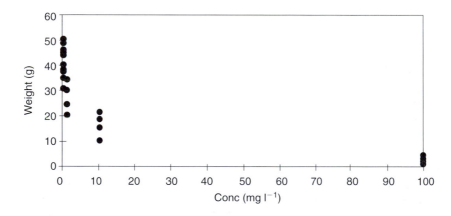

Fig. 7.7. Observed weights vs. concentrations for the standard substance data from Table 7.13a.

$$\mu(x) = \exp(\alpha + \beta x^\lambda) \qquad (x \geq 0) \qquad (18)$$

The logarithmic transformation $\eta = \log(\mu)$ yields

$$\eta(x) = \alpha + \beta x^\lambda \qquad (19)$$

which, for some λ, is a straight line model for η with intercept α and slope β. For a given λ, the exponential model (18) or (19) is also known as the log-linear model for μ. Since it is possible that at extremely high concentrations the insect response reaches a plateau above zero, the exponential or log-linear model can be extended by including an additional parameter, say γ ($\gamma \geq 0$), such that:

$$\mu(x) = \gamma + \exp(\alpha + \beta x^\lambda) \qquad (20)$$

In this model, $\mu(0) = \gamma + \exp(\alpha)$ and $\mu(\infty) = \gamma$.

2. The hyperbolic model is expressed as:

$$\mu(x) = \beta/(x + \alpha) \qquad (x \geq 0) \qquad (21)$$

where α and β are unknown positive parameters. The function $\mu(x)$ is a monotonically decreasing function of x. In this model, $\mu(0) = \beta/\alpha$ and $\mu(\infty) = 0$. If the concentration scale is changed, for instance, by taking a power transformation of x, the model is expressed as:

$$\mu(x) = \beta/(x^\lambda + \alpha) \qquad (x \geq 0) \qquad (22)$$

for some $\lambda > 0$. The reciprocal transformation $\eta = 1/\mu$ yields

$$\eta(x) = \alpha/\beta + x^\lambda/\beta \qquad (23)$$

which, for some λ, is a straight line model for $\eta = 1/\mu$, with intercept α/β and slope $1/\beta$. The hyperbolic model (23) can be extended by including a third parameter γ ($\gamma \geq 0$) such that

$$\mu(x) = \gamma + \beta/(x^\lambda + \alpha) \qquad (24)$$

In this case, $\mu(0) = \gamma + \beta/\alpha$ and $\mu(\infty) = \gamma \geq 0$.

Both the exponential or log-linear, and the hyperbolic models for describing the concentration–response relationship are non-linear in their parameters α, β and γ. Thus, for a given value of λ, three-parameter models are fitted to the data. Moreover, if the power λ is unknown, it should be estimated from the data. In this case, four-parameter models are fitted to the data. In the following, for simplicity, we shall assume that $\gamma = 0$ and λ is known. Thus, we shall focus on fitting the exponential hyperbolic models given in eqs. (18) and (22), respectively.

In order to get an impression of a plausible model that fits our standard sample weight data, we first experiment with plots of $\log(y)$ and $1/y$ vs. power transformations x^λ for various choices of the value of λ. Examining the plot of $\log(\text{weight})$ vs. $x^{0.4}$ (Figure 7.8) suggests that a straight-line relationship is reasonable for the standard sample data. Judging from

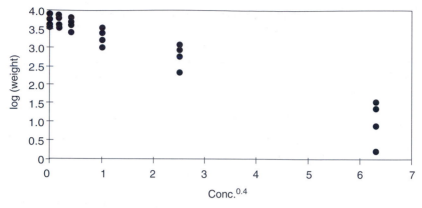

Fig. 7.8. Logarithm weight values, log(weight), vs. concentration power transformation $x^{0.4}$ for the standard substance data from Table 7.13a.

similar plots for the test sample data, it appears that a straight-line relationship between log(weight) and $x^{0.4}$ is also reasonable for the test sample data.

In order formally to fit the exponential model $\mu(x) = \exp(\alpha + \beta x^{0.4})$ to the standard sample data, we assume that the weight observations $\{y_{ij}\}$ ($i = 1, \dots, k$; $j = 1, \dots, r$) are independent and are taken from a specific family of probability distributions. For this family of distributions, there exists some specific function of the mean, say $V(\mu)$, such that $\text{var}(y_{ij})/V(\mu_i)$ is a constant, say ϕ, for all i ($i = 1, \dots, k$). For example, if observations $\{y_{ij}\}$ are normally distributed with means $\mu_i = \exp(\alpha + \beta x_i^{0.4})$ ($i = 1, \dots, k$) and common variance, then one can put $V(\mu_i) = 1$ or $\text{var}(y_{ij}) = \phi$, which corresponds to the classical non-linear regression.

Examining the means and the variances of the weight observations (Table 7.13a) suggests that the variances vary with the means and, indeed, each variance, s_i^2, is almost equal to the corresponding mean, $\bar{y}_i$ ($i = 1, \dots 6$). Thus, it is reasonable to assume that $V(\mu_i) = \mu_i$ or $\text{var}(y_{ij})/\mu_i = \phi$, where ϕ is an unknown dispersion parameter. The situation in which $V(\mu_i) = \mu_i$ and $\phi = 1$ corresponds to the Poisson parent distributions, which commonly arise with frequency count data. The variance function $V(\mu_i) = \mu_i$ can also occur with continuous non-negative response data (as in our example). It should be emphasized that if an inappropriate variance function $V(\mu)$ is used, then the resulting standard errors of the estimated model parameters will generally be incorrect. For example, if in this example we use a classical non-linear regression, $\mu(x) = \exp(\alpha + \beta x^{0.4})$ with the assumption $V(\mu)=1$ or $\text{var}(y_{ij}) = \phi$, the standard errors of parameter estimates might be misleading, resulting in possible incorrect inferences about the model parameters. So, we wish to fit the exponential model $\mu(x) = \exp(\alpha + \beta x^{0.4})$ to our data with the assumption that $\text{var}(y_{ij})= \phi\mu_i$, where $\mu_i = \mu(x_i)$.

When ϕ is unknown, the form of the distributions of $\{y_{ij}\}$ cannot be fully specified. The estimation procedure of the model parameters, α and β, is based on a quasi-likelihood method, as briefly described at the end of the section on model fitting in connection with the modelling of mortality data with overdispersion. More specifically, this method involves maximization of the log-likelihood function of independent Poisson variables ($\phi = 1$), which is given by:[5]

$$L(\mathbf{y};\boldsymbol{\mu}) = r\sum_{i=1}^{k} \mu_i + \sum_{i=1}^{k}\sum_{j=1}^{r} y_{ij}\log(\mu_i)$$

subject to the restrictions implied by the exponential $\mu(x)$ model. Twice the difference between the log-likelihood of a full model that places no restrictions on the means and that of the fitted model provides a comparative measure of the two models (when $\phi = 1$). The full model includes k unknown parameters, μ_i, that are estimated as the sample means, $\bar{y}_i$ ($i = 1$, ... ,k), whereas the exponential $\mu(x)$ model includes $q = 2$ parameters. Let $\hat{\mu}_i = \exp(\hat{\alpha} + \hat{\beta}x_i^{0.4})$ ($i = 1$, ... ,k) represent the estimated means obtained from the exponential model with $\phi = 1$. A related comparative measure of the two models is provided by Pearson's statistic, defined as:

$$X^2 = r\sum_{i=1}^{k} (\bar{y}_i - \hat{\mu}_i)^2/\hat{\mu}_i \qquad (25)$$

Since the dispersion parameter, ϕ is also unknown, it must be estimated from the data; it is commonly estimated as:

$$\hat{\phi} = [\sum_{i=1}^{k}\sum_{j=1}^{r} (y_{ij} - \hat{\mu}_i)^2/\hat{\mu}_i]/(kr - q) \qquad (26)$$

Under the assumption that $\text{var}(y_{ij}) = \phi\mu_i$, the estimated model parameters are equivalent to the estimates, $\hat{\alpha}$ and $\hat{\beta}$, obtained when $\phi = 1$, while the estimated variances and covariances are multiplied by $\hat{\phi}$. Moreover, the scaled Pearson's statistic, defined as $[r\sum_{i=1}^{k} (\bar{y}_i - \hat{\mu}_i)^2/\hat{\mu}_i]/\hat{\phi}$, may be used to assess goodness-of-fit of the exponential model. Under certain conditions, the significance of this statistic is judged in terms of a $\chi^2(k - q)$ distribution.

The SAS GENMOD procedure was used for fitting the exponential model $\mu_i = \exp(\alpha + \beta x_i^{0.4})$, with $\text{var}(y_{ij}) = \phi\mu_i$, to the data from the standard and from the test samples. This procedure fits generalized linear models, as defined by Nelder and Wedderburn (1972)[6] (see Appendix 7.1 for a concise discussion of the generalized linear models). The results of the analysis produced by the SAS GENMOD procedure are presented in Table 7.14. Note that in Table 7.14, X^2(fitted) is defined as $\sum_{i=1}^{k}\sum_{j=1}^{r} (y_{ij} - \hat{\mu}_i)^2/\hat{\mu}_i$, whereas the deviance represents a closely related measure of dispersion.

The test sample produces: $4\sum_{i=1}^{6}(\bar{y}_i - \hat{\mu}_i)^2/\hat{\mu}_i = 8.7284$ and $\hat{\phi} = 1.333$. The ratio $8.7284/1.333 = 6.548$ is not significant, and we therefore conclude that the non-linear model, $\mu(x) = \exp(\alpha + \beta x^{0.4})$, with $\text{var}(y_{ij}) = \phi\mu(x_i)$, is acceptable for the test sample. A similar conclusion is reached regarding the standard sample.

The predicted weights for the standard and test samples are:

$$\hat{\mu}(x) = \exp(3.7754 - 0.4112x^{0.4}) \qquad \text{for the standard sample}$$
$$\hat{\mu}(x) = \exp(3.7966 - 0.8664x^{0.4}) \qquad \text{for the test sample.}$$

Table 7.14. The exponential models $\mu(x) = \exp(\alpha + \beta x^{0.4})$ with $\text{var}(y_{ij}) = \phi\mu(x_i)$ fitted to the standard and the test sample data from Table 7.13.

(a) Standard sample

Criteria for assessing goodness of fit

Criterion	df	Value	Value/df
Deviance (fitted)	22	21.7190	0.9872
Pearson X^2 (fitted)	22	21.1541	0.9615

Analysis of parameter estimates

Parameter	Estimate	SE	P value
Intercept α_s	3.7754	0.0441	0.0001
Slope β_s	−0.4112	0.0354	0.0001
Scale $\phi^{1/2}$	0.9806		

(b) Test sample

Criteria for assessing goodness of fit

Criterion	df	Value	Value/df
Deviance (fitted)	22	27.6909	1.2587
Pearson X^2 (fitted)	22	29.1331	1.3331

Analysis of parameter estimates

Parameter	Estimate	SE	P value
Intercept α_T	3.7966	0.0584	0.0001
Slope β_T	−0.8664	0.0859	0.0001
Scale $\phi^{1/2}$	1.1546		

Table 7.15 presents the estimated mean weight, $\hat{\mu}(x)$, and the 95% confidence limits for the true weight, $\mu(x)$, for each of the standard and the test samples. The lower and upper confidence limits at a given concentration, x, are denoted by $\hat{\mu}_L(x)$ and $\hat{\mu}_U(x)$, respectively. These limits are based on the assumption that $\lambda = 0.4$.

We can now ask what is the effective concentration, EC_{50}, that causes a 50% reduction in insect weights, as compared with untreated controls. More generally, for a given π ($0 < \pi < 1$), $EC_{100\pi}$ represents the effective concentration which causes $100\pi\%$ reduction in insect weights, as compared with untreated controls.

For the exponential model, $EC_{100\pi}$ is obtained by solving the equation $\exp(\beta x^\lambda) = 1 - \pi$ for x. The solution is:

$$EC_{100\pi} = (\log(1 - \pi)/\beta)^{1/\lambda} \tag{27}$$

For a given π, the efficacy of a test substance relative to that of a standard, is defined by:

$$\rho_\pi = EC_{100\pi;S} / EC_{100\pi;T} = (\log(1 - \pi)/\beta_S)^{1/\lambda_S} / (\log(1 - \pi)/\beta_T)^{1/\lambda_T} \tag{28}$$

Table 7.15. Ninety-five per cent confidence limits for $\mu(x)$ obtained by fitting the exponential model $\mu(x) = \exp(\alpha + \beta x^{0.4})$ with $\mathrm{var}(y_{ij}) = \phi\mu(x_i)$ to the standard and the test sample data from Table 7.13.

(a) Standard sample

	Conc. (mg l^{-1})					
	0	0.01	0.10	1.0	10	100
$\hat{\mu}_L(x)$	40.002	37.672	34.322	26.745	13.378	2.18
$\hat{\mu}(x)$	43.614	40.862	37.028	28.909	15.524	3.25
$\hat{\mu}_U(x)$	47.553	47.553	39.947	31.248	18.015	4.85

(b) Test sample

	Conc. (mg l^{-1})					
	0	0.01	0.10	1.0	10	100
$\hat{\mu}_L(x)$	39.732	35.053	28.660	16.232	3.479	0.069
$\hat{\mu}(x)$	44.548	38.832	31.552	18.732	5.054	0.188
$\hat{\mu}_U(x)$	49.948	43.018	34.735	21.613	7.342	0.514

$\hat{\mu}(x)$ is the model-fitted weight at concentration level x; $\hat{\mu}_L(x)$ is the lower confidence limit; $\hat{\mu}_U(x)$ is the upper confidence limit.

where $EC_{100\pi;T}$ and $EC_{100\pi;S}$ denote the effective concentrations of the test and standard substances, respectively. A test preparation is more effective than a standard at a given weight reduction, π, if $\rho_\pi > 1$. If $\lambda_T = \lambda_S$, the relative efficacy becomes constant or $\rho = (\beta_T/\beta_S)^{1/\lambda}$, over all values of π. For our data, $\lambda_T = \lambda_S = 0.4$ while the estimated relative efficacy is: $\hat{\rho} = (0.8664/0.4112)^{1/0.4} = 6.444$. The calculations of the standard errors of the estimated effective concentrations and estimated relative efficacy are omitted.

Bioassays with Time–Mortality Data

The following is a constructed example of a bioassay with time–mortality data.

An experiment was conducted to study the effect of a nematode preparation at a given concentration upon the mortality of ticks. At the beginning of the experiment, a group of 60 similar ticks was exposed to the substance at a concentration of 50 nematodes per 0.5 ml water. The number of ticks dying between the beginning of the experiment (day 0) and the first day of exposure (day 1) was recorded at day 1 and was equal to 4. The dead ticks were removed and the study continued. Thus, 60 − 4 = 56 ticks were still alive after 1 day of exposure. The number of ticks dying between the first day and the second day of exposure (day 1 to day 2) was observed at day 2 and was equal to 6. Thus, 56 − 6 = 50 ticks were alive after 2 days of exposure. Again, dead ticks were removed from the experiment. Continuing in this way yielded the following observations taken at 1-day intervals over 12 days of exposure to the substance:

4, 6, 7, 7, 6, 6, 4, 4, 4, 3, 3, 1

The experiment terminated after 12 days. The daily cumulative total numbers of dead ticks observed over 12 days were:

4, 10, 17, 24, 30, 36, 40, 44, 48, 51, 54 and 55.

At the end of the study, 55 ticks were dead while five ticks were still alive; no information is available about the times of mortality of the 5 ticks which survived beyond the duration of the experiment. Mortality data of this form are known as grouped or interval data, with a single censored-time observation corresponding to the last time point (day 12) at which a record was taken.

Let $F(t)$ represent the theoretical cumulative proportion of ticks dying by time t, or the probability of a given tick dying by time t. The mortality function $F(t)$ is interpreted as a cumulative probability distribution function of a random variable, T, representing the lifetime of a tick under study or the time of exposure which, if exceeded, causes a tick to die. By definition, $F(t) = pr\{T \le t\}$, and $F(0) = 0$, where $t = 0$ corresponds to the time of starting the experiment. It may happen that at long exposure times 100% mortality is never reached, but a plateau below 100%, say γ, is attained. The term $S(t) = 1 - F(t)$ represents the probability of a given tick's survival beyond time t.

We wish to describe the mortality trend over the time of exposure, or to establish a time–mortality relationship. We first notice that since one group of 60 ticks was used to assess mortality at 1-day intervals over the duration of the experiment, the daily cumulative mortality proportions are correlated observations. Therefore, the statistical methods of modelling concentration–mortality data (e.g. probit analysis), described previously, are not valid for analysing the time–mortality data of our example. If different groups of ticks were used for assessing mortality at distinct times of exposure, the cumulative mortality observations would be independent and the data could then be analysed by the methods described above. However, since an experiment designed as a follow-up of one group of ticks is generally more economical than a setup involving independent groups under distinct times of exposure, a scientific approach would be to apply the former experimental design and to analyse the data by appropriate methods.

Statistical methods for analysing time–mortality data are described in statistical literature on survival analysis. Interested readers are referred to Gross and Clark (1975), Lawless (1982), Collett (1991) and Lee (1992) for detailed discussion and additional references. The statistical methods include analysis of various types of time–mortality data. For example, time–mortality data may consist of the precise times of death of subjects receiving a treatment, possibly with some censored observations that correspond to times at which several subjects are withdrawn or lost from the study for reasons other than dying as a result of the treatment. This form of data commonly arises in biological or medical follow-up studies. In this section we shall present a brief description of basic methods used for analysing interval time–mortality data with single censored-time observation, as in our example.

We first estimate F(t), which represents the theoretical cumulative mortality by time t, by a non-parametric method. This means that we assume that there is no underlying parametric function, such as probit or logit, for describing a mortality trend over time.

It is assumed that at the beginning of the experiment N ticks were exposed to the substance; during the study, mortality records were taken at M distinct times of exposure, say $0 < t_1 < t_2 < \ldots < t_M$. In our example, $N = 60$ and $t_j = j$ (days) for $j = 1, \ldots, 12$. Let d_j represent the number of ticks dying between t_{j-1} and t_j, which is recorded at time t_j ($j = 1, \ldots, M$). The total number of ticks dying by time t_j is: $D_j = (d_1 + \ldots + d_j)$ ($j = 1, \ldots, M$). At the end of the experiment $(N - D_M)$ ticks remain alive, and their mortality times are unknown. In its simplest form, the estimated probability of the total mortality by time t_j is the proportion of ticks dying by time t_j. This estimate is denoted by $\hat{\mathrm{F}}(t_j)$ and equals D_j/N ($j = 1, \ldots, M$). The estimated mortality function, F(t) (or survival function, S(t)) is formed by a step-function with changes occurring only at the times of observations. Thus,

$$\hat{\mathrm{F}}(t) = 0 \qquad \hat{\mathrm{S}}(t) = 1 \qquad \text{for } 0 \le t < t_1$$
$$\hat{\mathrm{F}}(t) = D_j/N \qquad \hat{\mathrm{S}}(t) = 1 - D_j/N \qquad \text{for } t_j \le t < t_{j+1} \;\; j = 1, \ldots, M-1$$
$$\hat{\mathrm{F}}(t) = D_M/N \qquad \hat{\mathrm{S}}(t) = 1 - D_M/N \qquad \text{for } t \ge t_M$$

Note that the survival estimate at time $t = t_M$ is $1 - D_M/N$; with no better survival estimate available from the data, we estimate the survival beyond time t_M also as $1 - D_M/N$. If all ticks were dead by time t_M, the estimated survival at $t = t_M$ would drop to zero, and would equal zero for all $t > t_M$.

For our tick mortality data, the estimated daily cumulative mortalities over 12 days of exposure are:

4/60, 10/60, 17/60, 24/60, 30/60, 36/60, 40/60, 44/60, 48/60, 51/60, 54/60 and 55/60.

The step-function of the estimated mortality function is presented in Fig. 7.9.

A summary measure of the cumulative probablity of mortality is the median lethal time, LT_{50}, required to kill 50% of the tick population. Let t_{j-1} and t_j denote consecutive time points satisfying the expressions: $\hat{\mathrm{F}}(t_{j-1}) = D_{j-1}/N < 0.50$, and $\hat{\mathrm{F}}(t_j) = D_j/N \ge 0.50$. Since mortality may vary with time of exposure, the estimated $\hat{\mathrm{LT}}_{50}$ can be obtained by linear interpolation using the following equation:

$$\hat{\mathrm{LT}}_{50} = t_{j-1} + (t_j - t_{j-1})[0.50 - \hat{\mathrm{F}}(t_{j-1})]/[\hat{\mathrm{F}}(t_j) - \hat{\mathrm{F}}(t_{j-1})] \qquad (29)$$

For our data, the length of the j-th interval is: $(t_j - t_{j-1}) = 1$ day, for $j = 1, \ldots, 12$. Since $\hat{\mathrm{F}}(t_5) = 0.50$, LT_{50} is estimated by: $\hat{\mathrm{LT}}_{50} = 5$ days. Thus, 5 days of exposure are required to produce 50% mortality of these ticks.

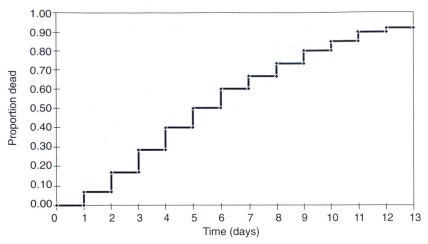

Fig. 7.9. Step-function of a non-parametric estimate of the mortality function, F(t), for the tick mortality data.

More generally, for a given value of π ($0 < \pi < 1$), the lethal time $\hat{\mathrm{LT}}_{100\pi}$ represents the time of exposure required to produce $100\pi\%$ mortality of the ticks. If t_{j-1} and t_j are time points satisfying the expressions: $\hat{F}(t_{j-1}) < \pi$ and $\hat{F}(t_{j-1}) \geq \pi$, then the estimated $\mathrm{LT}_{100\pi}$ is obtained by linear interpolation using the following equation:

$$\hat{\mathrm{LT}}_{100\pi} = t_{j-1} + (t_j - t_{j-1})[\pi - \hat{F}(t_{j-1})]/[\hat{F}(t_j) - \hat{F}(t_{j-1})] \qquad (30)$$

For our data, since $\hat{F}(t_8) = 0.7333$ and $\hat{F}(t_9) = 0.80$, the LT_{75} is estimated as 8.253 days. Moreover, since $\hat{F}(t_{11}) = 0.90$, the LT_{90} is estimated as 11 days. Notice that, since $\hat{F}(t) = 55/60$ for all $t \geq 12$, LT_{95} cannot be estimated from our data.

The SAS LIFETEST procedure calculates non-parametic estimates of mortality (or survival), with their corresponding estimated variances; it also provides estimates of $\mathrm{LT}_{100\pi}$ for various values of π, with corresponding approximate 95% confidence intervals.

A time–mortality study is usually repeated with at least three different groups of ticks, with mortalities being recorded for each group over distinct times. If, at each time point of observation, the variability among replicates is small, mortality data can then be pooled over replicates to produce a single mortality distribution over time. The SAS LIFETEST procedure produces the log-rank and Wilcoxon test statistics for testing homogeneity of several mortality distributions. Readers may refer to the statistical literature on survival analysis, mentioned above, for more details.

We now consider parametric estimation of the mortality function F(t), which represents the cumulative distribution function of a continuous random variable, T, defined as the lifetime of a tick under study. The function

F(t) is a monotonically increasing function of time, t, and is assumed to define a smooth, S-shaped curve. A widely used function for describing the trend of mortality over time is the Weibull distribution (Weibull, 1951), expressed as:

$$F(t) = 1 - \exp(-\alpha t^{\gamma}) \tag{31}$$

where α and γ are unknown positive parameters. Parameter α is viewed as a scale parameter, while γ is a shape parameter that determines the form and skewness of the distribution of T. The special case of $\gamma = 1$ corresponds to the family of exponential distributions.

Application of the complementary log–log (cll) transformation, defined by $\log[-\log(1-F(t))]$, yields:

$$\log[-\log(1-F(t))] = \log(\alpha) + \gamma\log(t) \tag{32}$$

Thus, the Weibull model assumes that the relationship between $\log[-\log(1-F(t))]$ and $\log(t)$ is linear, with intercept $\log(\alpha)$ and slope γ.

If $\hat{F}(t_j)$ represents the non-parametric estimate of the cumulative mortality by time t_j, then a scatter plot of $\log[-\log(1-\hat{F}(t_j))]$ against $\log(t_j)$ that resembled a straight-line relationship would suggest that the Weibull model fitted the time–mortality data. Figure 7.10 shows the scatter plot of $\log[-\log(1-\hat{F}(t_j))]$ against $\log(t_j)$ for our tick mortality data.

It can be seen from Fig. 7.10 that the Weibull distribution appears to be a reasonable choice for modelling our data. Thus, our data seem adequate for estimating the parameters of the Weibull distribution. Notice that for a case in which 100% mortality is never reached, but a plateau below 100% is attained at long exposure times, it would be inappropriate to fit a Weibull distribution to the time–mortality data.

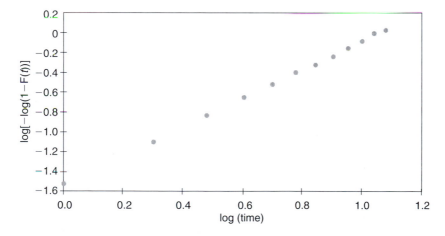

Fig. 7.10. Complementary-log-log transformation of $\hat{F}(t_j)$ plotted against $\log(t_j)$ for the tick mortality data.

In general, it is common to consider the cumulative probability distribution function of a random variable $W = \log(T)$ or the extreme-value distribution, rather than the Weibull distribution of T. The extreme-value distribution depends on two parameters: μ (intercept) and σ (scale), that are related to the parameters of the Weibull distribution by $\gamma = 1/\sigma$ and $\alpha = \exp(-\mu/\sigma)$.

For a given value of π ($0 < \pi < 1$), let $\log(LT_{100\pi})$ represent the logarithm of the time of exposure required to achieve a mortality of $100\pi\%$. It can be seen that $\log(LT_{100\pi})$ can be expressed as:

$$\log(LT_{100\pi}) = \mu + \sigma\log[-\log(1-\pi)] \tag{33}$$

The exponential function, $\exp(\log(LT_{100\pi}))$ produces $LT_{100\pi}$.

Statistical methods have been developed for estimating the unknown parameters of a Weibull distribution by the use of various types of time–mortality data (see references on survival analysis, mentioned above).

The SAS LIFEREG procedure was used for fitting a Weibull model to the time-interval data with single censored-time observation of our example. The dependent variable is the time interval between t_{j-1} and t_j, with a weight variable equal to the number of dead ticks, d_j recorded at time t_j. The right censored time is $t_M = 12$ (days) with weight corresponding to the number of ticks remaining alive. The estimated parameters of the extreme-value distribution of $\log(T)$ and their standard errors are: $\hat{\mu} = 1.8619$ (SE $= 0.0948$) and $\hat{\sigma} = 0.6850$ (SE $= 0.0780$). Thus, the parameter estimates of the Weibull distribution are: $\hat{\gamma} = 1/0.685 = 1.460$ and $\hat{\alpha} = \exp(-1.8619/0.685) = 0.066$.

For a given value of π ($0 < \pi < 1$), the estimated lethal time $LT_{100\pi}$ is obtained by substituting the estimated parameters, $\hat{\mu}$ and $\hat{\sigma}$, in eq. (33) and taking an exponential function. For example, $\hat{LT}_{50}$ is given by: $\hat{LT}_{50} = \exp(1.8619 + 0.685[\log(-\log(0.5))]) = 5.006$. Estimated lethal times, $\hat{LT}_{100\pi}$, for various values of π, their corresponding standard errors (SE), and lower and upper limits of the 95% confidence intervals ($\hat{LT}_L$ and $\hat{LT}_U$) are presented in Table 7.16. Note the similarity between the estimates obtained by both parametric and non-parametric methods.

Bioassays with Time–Concentration–Mortality Data

The following is a constructed example of a bioassay with time–concentration–mortality data.

An experiment was conducted to study the effects of various concentrations of nematodes on tick mortality over time of exposure. The experiment was designed with $k = 5$ concentrations, say: 25, 50, 100, 250 and 500 nematodes per 0.5 ml water. Five groups, each of 60 similar ticks, were exposed to the substance at the beginning of the experiment, with each group receiving a different nematode concentration. Mortality records were taken at 1-day intervals over a period of 12 days. Dead ticks were removed

Table 7.16. Estimated lethal times $\hat{LT}_{100\pi}$, standard errors, and lower and upper 95% confidence limits obtained by fitting the Weibull model to the tick mortality data.

π	$\hat{LT}_{100\pi}$	SE	$\hat{LT}_L$	$\hat{LT}_U$
0.25	2.7415	0.4111	2.0433	3.6781
0.50	5.0073	0.5250	4.0771	6.1497
0.75	8.0503	0.7458	7.5072	9.6532
0.90	11.3962	1.1696	10.5991	13.9355

π = cumulative probability of mortality; $\hat{LT}_{100\pi}$ = estimated lethal time of exposure required to achieve 100π% mortality; SE = standard error; $\hat{LT}_L$ = lower 95% confidence limit; $\hat{LT}_U$ = upper 95% confidence limit.

from the experiment upon detection. The experiment terminated after 12 days. Artificial mortality data for such an experiment, recorded at 1-day intervals, are presented in Table 7.17.

Let $F(t;x)$ represent the theoretical cumulative proportion of ticks exposed to concentration x dying by time t, or the probability of one of these ticks dying by time t. We wish to study the effects of various concentrations and distinct times of exposure upon tick mortality.

In general, it is assumed that k groups, each of N ticks, are randomly assigned to various concentrations, $(x_1, \ldots, x_k)$, with each exposed to a different concentration. Let $0 < t_1 < t_2 < \ldots < t_M$ represent M time points at which mortality records are taken over the duration of the experiment. For our data, $N = 60$, $k = 5$; $t_j = j$ days ($j = 1, \ldots, 12$). Let d_{ij} denote the number of ticks exposed to concentration x_i which died between t_{j-1} and t_j ($i = 1, \ldots, k; j = 1, \ldots, M$). The total number of ticks exposed to concentration x_i, which died by time t_j, is: $D_{ij} = (d_{i1} + \ldots + d_{ij})$. At the end of the experiment, that is, at time $t_M = 12$ (days), $(N - D_{iM})$ ticks exposed to concentration x_i were still alive.

We start with a non-parametric estimation of the cumulative mortality distribution function, separately for each concentration. Let $\hat{F}(t_j;x_i)$ represent the estimated cumulative proportion of ticks exposed to concentration x_i, which die by time t_j. By means of the SAS LIFETEST procedure, we obtain five step-functions of $\hat{F}(t;x_i)$, for $i = 1, \ldots, 5$ as described in the previous section. Let the lethal time $LT_{50}(x_i)$ represent the time required to kill 50% of the tick population exposed to concentration x_i ($i = 1, \ldots, 5$). The estimated lethal times $LT_{50}(x_i)$ are: 6.666, 4.625, 4.286, 3.75 and 3.2 (days), for exposure to concentrations of 25, 50, 100, 250 and 500 nematodes per 0.5 ml water, respectively. Thus, it is seen that the lethal time $LT_{50}(x)$ decreases as concentration x increases, or that the speed of killing increases as concentration increases. Tests of equality of the five mortality distributions (log rank and Wilcoxon tests) indicate significant differences among the five distributions ($P < 0.01$).

Table 7.17. Artificial mortality data for five groups of ticks, recorded daily over 12 days; each group consisted of 60 ticks that had been exposed to a given concentration at the beginning of the experiment.

Time (days)	Concentration (nematodes per 0.5 ml)				
	25	50	100	250	500
1	4	3	4	5	8
2	4	8	7	9	12
3	5	6	9	10	8
4	4	8	8	8	10
5	6	8	7	10	7
6	5	6	6	6	5
7	3	5	5	4	5
8	4	3	5	1	2
9	3	3	4	1	1
10	4	2	2	1	1
11	3	3	1	2	1
12	1	1	2		
Total mortality	46	56	60	57	60
Total survival	14	4	0	0	0

Time–concentration–mortality data can be modelled by parametric methods. One can fit Weibull distributions to the time–mortality data separately for each of the five concentrations. Indeed, scatter plots of $\log[-\log(1-\hat{F}(t_j;x_i))]$ against $\log(t_j)$ for each concentration show straight-line relationships. Similarly, one can model concentration–mortality relationships by fitting probit curves separately for each of the distinct times of exposure, t_j, $(j = 1, \ldots ,M)$. The statistical analysis based on fitting a time–mortality model separately for each concentration, and concentration–mortality model separately for each time of exposure, is inefficient, because not all the data are used in the estimation procedure. Furthermore, the activity of the substance at different concentrations and over the time of exposure may be difficult to describe.

Preisler and Robertson (1989) proposed complementary log–log models for describing time–concentration–mortality relationships (see also Robertson and Preisler, 1992). The models are presented as follows.

Let κ_{ij} represent the conditional probability of a tick exposed to concentration x_i dying by time t_j, given that it was alive at time t_{j-1} $(i = 1, \ldots ,k; j = 1, \ldots ,M)$. The conditional probabilities $\{\kappa_{ij}\}$ are modelled by:

$$\kappa_{ij} = 1 - \exp[-\exp(\gamma_j + \beta\log(x_i))] \tag{34}$$

where γ_j $(j = 1, \ldots ,M)$ and β are $M + 1$ unknown parameters, with each γ_j corresponding to its own time interval (t_{j-1}, t_j).

Application of the complementary log–log transformation produces:

$$\log[-\log(1 - \kappa_{ij})] = \gamma_j + \beta \log(x_i) \qquad (35)$$

The cumulative probability, $F(t_j; x_i)$ of a tick exposed to concentration x_i dying by time t_j, can be approximated by

$$F(t_j; x_i) = 1 - \prod_{k=1}^{j} (1 - \kappa_{ij}) = 1 - \exp[-\exp(\tau_j + \beta \log(x_i))] \qquad (36)$$

where the parameter, τ_j is defined as $\tau_j = \log[\exp(\gamma_1) + \dots + \exp(\gamma_j)]$.

Application of the complementary log–log transformation produces:

$$\log[-\log(1 - F(t_j; x_i))] = \tau_j + \beta \log(x_i) \qquad (37)$$

Thus, the complementary log–log models (35) and (37) describe a time–concentration–mortality relationship.

We first estimate the unknown parameters $\{\gamma_j\}$ and β of the cll model given in (35). Recall that d_{ij} denotes the number of ticks exposed to concentration x_i, which die between time t_{j-1} and time t_j and that $D_{ij} = (d_{i1} + \dots + d_{ij})$ represents the total number of these which die by time t_j. Let $N_{ij} = (N - D_{i,j-1})$ be the number of ticks that are alive at time t_{j-1}. It is reasonable to assume that observations d_{ij}s are independent and follow binomial distributions, $\text{Bin}(\kappa_{ij}, N_{ij})$, $(i = 1, \dots, k; j = 1, \dots, M)$, where each κ_{ij} denotes the conditional probability expressed in terms of the $(M + 1)$ parameters included in the cll model given in eq. (35). Note that, without the cll model constraints, each conditional probability κ_{ij} is estimated as the proportion d_{ij}/N_{ij} $(i = 1, \dots, k; j = 1, \dots, M)$. The maximum likelihood estimation procedure can be applied to the binomial data $\{d_{ij}, N_{ij}\}$ $(i = 1, \dots, k; j = 1, \dots, M)$, to estimate the unknown model parameters, γ_js and β.

The SAS GENMOD procedure was used for fitting the cll model (35) to the binomial data $\{d_{ij}, N_{ij}\}$. In all calculations, 58 observations were used for estimating $(12 + 1) = 13$ unknown parameters. The values of the Pearson and the deviance statistics used to assess goodness of fit are: 18.913 and 19.427, respectively. Under certain conditions, if the model is correct, each of the goodness-of-fit statistics approaches a χ^2 distribution with $58-13 = 45$ df. Thus, if the model is correct, the mean deviance should be close to 1. For our data, the mean deviance equals $19.427/45 = 0.432$, which is considerably smaller than unity. This indicates underdispersion of the binomial data $\{d_{ij}, N_{ij}\}$, which probably resulted from small sample sizes. The estimated cll model parameters and their standard errors are presented in Table 7.18. All parameters differ significantly from zero.

The estimated probability, $\hat{F}(t_j; x_i)$ of a tick exposed to concentration x_i dying by time t_j is obtained by substituting the estimated parameters $\{\hat{\gamma}_j\}$ and $\hat{\beta}$ in model (36). Thus,

$$\hat{F}(t_j; x_i) = 1 - \exp[-\exp(\hat{\tau}_j + \hat{\beta} \log(x_i))] \qquad (38)$$

where $\hat{\tau}_j = \log[\exp(\hat{\gamma}_1) + \dots + \exp(\hat{\gamma}_j)]$.

For our data, the estimated τ_js for $j = 1, \dots, 12$ are: -4.2094, -3.1420,

Table 7.18. Complementary-log-log (cll) model fitted to the conditional mortality probabilities, for time–concentration–mortality data from Table 7.17.

Parameter	Estimate	SE
Time γ_1	-4.2094	0.3547
Time γ_2	-3.5634	0.3281
Time γ_3	-3.4289	0.3274
Time γ_4	-3.2098	0.3241
Time γ_5	-2.9348	0.3194
Time γ_6	-2.9381	0.3291
Time γ_7	-2.8663	0.3372
Time γ_8	-2.9527	0.3617
Time γ_9	-2.9157	0.3815
Time γ_{10}	-2.8304	0.4005
Time γ_{11}	-2.4913	0.3960
Time γ_{12}	-2.9533	0.5462
Conc. β	0.3533	0.0571

-2.5820, -2.1542, -1.7771, -1.5046, -1.2765, -1.1050, -0.9535, -0.8111, -0.6402 and -0.5459, respectively. Thus, for instance, the estimated probability that a tick exposed to a concentration of $x = 250$ (nematodes 0.5 ml^{-1} water) will die by $t = 3$ days of exposure is:

$\hat{F}(3;250) = 1 - \exp[-\exp(\hat{\tau}_3 + \hat{\beta}\log(250))] = 1 - \exp[-\exp(-2.5820 + 0.3533\log(250))] = 0.4125$. Notice the similarity to the non-parametric estimate of 26/60 or 0.431. Figure 7.11 shows the fitted mortality probabilities for concentration x_i, by time of exposure $t_j (i = 1, \dots ,5; j = 1, \dots ,12)$.

For a given π $(0 < \pi < 1)$ and time t_j, the estimated logarithm of the lethal concentration $\log(\hat{LC}_{100\pi}(t_j))$ that causes a total mortality of $100\pi\%$ by time t_j is obtained by solving the following equation for $\log(x)$:

$$\hat{F}(t_j;x) = 1 - \exp[-\exp(\hat{\tau}_j + \hat{\beta}\log(x))] = \pi$$

The solution is

$$\log(\hat{LC}_{100\pi}(t_j)) = (\log[-\log(1-\pi)] - \hat{\tau}_j)/\hat{\beta} \qquad (39)$$

Use of an exponential function leads to the estimated lethal concentration $\hat{LC}_{100\pi}(t_j)$ that causes a total mortality of $100\pi\%$ at exposure time, t_j. For example, the logarithm of the median lethal concentration that causes a total mortality of 50% at 4 days of exposure is $\log(\hat{LC}_{50}(4)) = (\log[-\log(.50)]+2.1542)/0.3533 = 5.06$. Thus, the corresponding median lethal concentration at 4 days of exposure is $\hat{LC}_{50}(4) = \exp(5.06) = 157.58$ nematodes per 0.5 ml water. Table 7.19 shows the estimated $\hat{LC}_{50}(t_j)$ and $\hat{LC}_{90}(t_j)$ for $t_j = 4, \dots ,11$. Each lethal concentration decreases as the time of exposure increases.

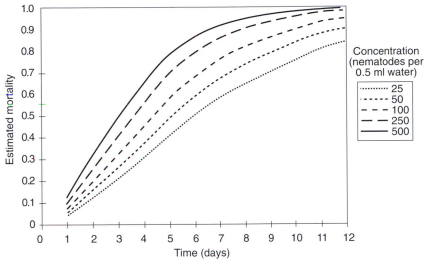

Fig. 7.11. Fitted mortality probabilities over time of exposure to the substance, at various concentrations, for the time–concentration–mortality data from Table 7.17.

Table 7.19. Estimated lethal concentrations $\hat{L}C_{50}(t_j)$ and $\hat{L}C_{90}(t_j)$ (nematodes per 0.5 ml water) at $t_j = 4, \ldots, 11$ days of exposure for the data from Table 7.17.

	Time (days)							
	4	5	6	7	8	9	10	11
$\hat{L}C_{50}(t_j)$	157	54	25	13	8	5	4	2
$\hat{L}C_{90}(t_j)$	4709	1619	749	392	241	157	105	65

Nowierski *et al.* (1996) used the cll models (35) and (37) to describe time–dose–mortality relationships for bioassays of fungal isolates. These authors also suggested the use of linear interpolation, as in eq. (29), with $\hat{F}(t_j)$ replaced by $\hat{F}(t_j;x_j)$, to estimate $LT_{50}(x_i)$, the time needed to kill 50% of the ticks exposed to a concentration x_i. For example, for the data from Table 7.17, the predicted mortalities at 4 and 5 days of exposure to a concentration of 100 (nematodes 0.5 ml^{-1} water) are 0.4458 and 0.5771, respectively. Thus, the time it takes for 50% of the ticks exposed to a concentration of 100 (nematodes 0.5 ml^{-1} water) to be killed, is estimated as $\hat{L}T_{50}(100) = 4 + 0.0548/0.1313 = 4.417$ days. It can be seen that, for a given π $(0 < \pi < 1)$, the lethal time, $\hat{L}T_{100\pi}(x)$ decreases as the concentration x increases.

In the cll models (35) and (37), presented above, it is assumed that the mortality of the untreated control ticks (at zero concentration) is zero over the duration of the experiment. In fact, the untreated control or background mortality may either vary or be constant over time. Nowierski *et al.* (1996)

considered the case in which background mortality changes with time. Let c_j denote the cumulative background mortality at time t_j. Let $F_{adj}(t_j;x_i)$ represent the true or control-adjusted mortality, caused by a concentration x_i of the microbial agent alone at time t_j ($j = 1, \ldots ,M;\ i = 1, \ldots ,k$). The cumulative mortality probability for concentration x_i by time t_j, or $F(t_j;x_i)$ can be written as:

$$F(t_j;x_i) = c_j + (1-c_j)F_{adj}(t_j;x_i) \tag{40}$$

Note that a similar expression was used in the control-adjusted model given in eq. (4) or (5). The maximum likelihood estimation procedure can be used to fit the cll models, given in (35) and (37) with slope $\beta = 0$, to the background mortality data ($j = 1, \ldots ,M$). The estimated background mortality at time t_j is denoted by $\hat{c}_j$ ($j = 1, \ldots ,M$) and the cll models (35) and (37) are then fitted to the adjusted mortality data. Furthermore, it can be seen that, for a given π ($0 < \pi < 1$), the estimated logarithm of the lethal concentration $\log(\hat{LC}_{100\pi}(t_j))$ that causes a total mortality of $100\pi\%$ by time t_j is obtained from

$$\log(\hat{LC}_{100\pi}(t_j)) = (\log[-\log(1-\pi_{adj}(t_j))] - \hat{\tau}_j)/\hat{\beta} \tag{41}$$

where $\pi_{adj}(t_j) = (\pi - \hat{c}_j)/(1 - \hat{c}_j)$.

Finally, we note that time–concentration–mortality experiments are usually designed such that mortality records of ticks exposed to a specified concentration are taken sufficiently often over the duration of the experiment; the experiment must not terminate too early. At least five concentrations should be used, as discussed in 'Analysis and comparison of the efficacies of two microbial agents' (see p. 265). Each concentration–time–mortality experiment must be repeated at least three times. If extra-binomial variation exists, because of changes among replications in the conditional mortality probabilities $\{\kappa_{ij}\}$, the data can be modelled by including an overdispersion parameter, as described at the end of 'Model fitting' (see p. 254) (see also Nowierski *et al.*, 1996).

Summary

In this chapter, basic concepts and selected non-linear models for the analysis of bioassay data have been introduced. Probit regression models, including the control-adjusted and the power models, are used for analysing bioassays which yield concentration–mortality data. Measures of the efficacy of microbial agents are introduced, including those based on the median lethal concentration (LC_{50}), or more generally, the lethal concentrations $LC_{100\pi}$, for $0 < \pi < 1$. Some guidelines and recommendations for designing bioassays and analysing concentration–mortality data have been provided.

In bioassays which record sublethal effects, the relationship between a positive continuous response and concentration was established; it uses exponential or log-linear models. Measures of the efficacy of microbial agents are based on the median effective concentration (EC_{50}) or, more generally, on effective concentrations $EC_{100\pi}$ for $0 < \pi < 1$.

The complementary log–log model is used to analyse the relationship between the exposure time and the mortality/survival data. The efficacy measures of microbial agents are based on the median lethal time (LT_{50}) or, more generally, on lethal times $LT_{100\pi}$ for $0 < \pi < 1$. Probit, log-linear and complementary log–log regression models are special cases of a wider class of generalized linear models.

Finally, the analysis of concentration–time–mortality data, using complementary log–log regression models, was considered. The efficacy measures of microbial agents are based on median lethal concentration at exposure time t ($LC_{50}(t)$) or, more generally, on lethal concentrations $LC_{100\pi}(t)$ for $0 < \pi < 1$. The relationship between lethal concentration $LC_{100\pi}(t)$ and time of exposure was described, and it was shown that for a given value of π, $LC_{100\pi}(t)$ decreases as time of exposure t increases.

Acknowledgements

We wish to thank Professor Bill Venables of the University of Adelaide, Australia, and Dr Andre Levesque of Agriculture and Agri-Food, Canada, for their criticism of a previous version of this manuscript. Their valuable comments and suggestions led to improvement in the presentation of this chapter.

Notes

1 Log to the base 10, $\log_{10} = \log/\log(10)$, is also of common use.
2 It can be shown that when $\gamma = 0$ the control-adjusted model is also the extreme point with $\lambda = 0$ in the power family of curves. This becomes apparent when we notice that $\alpha + \beta (x^\lambda - 1)/\lambda$ produces the same model as $\alpha' + \beta'x^\lambda$ with $\alpha' = \alpha - \beta/\lambda$ and $\beta' = \beta/\lambda$, and that $(x^\lambda - 1)/\lambda$ tends to $\log(x)$ as λ approaches zero.
3 Terms involving the binomial coefficients were omitted from the log likelihood function since they do not affect parameter estimates or standard errors.
4 See a similar study in Robertson and Preisler (1992), which uses the DOSESCREEN program, with a logit analysis.
5 Quantities involving factorial terms of the observations are omitted from the log-likelihood function, since they do not affect parameter estimates and their SES.
6 The SAS GENMOD procedure can also be used to analyse concentration–mortality data. However, it does not provide confidence limits for lethal concentrations, $LC_{100\pi}$, as the SAS PROBIT procedure does.

References

Abbott, W.S. (1925) A method of computing the effectiveness of an insecticide. *Journal of Economic Entomology* 18, 265–267.

Cochran, W.G. (1954) Some methods for strengthening the common χ^2 tests. *Biometrics* 10, 417–451.

Cody, R.P. and Smith, J.K. (1991) *Applied Statistics and SAS Programming Language*, 3rd edn. New Holland, New York.

Collett, D. (1991) *Modelling Binary Data*. Chapman and Hall, New York.

Eaves, D.M. and Marcus, R. (1997) A computational approach to bioassay. Unpublished manuscript.

Fieller, E.C. (1940) The biological standardization of insulin. *Journal of the Royal Statistical Society*, Series B, 7, 1–53.

Finney, D.J. (1971) *Probit Analysis*, 3rd edn. Cambridge University Press, London.

Finney, D.J. (1978) *Statistical Methods in Biological Assay*, 3rd edn. Macmillan, New York.

Gross, A.J. and Clark, V.A. (1975) *Survival Distributions: Reliability Applications in the Biomedical Sciences*. John Wiley & Sons, New York.

Lawless, J.F. (1982) *Statistical Models and Methods for Lifetime Data*. John Wiley & Sons, New York.

Lee, E.T. (1992) *Statistical Methods for Survival Data Analysis*, 2nd edn. John Wiley & Sons, New York.

McCullagh, P. and Nelder, J.A. (1989) *Generalized Linear Models*, 2nd edn, Chapman & Hall, London.

Nelder, J.A. and Wedderburn, R.W.M. (1972) Generalized linear models. *Journal of the Royal Statistical Society*, Series A, 135, 370–384.

Nowierski, R.M., Zeng, Z., Jaronski, S., Delgano, F. and Swearingen, W. (1996) Analysis and modelling of time–dose–mortality of *Melanoplus sanguinipes*, *Locusta migratoria migratorioides*, and *Schistocerca gregaria* (Orthoptera: Acrididae) from *Beauveria*, *Metarhizium*, and *Paecilomyces* isolates from Madagascar. *Journal of Invertebrate Pathology* 67, 236–252.

Preisler, H.K. (1988) Assessing insecticide bioassay data with extra-binomial variation, *Journal of Economic Entomology* 81, 759–765.

Preisler, H.K. and Robertson, J.L. (1989) Analysis of time–dose–mortality data. *Journal of Economic Entomology* 82, 1534–1542.

Robertson, J.L. and Preisler, H.K. (1992) *Pesticide Bioassay with Arthropods*. CRC Press, Boca Raton, Florida.

SAS Institute Inc. (1990) *SAS/STAT User's Guide, Version 6*, 4th edn. SAS Inst., Cary, North Carolina.

SAS Institute Inc. (1996) *SAS/STAT Software: Changes and Enhancements through Release 6.11*. SAS Inst., Cary, North Carolina.

Weibull, W. (1951) A statistical distribution function of wide applicability. *Journal of Applied Mechanics* 18, 293–297.

Appendix 7.1: Generalized Linear Models

Generalized linear models (GLMs) introduced by Nelder and Wedderburn (1972) form an extension of classical linear models. Refer to the book of McCullagh and Nelder (1989) for a thorough discussion of statistical modelling using GLMs. More specifically, let $\mathbf{y} = [y_1, \ldots ,y_n]'$ represent a vector of numerical observations. Each element y_i is observed under set #i of experimental conditions, where 'set #i' refers, for example, to dose level or microbial concentration #i. We write $\boldsymbol{\mu} = [\mu_1, \ldots ,\mu_n]'$ for the vector of their means and $\boldsymbol{\varepsilon} = \mathbf{y} - \boldsymbol{\mu} = [\varepsilon_1, \ldots ,\varepsilon_n]'$ for the vector of their random errors. A general linear model assumes that the mean, μ_i is of the form

$$\mu_i = \sum_{j=1}^{p} x_{ij}\beta_j$$

where x_{ij} is the specified value of the jth covariate or explanatory variable for observation #i, and $\beta_1, \ldots ,\beta_p$ are unknown parameters. In matrix notation we may write $\boldsymbol{\mu} = \mathbf{X}\boldsymbol{\beta}$ where $\mathbf{X}$ is an nxp matrix called the regression matrix or design matrix and $\boldsymbol{\beta} = [\beta_1, \ldots ,\beta_p]'$ is a vector of unknown regression parameters which must be estimated from the data. This is the systematic part of the model. For example, we may write $\mu_i = \beta_1 + \beta_2 x_i$ where μ_i is the mean response of insects exposed to an entomopathogenic substance at concentration x_i. In this case $\mathbf{X} = [\mathbf{1},\mathbf{x}]$ consists of just two columns $\mathbf{1}$ and $\mathbf{x}$ which contain, respectively, all 1s, and concentrations x_i. Sometimes, a transformation of concentration x_i, such as $\log(x_i)$ or x_i^λ for some power λ, is considered. For the random part, it is assumed that the elements ε_i ($i = 1, \ldots ,n$) are independent normal variables with zero means and a common variance σ^2.

The extension to a generalized linear model consists of the following elements:

A monotonic function $g(\mu)$, called a link function or response metameter is specified. The systematic part is written in the form

$$\eta_i = g(\mu_i) = \sum_{j=1}^{p} x_{ij}\beta_j$$

for $i = 1, \ldots ,n$. The η_is are called linear predictors. Thus, the link function g describes how the mean μ_i is related to the linear predictor η_i.

For the random part, it is assumed that the observations y_i are independent and follow one of the exponential family of distributions. This implies that the variance of each observation y_i depends on the mean, μ_i through a variance function $V(\mu_i)$, such that $\text{var}[y_i]/V(\mu_i) = \phi$ is a constant for all i. The constant ϕ is called a dispersion parameter; it is either known or must be estimated. For specific distribution families, such as binomial or Poisson families, the dispersion parameter is known ($\phi = 1$), whereas for the normal or the gamma distributions, ϕ is unknown.

The combination of the identity link $g(\mu) = \mu$ and the normal distribution associated with $V(\mu) = 1$ and unknown dispersion (thus variance) is the general linear model, which includes multiple regression, analysis of variance and analysis of covariance. Other popular link functions are: natural logarithm or log(μ), reciprocal or $1/\mu$, logit or log($\mu/(1-\mu)$) for $0 \leq \mu \leq 1$, and probit or $\Phi^{-1}(\mu)$, where Φ^{-1} is the inverse of the standard normal cumulative distribution function. Commonly used distribution functions and associated variance functions are: Poisson, with $V(\mu) = \mu$; binomial, with $V(\mu) = \mu(1-\mu)$ for $0 \leq \mu \leq 1$; and gamma, with $V(\mu) = \mu^2$. The analysis of dichotomous data tends to use logit and probit links and the binomial distribution almost exclusively. The analysis of frequency count data often tends to use the combination of $g(\mu) = \mu$ and the Poisson distribution. The gamma distribution implies that the coefficient of variation $\phi^{1/2} = (\text{var}[y_i])^{1/2}/\mu_i$ is constant; it is often used for analysing positive continuous response data.

Once a particular link function and a probablity distribution family have been chosen, the fitting of the generalized linear model proceeds similarly to that of the general linear model, with the usual blend of graphics and formal hypothesis testing associated with general linear models. More specifically, the maximum likelihood estimation (mle) method is used for estimating the model parameters. The likelihood function is the joint probability of all observations y_i with means μ_i expressed in terms of their unknown parameters. The mle procedure is based on maximization of the likelihood function or equivalently the log-likelihood function, $L(\mathbf{y};\boldsymbol{\mu})$, with respect to the unknown parameters that express the μ_is. Substitution of the maximum likelihood estimators of the model parameter yields the estimated vector of the means, which is denoted by $\hat{\boldsymbol{\mu}}$. The maximum value of the log likelihood function is denoted by $L(\mathbf{y}; \hat{\boldsymbol{\mu}})$.

For specific distribution families with $\phi = 1$ (e.g. the binomial and Poisson families), a deviance function is defined as

$$\text{Dev}(\mathbf{y};\boldsymbol{\mu}) = 2[L(\mathbf{y};\mathbf{y}) - L(\mathbf{y};\boldsymbol{\mu})]$$

Note that the term $L(\mathbf{y};\mathbf{y})$ is the maximum achievable log likelihood obtained when all μ_is are allowed to be distinct, without any constraints and it is, therefore, estimated as the y_is. Maximizing $L(\mathbf{y};\boldsymbol{\mu})$ is equivalent to minimizing $\text{Dev}(\mathbf{y};\boldsymbol{\mu})$, subject to the constraints implied by the model. The deviance of the fitted model is denoted by $\text{Dev}(\text{fitted}) = \text{Dev}(\mathbf{y};\hat{\boldsymbol{\mu}})$, where $\hat{\boldsymbol{\mu}}$ represents the estimated vector of the means for a given model. Thus, $\text{Dev}(\text{fitted})$ measures the difference between the maximum achievable and the fitted model log likelihood values. It can be shown that under certain conditions $\text{Dev}(\text{fitted})$ tends to be approximately equal to Pearson's statistic defined as

$$X^2(\text{fitted}) = \Sigma(y_i - \hat{\mu}_i)^2/V(\hat{\mu}_i)$$

The $X^2(\text{fitted})$ is the sum of the squared differences $(y_i - \hat{\mu}_i)^2$ between the observations y_i and the estimated means $\hat{\mu}_i$, weighted by taking into account the relative reliabilities of the y_is. Both $\text{Dev}(\text{fitted})$ and $X^2(\text{fitted})$ measure how poorly the model fits the data. Under certain conditions, if sample sizes are large and if the model is correct, the distribution of each of the Dev(fitted) and X^2(fitted) statistics tends to a χ^2 distribution with $n - q$ df, where n is the number of observations and q is the number of parameters included in the model. Thus, a Dev(fitted) or X^2 (fitted) that seems too large in terms of the χ^2 distribution with $n - q$ df is considered to be evidence against the model.

Given observations $\mathbf{y} = [y_1, \ldots, y_n]'$, we can fit various models for the true means, $\boldsymbol{\mu} = [\mu_1, \ldots, \mu_n]'$. Depending on the covariates which are expressed by $\mathbf{X}$, the estimated means in $\hat{\mu}_i$s might all be the same (the null model) or, at the other extreme, they might include as many unrelated distinct values as there are distinct covariate patterns $\mathbf{x}_i = [x_{i1}, \ldots, x_{ip}]$ for $i = 1, \ldots, n$ (the full model). Suppose we obtain a vector, $\hat{\boldsymbol{\mu}}_A$ of estimates from a model with q_A parameters known to be true, and $\hat{\boldsymbol{\mu}}_B$ from a submodel with q_B parameters ($q_B < q_A$). For distribution families with $\phi = 1$, the lack-of-fit of the submodel B is tested by means of the log-likelihood ratio statistic, expressed as:

$$2[L(\mathbf{y}; \hat{\boldsymbol{\mu}}_A) - L(\mathbf{y}; \hat{\boldsymbol{\mu}}_B)] = \text{Dev(fitted B)} - \text{Dev(fitted A)}$$

or, equivalently, by a corresponding Pearson's statistic. These statistics measure how far wrong is model B. Under certain conditions, if the sample sizes are large and if submodel B is correct, the distribution of each of these test statistics approaches a $\chi^2(q_A - q_B)$ distribution. Thus, large values in terms of the $\chi^2(q_A - q_B)$ form evidence against model B.

For distribution families with unknown ϕ (e.g. normal and gamma), the dispersion parameter ϕ must be estimated from the data. Common estimates of ϕ are: $\hat{\phi} = \text{Dev(fitted B)}/(n - q_B)$ or $X^2\text{(fitted B)}/(n - q_B)$ or the maximum likelihood estimate. In this case, the lack-of-fit of model B is judged in terms of the scaled deviance, defined as $[\text{Dev(fitted A)} - \text{Dev(fitted B)}]/\hat{\phi}$ or in terms of a corresponding scaled Pearson's statistic. Under certain conditions, if sample sizes are large and if submodel B is correct, the significance of each of these statistics is judged in terms of the χ^2 distribution with $q_A - q_B$ df.

In some situations, the form of the likelihood function for the observed data is not completely specified. In such a case, the maximum quasi-likelihood estimation procedure is used for estimating the model parameters. Most of the relevant theory appears in McCullagh and Nelder (1989). For example, proportion data with extra-binomial variation or count data with extra-Poisson variation may be analysed by means of quasi-likelihood theory.

Legislation Affecting the Collection, Use and Safe Handling of Entomopathogenic Microbes and Nematodes

8

D. Smith

CABI Bioscience UK Centre (Egham), Egham, Surrey, UK

Introduction

Collection, isolation, handling and maintenance of organisms are subject to safety regulations and legal obligations. Legislation continues to develop and change (Smith, 1996), which undoubtedly affects biologists who collect and characterize organisms in their goal to further scientific knowledge. In particular, the Convention on Biological Diversity (CBD), which was signed at Rio de Janeiro in 1992 and came into force in December 1993 (CBD, 1992) and has now been ratified by more than 140 countries, controls access to *in situ* organisms. The CBD gives sovereign rights over genetic resources to the country of origin. In the simplest of terms, the CBD requires a biologist who wishes to collect genetic resources to seek prior informed consent from relevant authorities and negotiate fair and equitable sharing of benefits that may arise from their use before access can be granted. The Convention and national legislation on access to genetic resources place an enormous duty on the shoulders of the collector, but are not intended to prevent the advancement of science.

Organisms of hazard groups 2, 3 and 4 (see p. 306 for definitions of hazard groups) are hazardous substances under the UK Control of Substances Hazardous to Health (COSHH) legislation. They fall under the EU Biological Agents Directives and are dangerous goods as defined by the International Air Transport Association (IATA) Dangerous Goods Regulations (IATA, 1998), where requirements for their packaging are defined. Further, there are restrictions on distribution imposed by National Postal Authorities, according to which more and more countries prohibit receipt of Infectious, Perishable Biological Substances (IPBS) and, in some cases, Non-infectious

© CAB *International* 2000. *Bioassays of Entomopathogenic Microbes and Nematodes* (eds A. Navon and K.R.S. Ascher)

Perishable Biological Substances (NPBS), including hazard group 1 organisms. Whether organisms are shipped by mail, courier or by hand and whether between or within countries, thought must be given to the regulations that control these matters. The EC Directives 93/88/EEC and 90/679/EEC on Biological Agents set mandatory control measures for laboratories requiring that risk assessments are carried out on all organisms handled. This necessitates the assignment of each strain to a hazard group following a thorough risk assessment including a positive inclusion into hazard group 1 when they are not categorized in hazard groups 2, 3 or 4. The risk assessment should include an assessment of all hazards involved, including the production of toxic metabolites and the ability to cause allergic reactions. Most entomopathogens, in particular those currently used for pest control, are classified as hazard group 1, but care must be taken as they may produce toxic metabolites.

The implications of a laboratory's health and safety procedures stretch beyond the laboratory to all those who may come in contact with substances and products from that laboratory. An organism in transit will potentially put carriers, postal staff, freight operators and recipients at risk. It is essential that safety and shipping regulations be followed to ensure safe transit. More stringent shipping regulations have evolved because of increasing cases of careless and negligent handling. Sound packaging and correct labelling and information must be used to minimize risk.

This chapter draws attention to legislation and requirements relevant to collecting and handling biological specimens. Many countries do not have health and safety or access to genetic resources legislation and in such cases it is recommended that as a guide the regulations of other countries can be followed. Attention is drawn to legislation that could be followed. Ownership of intellectual property rights is discussed, in particular, how this relates to patenting living organisms and the CBD. The importance of health and safety legislation in handling, storage and supply of organisms is raised and their classification on the basis of hazard and associated risk are discussed. Regulations affecting the distribution of organisms covered here concern postal, shipping, packaging, quarantine and control of dangerous pathogens and safety information provision. Sources for further information are provided and some suggestions on sound practices are offered.

Ownership of Intellectual Property Rights

Organisms can be collected from different habitats all over the world. This begs the question: Who owns these organisms and the intellectual property rights associated with them? The CBD bestows sovereign rights over genetic resources to the country where they arise. The landowner and the national government are the first stakeholders, followed by the collector, those involved in purification and growing the organism, the discoverer of intellectual property, depositor, collection owner and the developer of any process. It is clear that they do not all have an equal stake and this will depend upon their input to discovery or process. This has implications for the sharing of benefits arising from exploitation of the genetic resource. These are amongst the issues that are being discussed by delegates from the countries signatory to the CBD who meet at the Conference of the Parties and their workgroups. Information on the progress of these discussions can be found on the CBD web site (http://www.biodiv.org/).

Patents including living organisms

The general principles of international patent law require that details of an invention must be fully disclosed to the public. Inventions involving the use of organisms present problems of disclosure as a patented process often cannot be tested following the publication of a written description alone. If a process involving an organism has novelty, inventiveness, utility or application and sufficient disclosure, it can be subject to patent (Kelley and Smith, 1997). Organisms are not patentable in their natural state or habitat, new species are discoveries, not inventions. The invention of a product, a process of manufacture or a new use for a known product is an intellectual property owned by the inventor whether it involves an organism or not. It is often difficult to patent organisms as products themselves, although genetically manipulated microorganisms are usually considered as a human invention and are therefore patentable. The situation is less clear in the case of spontaneous or induced mutants of naturally occurring organisms, which fall midway between the natural and the artificial.

In many cases the organism involved must be part of the disclosure. This reasoning has led to an increasing number of countries either requiring by law or recommending to inventors that the written disclosure of an invention involving the use of organisms be supplemented by the deposit of the organism in a recognized culture collection. It is recommended by most patent lawyers that the organism is deposited, regardless of it being a requirement, rather than running the risk of the patent being rejected. To remove the need for inventors to deposit their organisms in a collection in every country in which they intend to seek patent protection, the 'Budapest Treaty on the International Recognition of the Deposit of Micro-organisms for the Purpose of Patent Procedure' was concluded in 1977 and came into force towards the end of 1980. This recognizes named culture collections as 'International Depositary Authorities' (IDA) and a single deposit made in any one is accepted by every country party to the treaty. Any collection can become an IDA providing it has been formally nominated by a contracting state and meets certain criteria. The CABI Genetic Resource Collection is one of 29 IDAs around the world accepting patent deposits of fungi, including yeasts, phytopathogenic bacteria and nematodes. There are six other

IDAs in the UK and many of the others are collections belonging to the European Culture Collection Organization (Anon, 1995). The Budapest Treaty provides an internationally uniform system of deposit and lays down the procedures which depositor and depository must follow, the duration of deposit and the mechanisms for the release of samples. Thirty-six states and the European Patent Office are now party to the Budapest Treaty.

The World Intellectual Property Organization (WIPO) publishes data on the numbers of microorganisms deposited in collections under the terms of the Budapest Treaty (1977). Since the treaty's inception, there were 24,712 deposits up to December 1994 (Anon, 1996a). Patent protection is covered in Article 16 of the CBD under which parties must cooperate, subject to national legislation and international law, to ensure patents and other intellectual property rights are supportive of, and do not contravene, the objectives of the Convention (Fritze, 1994). This remains an area of dispute as the Article leaves open the possibility that the CBD takes priority over national patent law. Patent law and the CBD are generally compatible but can conflict in cases where exploitation may endanger the resource. In many cases where organisms are grown artificially there is no threat to the existence of the species. Details of the requirements for a collection which relate to the deposit of an organism can be obtained directly from IDA collections and are summarized by Kelley and Smith (1997).

It is quite clear that every intermediary in an improvement or development process is entitled to a share of the IPR, which adds another dimension to ownership. It is therefore critical that clear procedures on access, mutually agreed terms on fair and equitable sharing of benefits and sound material transfer agreements are in place to protect interested parties.

The Convention on Biological Diversity

The CBD aims to encourage the conservation and sustainable utilization of the genetic resources of the world and has a number of articles that affect biologists. These cover:

- Development of national strategies for the conservation and sustainable use of biological diversity.
- Identification, sampling, maintenance of species and their habitats and the production of inventories of indigenous species.
- Encouraging *in situ* and in-country *ex situ* conservation programmes.
- Adoption of economically and socially sound measures to encourage conservation and sustainable use of genetic resources.
- Establishment of educational and training programmes and the encouragement of research.
- Commitment to allow access to genetic resources for environmentally sound uses on mutually agreed terms and with prior informed consent.

- Fair and equitable sharing of benefits and transfer of technology resulting from exploitation of genetic resources.
- Exchange of information.
- Promotion of technical and scientific cooperation.

The principles should not affect the development of science, but unfortunately some countries' legislation is placing obstacles in the way. It is essential that biologists continue to lobby their country representatives to ensure that science is not impeded whilst the principles of the CBD are being implemented. For example, Brazil is currently developing a new intellectual property law, which includes an article protecting traditional and ethnic knowledge. However, scientists fear that if the law is too restrictive and, by making it a criminal offence to remove genetic material from the country, it will restrict collaboration with foreign researchers (Neto, 1998). Currently, any foreign scientist wishing to work with biodiversity in Brazil must be accompanied by researchers from Brazil and confer co-authorship on subsequent publication of results.

The CBD requires that prior informed consent (PIC) be obtained in the country where organisms are to be collected before access is granted. Terms on which any benefits will be shared must be agreed in advance. The benefit sharing may include monetary elements but may also include information, technology transfer and training. If the organism is passed to a third party it must be under terms agreed by the country of origin. This will entail the use of material transfer agreements between supplier and recipient to ensure benefit sharing with, at least, the country of origin. Many biological resource centres or culture collections have operated benefit sharing agreements since they began, giving organisms in exchange for deposits and re-supplying the depositor with the strain if a replacement is required. However, huge rewards that may accompany the discovery of a new drug are illusory as the hit rate is often reported as less than 1 chance in 250,000. In the meantime, access legislation and the hope for substantial financial returns from isolated strains are restricting the free deposit in public service collections and the legitimate free movement of strains. An EU DG XII project, Micro-organisms Sustainable Use and Access Regulation International Code of Conduct (MOSAICC) is developing mechanisms to allow traceability and enable compliance with the spirit of the CBD and with national and international laws governing the distribution of microorganisms, whilst not restricting scientific goals (Davison *et al.*, 1998).

Perspectives on the Convention

There have been several interpretations on how the Convention affects the microbiologist, but discussions continue to leave unresolved problems. Six years on from the Earth Summit in Rio de Janeiro, yet more problems were raised at the fourth Conference of the Parties held in Bratislava in 1998. It is difficult for biologists to know what is required of them, particularly where there is often no country law and codes of practice for them to follow. Whilst the debate continues, there is a danger that biologists will ignore their responsibilities under the CBD and thus compound the problems. A few simple steps to allow traceability of organisms can ensure compliance with legislation. Interested parties have expressed their views, some thinking the Convention impractical and impossible to implement, others seeing the opportunity for the protection of biodiversity and its sustainable use. Some of these points of view are expressed below (Kirsop, 1993; Sands, 1994; Kelley, 1995).

- Total opposition to intellectual property rights on life forms including humans, animals, plants, microorganisms or their genes, cells or other parts.
- Organisms are the heritage of humankind and consequently should be available without restriction. This view was expressed by the signatories of the 'Thammasat Resolution', *Building and Strengthening Our sui genesis Rights*, the result of negotiation between 45 representatives of non-government and government agencies, academia and others from 19 countries (Internet communication: 1997; Genetic Resources Action International@igc.org).
- Developing countries see the Convention as a means for financing development.
- Developed countries hoped it would help to preserve threatened environments and biodiversity.
- There is an opportunity for individuals, organizations and countries to profit from equitable sharing of the benefits.
- There is an opportunity to harness the environment as a source of genetic material, rather than of renewable or non-renewable materials, by giving countries a financial interest in its maintenance and preservation and thus allowing the international community as a whole to benefit.
- The Convention impedes the progress of science, and could have an effect on the continuance of some service collections.

It is clear that many concerns exist and these will take time to resolve. In the meantime, countries are developing legislation to control access to their genetic resources and biologists are struggling to comply. The IUCN has produced a guide to designing legal frameworks to determine access to genetic resources (Glowka, 1998) which examines the Convention and national access legislation. In the Philippines, Executive Order 247 puts in place a mechanism to ensure it controls access to and use of its genetic resources. The Andean countries are also developing their own regulations and procedures. The CBD Secretariat offers information on developments to attain workable regulations (http://www.biodiv.org/).

The role of public service collections

Public service culture collections are charged with several tasks, which are influenced by access legislation. They are in a unique position as custodians of *ex situ* genetic resources and therefore have a key role to play in the conservation of genetic resources (Kirsop and Hawksworth, 1994). Biologists who collect organisms for their research and publish information on them should make their most important strains available for confirmation of results and future use by depositing them in public service collections. This will aid collections in their major roles:

- The *ex situ* conservation of organisms.
- Custodians of a national resource.
- Provide a living resource to underpin the science base.
- Receive deposits subject to publication.
- Offer safe, confidential and patent deposit services.

The Convention should not affect these functions and will increase the importance and extent of the collection's role. However, depositors are increasingly concerned about who the customers are, and if their rights as a stakeholder are protected.

Approaches taken by collections to comply with the CBD

To date little guidance has been given to collections to determine actions necessary to comply with the CBD. Collections have therefore developed several approaches independently.

- Statements are prominently displayed on accession forms and on information accompanying delivery of strains explaining the implications of, and requesting compliance with, the Convention. This draws attention to the requirements but does not protect the sovereign rights of the country of origin nor any other stakeholder.
- A requirement for depositors to declare in writing that PIC has been obtained and that this includes unrestricted distribution of the materials to third parties or has clearly defined conditions on distribution.
- A requirement for a signed material transfer agreement on supply of material including mutually agreed terms.

These are minimum requirements and should be followed by all. The difficulty lies in defining the beneficiaries and what is a fair and equitable sharing of benefits.

Development of policies on intellectual property rights (IPR) and compliance with the CBD

Several organizations have addressed these issues and have developed and published their policies. These organizations include large national collections, international organizations and industrial companies. CAB International (CABI) is an intergovernmental organization established by treaty, dedicated to improving human welfare through the application of scientific knowledge in support of sustainable development worldwide, with emphasis on agriculture, forestry, human health and conservation of natural resources, and with particular attention to the needs of developing countries (http://www.cabi.org). The CABI Genetic Resources Collection (GRC) is based at the CABI Bioscience UK Centre (Egham) and is tasked with the collection of organisms to provide a resource for the scientific programmes of CABI and to underpin biotechnology, conservation and science in its member countries. CABI maintains extensive collections that originate from many different countries and has introduced policy and procedures to ensure compliance with the requirements of the CBD. This policy was agreed by member country representatives and published in the 13th Review Conference Proceedings (Anon, 1996b).

The CABI policy offers an example of a mechanism to enable compliance with the CBD. CABI keeps the rapidly changing situation under review and will adopt procedures required to keep its operations within the spirit of the Convention. CABI complies with national legislation of member country governments concerning rights over natural resources and access to genetic material, and interprets its policies in a manner consistent with the CBD. CABI treats all its living material holdings as subject to the sovereign rights of the country of origin. In considering any activity relating to the possible exploitation of biodiversity, CABI will seek to protect the interests of the source country of each element of biodiversity. CABI adds value to living or dead material it receives and collects particularly by ensuring authoritative identifications. It makes its reference collections and the information on them available to institutions in the countries of origin and the scientific community in general.

Supply of living strains from the CABI Collection requires a signed declaration from the recipient undertaking not to exploit the organism or related information. Recipients who wish to exploit materials are requested to negotiate terms through CABI. New deposits are equally controlled. Before CABI can consider the acceptance of strains into its collections confirmation is required from the depositor that the collector has made reasonable efforts to obtain PIC to collect the organisms and also has permission to deposit the strains in a public service culture collection. CABI also needs to know if there is any restriction on further distribution and if there are conditions that must be included in any material transfer agreement that may accompany the samples when they are passed to a third person.

CABI needs to have such information from all depositors regardless of the country of origin of the material or the collector. This is often difficult to enforce, as in most countries a PIC authority has not been appointed. In such cases, proof is required that a depositor has made reasonable efforts to get permission to collect from landowners and a government office.

Biological resource collections, such as public service collections, like the CABI Collection, often add value to received and collected biological material. This is done through purification, expert preparation, authoritative identification, description, determination of biochemical and other characteristics, comparison with related material, safe and effective storage/preservation, evaluation of value for specified uses, and indication of importance of beneficial and detrimental attributes. They often provide samples of deposited organisms free of charge to the depositor and participate in capacity building projects to help establish facilities and expertise in-country to maintain *ex situ* collections. This plays a role in the utilization of genetic resources and defines a collection as a stakeholder.

Suggested code of conduct for collection of biodiversity

Biologists collect entomopathogenic nematodes and microorganisms all over the world, not just in their own country. No matter where they collect they must do so legally, following national and international law. The following principles should be borne in mind:

- Do not collect material from any country without prior informed consent (PIC).
- Routinely seek documentation through standard identification or submission forms to clarify the status of the material received. Do not make the assumption that a sender of material has the authority to allow you to use and dispose of material as you deem fit.
- Do not make material available to third parties for the development of commercial products unless you have been given the authority to do so by the source country, or the recipient agrees in writing not to exploit such material without negotiating an agreement to do so with the source country and stakeholders.
- All material can be used for scientific research but the country of origin should benefit from receipt of published information.
- Collections should be prepared to loan dead material to scientists at government, university and research institutes in all countries of the world for research and identification purposes.
- The depositor of a culture into a collection should be given the right to the return of a replicate of their deposit on request at reasonable intervals without any charge.
- As far as practical, ensure that type material based on specimens submit-

ted is deposited in a recognized reference collection in accordance with any legislation of the country of origin. Where such legislation is not in existence, the material should be deposited in accordance with internationally accepted principles for stewardship of such material and the interests of the international scientific community as a whole.

Problems to be resolved

There are several problems that can impede the development of procedures for compliance with the CBD and these will need some time to resolve.

- Clarification is required on ownership, intellectual property rights and benefit sharing.
- Identification of country authorities who can grant prior informed consent.
- Identification of stakeholders and assessment of the value of their input.
- Establishing a clear, simple and flexible approach that avoids impractical bureaucracy.
- Monitoring and enforcement of procedures put in place.
- Keeping up to date with country legislation.

The CBD is not an opportunity for all to benefit financially, and prospects of accruing huge profits from exploiting an organism for the country of origin are very slim. Additionally, the process from sampling to market can take from 8–15 years, therefore nothing will be achieved quickly, and is likely to require considerable investment. The CBD was negotiated to protect genetic resources and ensure their sustainable use.

The agreement on Trade-related Aspects of Intellectual Property Rights (TRIPs) is thought to conflict with the CBD where there is concern that developing countries are required to allow companies to take out patents on products and processes of biotechnology. There are several forms of intellectual property rights that are relevant to the Convention in addition to patents, for example copyright, trade secrets and plant breeder's rights. The CBD requires that terms for technologies subject to IPR protection are to recognize and be consistent with adequate and effective protection of IPR (Glowka, 1998). In reality, so long as there is an agreement on mutually agreed terms for benefit sharing with the country of origin, the TRIPs agreement and patenting do not run contrary to the CBD.

Health and Safety

Organisms can present several challenges to health and safety including infection, poisoning and allergies (Anon, 1993; Stricoff and Walters, 1995). Handling, distribution and use of organisms are controlled by regulations.

Whether it is compliance with the law or duties of a caring employer, the basic requirements to establish a safe workplace are:

- Adequate assessment of risks.
- Provision of adequate control measures.
- Provision of health and safety information.
- Provision of appropriate training.
- Establishment of record systems to allow safety audits to be carried out.
- Implementation of good working procedures.

Good working practice requires assurance that correct procedures are being followed and this requires a sound and accountable safety policy.

The UK Management of Health and Safety at Work (MHSW) Regulations 1992 (Anon, 1992) are all encompassing and general in nature but overlap and lead into many specific pieces of legislation. The Control of Substances Hazardous to Health (COSHH) regulations require that every employer makes a suitable and sufficient assessment of the risks to health and safety to which any person, whether employed by them or not, may be exposed through their work (Anon, 1996d). These assessments must be reviewed regularly and must be recorded when the employer has more than five employees. The distribution of microorganisms to others outside the workplace extends these duties to protect others. Such assessments of risk are extended to other biological agents, such as entomopathogenic nematodes, through EC Council Directives on Biological Agents (90/679/EEC; 93/88/EEC).

The effect of some safety regulations on culture storage and supply

The COSHH regulations (1988) require a suitable risk assessment for all work that is liable to expose an employee to any substance that may be hazardous to health. This UK legislation has equivalents in other countries and at the European level. Organisms present different levels and kinds of hazard, evaluation of which represents an enormous, but necessary, task for biologists. A risk assessment, for example, must take into account the production of potentially hazardous toxins. Ultimately, a safe laboratory is the result of applying good techniques, a hallmark of technical excellence. Containment level 2 (Anon, 1996c) is easily achievable and should be standard practice in all laboratories handling organisms that present a risk of infection or of causing other harm. Good aseptic techniques applied by well-trained personnel will ensure pure and clean cultures and will minimize contact with the organism. However, accidents must also be taken into account when assessing risks. The employment of good laboratory practice and good housekeeping, workplace and equipment maintenance, together with ensuring that staff have relevant information and training, will minimize the risk of accidents. The establishment of emergency procedures to reduce potential harm is an additional and sensible precaution.

Classification of organisms on the basis of hazard

Various classification systems exist, including those of the World Health Organization (WHO), United States Public Health Service (USPHS), Advisory Committee on Dangerous Pathogens (ACDP), European Forum for Biotechnology (EFB) and the European Commission (EC). In Europe, the EC Council Directive (93/88/EEC) on Biological Agents sets a common base line which has been strengthened and expanded in many of the individual member states. In the UK, the definition and minimum handling procedures for pathogenic organisms are set by the ACDP, who list four hazard groups with corresponding containment levels (Anon, 1996c).

Group 1   A biological agent that is most unlikely to cause human disease.

Group 2   A biological agent that may cause human disease and which might be a hazard to laboratory workers but is unlikely to spread in the community. Laboratory exposure rarely produces infection and effective prophylaxis or treatment is available.

Group 3   A biological agent that may cause severe human disease and present a serious hazard to laboratory workers. It may present a risk of spread in the community but there is usually effective prophylaxis or treatment.

Group 4   A biological agent that causes severe human disease and is a serious hazard to laboratory workers. It may present a high risk of spread in the community and there is usually no effective prophylaxis or treatment.

Risk assessment

The species of bacteria, fungi and parasites falling into hazard groups 2 and 3 have been defined (Anon, 1996c). Similarly, all bacteria from the approved list have been assigned to an appropriate hazard group in Germany (Anon, 1997a, b, 1998). However, species of fungi have not been assigned to hazard group 1 (Anon, 1996c, d). Medically important fungi have been categorized into relevant hazard groups by de Hoog (1996). To meet the UK and European legislation, all microbiologists will have to make a risk assessment on the organisms with which they work or hold in collections. In the case of fungi, it is recognized that many may infect following traumatic inoculation through the skin, or infect a compromised patient, but do not infect healthy individuals. Most fungi from clinical specimens require Containment Level 2 (Anon, 1996c), although a higher degree of containment is specified for a few.

The COSHH regulations work well and can be easily applied in establishments with designed laboratories, but may not work as well in an industrial environment where very large volumes and more hazardous techniques may be used. Total containment is rarely applicable.

Assessment of the risks involved in handling organisms

Compared with chemicals, organisms are more difficult to name, less predictable and more difficult to enumerate or measure. Virulence and toxicity may vary from strain to strain and additional hazards, such as mycotoxin production and allergenicity, must be considered. To meet biological agents legislation, a step by step evaluation of a laboratory procedure or an industrial process must be carried out. The assessment must cover the procedure from the original inoculum or seed culture to the final product or the point where the organism is killed and disposed of. It must be noted that individuals may respond differently to exposure, with some being more sensitive than others. It is therefore critical that the full hazard potential of the organism is considered and that this is related to effects it may have on the particular individual carrying out the work.

Regulations Governing Distribution of Cultures

The distribution of organisms is controlled by numerous regulations. Some are discussed below and include postal and shipping regulations, requirements for packaging aimed at protecting handlers and recipients of organisms, and quarantine legislation to protect plant health. Countries have their own regulations governing the packaging and transport of biological material in their domestic mail. International Postal Regulations regarding the transport of human and animal pathogens are very strict because of the safety hazard they present. There are several organizations that set regulations controlling the international transfer of such material. These include the International Air Transport Association (IATA), International Civil Aviation Organization (ICAO), United Nations Committee of Experts on the Transport of Dangerous Goods, the Universal Postal Union (UPU) and the World Health Organization (WHO).

Postal regulations

It is common to send microorganisms by post, as this is more convenient and less expensive than airfreight (Rohde *et al.*, 1995). However, many countries prohibit the movement of biological substances through their postal services. The International Bureau of the UPU in Berne publishes all import and export restrictions for biological materials by national postal services (UPU, 1998). The UK Post Office leaflet on 'Infectious and non-infectious perishable biological substances in the overseas post' is available from the Post Office (Corporate Headquarters, 30 St James Square, London SW1 4PY. Tel: +44 171 490 2888; Fax: +44 181 681 9387) and provides relevant information. A list, which changes from time to time, of countries that will not accept human pathogens through the post can also be obtained from the Post Office (also see Anon, 1998; Smith, 1996).

Shipping regulations

The IATA Dangerous Goods Regulations require that shippers of microorganisms of hazard groups 2, 3 or 4 must be trained by IATA certified and approved instructors. They also require shippers' declaration forms, which should accompany the package in duplicate, and specified labels are used for organisms in transit by air (IATA, 1998). There are several other regulations that impose export restrictions on the distribution of microorganisms. These include control of distribution of agents that could be used in biological warfare and EU Council Regulation (3381/94/EEC) on the control of export of dual-use goods. More generally, countries are currently implementing Access Regulations to Genetic Resources under the Convention on Biological Diversity. It is critical that microbiologists are aware of, and follow, such legislation. Details can be found in Alexander and Brandon (1986), *Shipping of Infectious, Non-infectious and Genetically Modified Biological Material, International Regulations* (Anon, 1998) and IATA Dangerous Goods Regulations (IATA, 1998).

In Europe, class 6.2 Dangerous Goods are transported by road, packed according to EN 829 requirements. Transport by road is regulated by the *Accord Européen Relatif au Transport International des Merchandises Dangereuses par Routes* (ADR). This clearly separates class 6.2 into two subclasses: A, highly infectious material (hazard groups 3 and 4), and B, other infectious material. These two groups, A and B, have different packaging requirements. However, currently there are no manufacturers producing these different shipping containers so that the UN specification containers for class 6.2 materials must be used for both subclasses. The EU has made an attempt to coordinate Member State laws on transport of dangerous goods by road with the 'ADR-Directive' (EC Council Directive (94/55/EC) of 21 November 1994) *on the Approximation of the Laws of the Member States on the Transport of Dangerous Goods by Road* and its annexes (EC, 1996).

The basis for all regulations governing the safe transport of goods for all carriers are laid down in the *Orange Book, Recommendations on the Transport of Dangerous Goods, Tests and Criteria* (Anon, 1997c).

Packaging

IATA Dangerous Goods Regulations (DGR) require that packaging used for the transport of hazard group 2, 3 or 4 must meet defined standards according to IATA packing instruction 602 (class 6.2) (IATA, 1998). The DSMZ collects all relevant guidelines for the shipping of microorganisms and updates it on a regular basis (Anon, 1998). It is also available on the DSMZ web site (http://www.gbf.de/dsmz/shipping/shipping.htm). Packaging must meet EN 829 triple-containment requirements for hazard group 1 organisms. However, microorganisms that qualify as dangerous goods (class 6.2) and are sent by air must be in UN-certified packages. These packages must be sent by airfreight if the postal services of the countries through which it passes do not allow the organisms in their postal systems. They can only be sent airmail if the national postal authorities accept them. There are additional costs above the freight charges and package costs if the carrier does not have its own fleet because the package and documentation will need to be checked at the airport DGR centre.

Quarantine regulations

Clients in the UK who wish to obtain cultures of non-indigenous plant pathogens must first obtain a MAFF Plant Health Licence and provide a letter of authority. Such licences can be applied for in England and Wales from the Ministry of Agriculture, Fisheries and Food, Room 340, Foss House, Kings Pool, 1–2 Peace Holme Green, York YO1 2PX, and in Scotland from the Plant Health Section, Agricultural Science Agency, East Craigs, Edinburgh EH12 8NJ. Non-indigenous tree pathogens can only be supplied if the customer holds a current permit issued by The Forestry Commission, Forestry Commission Headquarters, 231 Corsthorphine Road, Edinburgh EH12 7AP.

All shipments to Canada and the USA of plant pathogens must be accompanied by import mailing labels, without which entry of cultures to these countries is refused. Applications for these labels, stating the names of the organisms and the purpose for which they are required, should be made for Canada to the Chief of the Plant Protection Division, Agriculture Canada Science Division, Science Service Building, Ottawa, Ontario, Canada K1AS 0C5, and for the USA to USDA Agricultural Research Service, Plant Protection and Quarantine, Room 764, 6505 Belcrest Road, Hyattsville, MD 20782, USA.

Information on the transport of plant pathogens throughout Europe can be obtained from the European and Mediterranean Plant Protection Organization (EPPO), 1 rue le Nôtre, 75016 Paris, France. EC Council Directive (77/93/EEC), on protective measures against the introduction of harmful organisms and of plant or plant products, also provides useful information.

Control of dangerous pathogens

There is considerable concern over the transfer of selected infectious agents capable of causing substantial harm to human health. There is potential for such organisms to be passed to parties not equipped to handle them or to persons who may make illegitimate use of them. Of special concern are pathogens and toxins causing anthrax, botulism, brucellosis, plague, Q fever, tularemia and all agents classified for work at Biosafety Level 4 (hazard group 4). The Australia Group Countries have strict controls for movement outside their group of countries but has lower restrictions within. The UK National Culture Collections have implemented a system involving the registration of customers to ensure bona fide supply (see http://www.ukncc.co.uk). The USA have rules that include a comprehensive list of infectious agents, registration of facilities that handle them and requirements for transfer, verification and disposal. Contravention of the rules entails criminal and civil penalties. In the UK, all facilities handling hazard groups 2, 3 or 4 must be registered. Strict control of hazard group 3 and 4 organisms is in place.

Safety Information Provided to the Recipient of Microorganisms

A safety data sheet must be despatched with an organism, indicating to which hazard group it belongs and the containment and disposal procedures required. In the UK, microorganisms are covered by the Control of Substances Hazardous to Health (COSHH) regulations (1988), HSW Act (Anon, 1974) s.6(4)(c) and subject to the Approved Code of Practice Biological Agents (Anon, 1996d). *Substances for Use at Work: the Provision of Information* (1985) provides details of the safety data that must be provided. Article 10 of the EC Council Directive (90/679/EEC) dictates that manufacturers, importers, distributors and suppliers must provide safety data sheets in a prescribed format. A safety data sheet accompanying a microorganism must include:

- The hazard group of the organism being despatched as defined by EC Directive 90/679/EEC *Classification of Biological Agents* and by the national variation of this legislation; for example, in the UK, as defined in the Advisory Committee on Dangerous Pathogens (ACDP) *Categorization of Biological Agents*, 4th edition, and the Approved Code of Practice (ACOP) for Biological Agents (Anon, 1996c).
- A definition of the hazards and assessment of the risks involved in handling the organism.
- Requirements for the safe handling and disposal of the organism:
    containment level,
    opening cultures and ampoules,
    transport,
    disposal,
    procedures in case of spillage.

Such information is absolutely essential to enable the recipient of organisms to handle and dispose of the organisms safely.

Summary

Legislation controls the safe handling and use of organisms, and biologists must ensure they keep abreast of existing, new and changing regulations. Misuse and abuse of rules will inevitably result in even more restrictive legislation that will make the exchange of organisms for legitimate use even more difficult. Health and safety, packaging and shipping, and controlled distribution legislation may be extensive and sometimes cumbersome but it is there to protect us and must be followed. Biologists wishing to collect organisms, characterize them and investigate their roles in nature must remember that many rules and regulations govern their actions. If the organisms or their products are to be exploited, then the country of origin must be taken into account. If agreements are in place, including permission to collect and how the organism may be used, and a suitable risk assessment is completed as soon as practicable, the process of compliance with the law is made much simpler. In the interests of the progress of science, biologists must be able to exchange the organisms upon which their hypotheses and results are based, but they must do this in a way that presents minimum risk to those who come into contact with the organism. Further information can be found in a paper published on the internet on the Society for General Microbiology web site (http://www.socgenmicrobiol.org.uk).

References

Alexander, M.T. and Brandon, B.A. (eds) (1986) *Packaging and Shipping of Biological Materials at ATCC.* American Type Culture Collection, Rockville, Maryland.

Anon (1974) *Health and Safety at Work etc. Act 1974.* HMSO, London, 117 pp.

Anon (1992) *The UK Management of Health and Safety at Work (MHSW) Regulations.* Health and Safety Executive, HSE Books, Sudbury.

Anon (1993) *Laboratory Biosafety Manual,* 2nd edn. World Health Organization, UK.

Anon (1995) *European Culture Collections: Microbial Diversity in Safe Hands, Information on Holdings and Services.* European Culture Collections' Organization (ECCO). Information Centre for European Culture Collections, Braunschweig, Germany, 48 pp.

Anon (1996a) *Industrial Property Statistics, 1994, Part II.* World Intellectual Property Organization (WIPO), Geneva.

Anon (1996b) *CAB INTERNATIONAL 13th Review Conference Proceedings.* CAB International, Wallingford, UK.

Anon (1996c) *Categorization of Pathogens According to Hazard and Categories of Containment,* 4th edn. Advisory Committee on Dangerous Pathogens (ACDP). HMSO, London.

Anon (1996d) *COSHH (General ACOP), Control of Carcinogenic Substances, Biological Agents: Approved Codes of Practice (1996)*. HSE Books, London.

Anon (1997a) Sichere Biotechnologie, Eingruppierung biologischer Agenzien: Bakterien. *Berufsgenossenschaft der Chemischen Industrie. Merkblatt B 006, 2/97, ZH 1/346.* Jerdermann-Verlag, Heidelberg.

Anon (1997b) Sichere Biotechnologie, Eingruppierung biologischer Agenzien: Fungi. *Berufsgenossenschaft der Chemischen Industrie. Merkblatt B 006, 2/97, ZH 1/346.* Jerdermann-Verlag, Heidelberg.

Anon (1997c) *Orange Book, Recommendations on the Transport of Dangerous Goods, Tests and Criteria*, 10th edn. UNO, New York.

Anon (1998) *Shipping of Infectious, Non-infectious and Genetically Modified Biological Materials, International Regulations*. DSMZ – Deutsche Sammlung von Mikroorganismen und Zellkulturen GmbH, Braunschweig, Germany.

CBD (1992) *The Convention on Biological Diversity* 31 I.L.M. 822.

COSHH (1988) *Statutory Instruments 1994 No. 3246. Health and Safety: the Control of Substances Hazardous to Health (Amendment) Regulations 1994.* HMSO, London.

Davison, A., Brabandere, J. de and Smith, D. (1998) Microbes, collections and the MOSAICC approach. *Microbiology Australia* 19, 36–37.

de Hoog, G.S. (1996) Risk assessment of fungi reported from humans and animals. *Mycoses* 39, 407–417.

EC Council Directive 77/93/EEC (1997) On protective measures against the introduction into the Member States of harmful organisms of plant or plant products. *Official Journal of the European Communities* 20, 20–54.

EC Council Directive 90/679/EEC (1990) Protection of workers from risks related to biological agents. *Official Journal of the European Communities* L374, 31.

EC Council Directive 93/88/EEC (1993) Directive amending Directive 90/679/EEC on the protection of workers from risks related to exposure to biological agents. *Official Journal of the European Communities* L307.

EC Council Directive 94/55/EC (1994) On the approximation of the laws of Member States on the transport of dangerous goods by road. *Official Journal of the European Communities* L319, 7.

EC Council Regulation 3381/94/EEC (1994) On the control of export of dual-use goods. *Official Journal of the European Communities* L 367, 1.

EC (1996) Annexes of EC Council Directive 94/55/EC on the approximation of the laws of Member States on the transport of dangerous goods by road. *Official Journal of the European Communities* L275.

Fritze, D. (1994) Patent aspects of the Convention at the microbial level. In: *The Biodiversity of Micro-organisms and the Role of Microbial Resource Centres*. World Federation for Culture Collections (WFCC), Germany, pp. 37–43.

Glowka, L. (1998) *A Guide to Designing Legal Frameworks to Determine Access to Genetic Resources*. IUCN, Gland, Switzerland, 98 pp.

IATA (1998) *Dangerous Goods Regulations*, 39th edn. IATA, Geneva.

Kelley, J. (1995) Micro-organisms, Indigenous Intellectual Property Rights and the Convention on Biological Diversity. In: Allsopp, D., Colwell, R.R. and Hawksworth, D.L. (eds) *Microbial Diversity and Ecosystem Function*. CAB International, Wallingford, UK, pp. 415–426.

Kelley, J. and Smith, D. (1997) Depositing micro-organisms as part of the patenting process. In: *European BioPharmaceutical Review,* 2, Ballantyne Ross Ltd, London, pp. 52–56.

Kirsop, B. (1993) *The Biodiversity Convention: some Implications for Microbiology and Microbial Resource Centres*. World Federation for Culture Collections (WFCC), Braunschweig, Germany.

Kirsop, B. and Hawksworth, D.L. (1994) *The Biodiversity of Micro-organisms and the Role of Microbial Resource Centres*. World Federation for Culture Collections (WFCC), Braunschweig, Germany.

Neto, R.B. (1998) Briefing bioprospecting: Brazil's scientists warn against 'nationalistic' restrictions. *Nature* 392, 538.

Rohde, C., Claus, D. and Malik, K.A. (1995) Technical Information Sheet No. 14: Packing and shipping of biological materials: some instructions, legal requirements and international regulations. *World Journal of Microbiology and Biotechnology* 11, 706–710.

Sands, P. (1994) Microbial Diversity and the 1992 Convention on Biological Diversity. In: Kirsop, B. and Hawksworth, D.L. (eds) *The Biodiversity of Micro-organisms and the Role of Microbial Resource Centres*. World Federation for Culture Collections, Braunschweig, Germany, pp. 9–27.

Smith, D. (ed.) (1996) *Committee on Postal, Quarantine and Safety Regulations Report 1996, Postal, Quarantine and Safety Regulations: Status and Concerns*. World Federation for Culture Collections (WFCC), Braunschweig, Germany, 39 pp.

Stricoff, R.S. and Walters, D.B. (1995) *Handbook of Laboratory Health and Safety*, 2nd edn. John Wiley & Sons, New York, 462 pp.

UPU (1998) *Universal Postal Convention, Compendium of Information*. Edition of 1 January 1996 – last update 15.06.98. Universal Postal Union, Berne (International Bureau).

The EU/EC Directives are available from the Office for Official Publications of the European Communities, 2 rue Mercier, L-2985 Luxembourg.

Appendix 8.1: List of Abbreviations Used

ACDP	Advisory Committee on Dangerous Pathogens
ACOP	Approved Code of Practice
ADR	Accord Européen Relatif au Transport International des Merchandises Dangereuses par Routes
CABI	CAB International
CBD	Convention on Biological Diversity
COSHH	Control of Substances Hazardous to Health
DGR	Dangerous Goods Regulations
DSMZ	Deutsche Sammlung von Mikroorganismen und Zellkulturen
EC	European Commission
EFB	European Forum for Biotechnology
EPPO	European and Mediterranean Plant Protection Organization
GRC	Genetic Resources Collection
IATA	International Air Transport Association
ICAO	International Civil Aviation Organization
IDA	International Depositary Authorities
IPBS	Infectious, Perishable Biological Substances
IPR	Intellectual Property Rights
MHSW	Management of Health and Safety at Work
MOSAICC	Micro-organisms Sustainable Use and Access Regulation International Code of Conduct
NPBS	Non-infectious Perishable Biological Substances
PIC	Prior Informed Consent
UPU	Universal Postal Union
USPHS	United States Public Health Service
WHO	World Health Organization
WIPO	World Intellectual Property Organization

Index

Figures in **bold** indicate major references. Figures in *italic* refer to diagrams, photographs and tables.